Branching Processes: Variation, Growth, and Extinction of Populations

Biology takes a special place among the other natural sciences because biological units, be they pieces of DNA, cells, or organisms, reproduce more or less faithfully. As for any other biological process, reproduction has a large random component. The theory of branching processes was developed especially as a mathematical counterpart to this most fundamental of biological processes. This active and rich research area allows us to make predictions about both extinction risks and the development of population composition, and also uncovers aspects of a population's history from its current genetic composition. Branching processes play an increasingly important role in models of genetics, molecular biology, microbiology, ecology, and evolutionary theory. This book presents this body of mathematical ideas for a biological audience, but should also be enjoyable to mathematicians, if only for its rich stock of realistic biological examples. It can be read, in full, by anyone with a basic command of calculus, matrix algebra, and probability theory. Any more advanced standard facts needed from basic probability theory are treated in a special appendix.

PATSY HACCOU is Associate Professor of Mathematical Biology at the Institute of Biology, Leiden University. She is co-author of *Statistical Analysis of Behavioural Data*.

PETER JAGERS is Professor of Mathematical Statistics at the Chalmers University of Technology and Gothenburg University. He is a member of the Swedish Academy of Sciences. He is author of *Branching Processes with Biological Applications* and co-editor of *Classical and Modern Branching Processes*.

VLADIMIR VATUTIN is Leading Researcher at the Steklov Mathematical Institute, Moscow, Russia. He is co-author of *Probabilistic Methods in Physical Research*.

Cambridge Studies in Adaptive Dynamics

Series Editors

ULF DIECKMANN
Adaptive Dynamics Network
International Institute for
Applied Systems Analysis
A-2361 Laxenburg, Austria

JOHAN A.J. METZ
Institute of Biology
Leiden University
NL-2311 GP Leiden
The Netherlands

The modern synthesis of the first half of the twentieth century reconciled Darwinian selection with Mendelian genetics. However, it largely failed to incorporate ecology and hence did not develop into a predictive theory of long-term evolution. It was only in the 1970s that evolutionary game theory put the consequences of frequency-dependent ecological interactions into proper perspective. Adaptive Dynamics extends evolutionary game theory by describing the dynamics of adaptive trait substitutions and by analyzing the evolutionary implications of complex ecological settings.

The *Cambridge Studies in Adaptive Dynamics* highlight these novel concepts and techniques for ecological and evolutionary research. The series is designed to help graduate students and researchers to use the new methods for their own studies. Volumes in the series provide coverage of both empirical observations and theoretical insights, offering natural points of departure for various groups of readers. If you would like to contribute a book to the series, please contact Cambridge University Press or the series editors.

1. *The Geometry of Ecological Interactions: Simplifying Spatial Complexity*
   Edited by Ulf Dieckmann, Richard Law, and Johan A.J. Metz

2. *Adaptive Dynamics of Infectious Diseases: In Pursuit of Virulence Management*
   Edited by Ulf Dieckmann, Johan A.J. Metz, Maurice W. Sabelis, and Karl Sigmund

3. *Adaptive Speciation*
   Edited by Ulf Dieckmann, Michael Doebeli, Johan A.J. Metz, and Diethard Tautz

4. *Evolutionary Conservation Biology*
   Edited by Régis Ferrière, Ulf Dieckmann, and Denis Couvet

5. *Branching Processes: Variation, Growth, and Extinction of Populations*
   Patsy Haccou, Peter Jagers, and Vladimir A. Vatutin

In preparation:

Fisheries-induced Adaptive Change
Edited by Ulf Dieckmann, Olav Rune Godø, Mikko Heino, and Jarle Mork

Elements of Adaptive Dynamics
Edited by Ulf Dieckmann and Johan A.J. Metz

Branching Processes: Variation, Growth, and Extinction of Populations

Patsy Haccou, Peter Jagers, and Vladimir A. Vatutin

CAMBRIDGE UNIVERSITY PRESS
Cambridge, New York, Melbourne, Madrid, Cape Town, Singapore, São Paulo

Cambridge University Press
The Edinburgh Building, Cambridge CB2 8RU, UK

Published in the United States of America by Cambridge University Press, New York

www.cambridge.org
Information on this title: www.cambridge.org/9780521832205

First published 2005

A catalogue record for this publication is available from the British Library

ISBN 978-0-521-83220-5 hardback

Transferred to digital printing 2007

Cover and illustrations by
Eva Engstrand

Contents

Authors

Main authors

Patsy Haccou (haccou@rulsfb.leidenuniv.nl) Institute of Biology, Leiden University, NL-2311 GP Leiden, The Netherlands

Peter Jagers (jagers@math.chalmers.se) Department of Mathematical Sciences, Chalmers University of Technology and Gothenburg University, SE-412 96 Gothenburg, Sweden

Vladimir A. Vatutin (vatutin@mi.ras.ru) Department of Discrete Mathematics, Steklov Mathematical Institute, 119991 Moscow, Russia

Contributors

Marina Alexandersson (marina.alexandersson@fcc.chalmers.se) Fraunhofer-Chalmers Centre SE-412 88 Gothenburg, Sweden

Gerold Alsmeyer (gerolda@math.uni-muenster.de) Institute of Mathematical Statistics, Faculty of Mathematics and Informatics, D-48149 Münster, Germany

A.D. Barbour (adb@amath.unizh.ch) Angewandte Mathematik, Universität Zürich, Winterthurerstrasse 190, CH-8057 Zürich, Switzerland

Michel Durinx (durinx@rulsfb.leidenuniv.nl) Institute of Biology, Leiden University, NL-2311 GP Leiden, The Netherlands

Mats Gyllenberg (mats.gyllenberg@helsinki.fi) Department of Mathematics and Statistics, University of Helsinki, FIN-20014 Helsinki, Finland

Göran Högnäs (goran.hognas@abo.fi) Department of Mathematics, Åbo Akademi University, FIN-20 500 Åbo, Finland

Vincent A.A. Jansen (vincent.jansen@rhul.ac.uk) School of Biological Sciences, Royal Holloway, University of London, Egham, Surrey TW20 0EX, United Kingdom

Fima Klebaner (fima.klebaner@sci.monash.edu.au) School of Mathematical Sciences Building 28M, Monash University, Clayton VIC 3800 Australia

Marek Kimmel (kimmel@stat.rice.edu) Department of Statistics, Rice University, MS-138 Houston, TX 77005, USA

Thomas G. Kurtz (kurtz@math.wisc.edu) Departments of Mathematics and Statistics, University of Wisconsin, Madison, WI 53706, USA

Johan A.J. Metz (metz@rulsfb.leidenuniv.nl) Institute of Biology, Leiden University, NL-2311 GP Leiden, The Netherlands

Peter Olofsson (olofsson@stat.rice.edu) Department of Statistics, Rice University, MS-138 Houston, TX 77005, USA

Serik Sagitov (serik@math.chalmers.se) Department of Mathematical Sciences, Chalmers University of Technology and Gothenburg University, SE-412 96 Gothenburg, Sweden

Nico Stollenwerk (nks22@cam.ac.uk) School of Biological Sciences, Royal Holloway, University of London, Egham, Surrey TW20 0EX, United Kingdom

Simon Tavaré (stavare@usc.edu) Program in Molecular and Computational Biology, University of Southern California, DRB 155 Los Angeles, CA 90089, USA

Acknowledgments

Natural science has always had two equally important sides: careful observation of facts and the creation of theories to explain what knowledge has been gathered. This is a triviality, and the difference is not so clear-cut; nobody makes observations without brooding over them (at least, we hope this is the case) and nobody creates and analyzes theories that do not explain the facts (at least, so it ought to be). Still, in task and attitude there is an abyss between the fact-gathering Linnaean entomologists in the field and the theoretical physicists at their desks, deducing properties of a model universe.

Even though there has been a connection between biology and mathematics ever since Darwin – his cousin Galton was one of the creators of the particular theory that is the topic of this book – biology has long been dominated by fact finding. In spite of many isolated earlier contributions, it is only recently, through the advances of theoretical and molecular biology, and (of course) the advent of the computer, that general conceptual thinking in mathematical terms, modeling and analysis, has assumed a more substantial role in the life sciences.

This book is a contribution to this development. As ought to be the case, it is a collective venture. The main parts are written by one biologist and two mathematicians. Probably, the three of us know who wrote what first, but the parts in the main jointly authored by us (Chapters 1–3 and 5–6) have been written and rewritten so often that we must all be held responsible for them. However, we have also had the privilege of contributions from others. The authorship of individual contributions is given in the text, as opposed to that of the main body of the book.

The idea of writing this book has its roots in a Population Dynamics Meeting in Gothenburg in 1998. One of the participants was Hans Metz, and another Ulf Dieckmann. Through their Adaptive Dynamics Network at the International Institute for Applied Systems Analysis (IIASA) at Schloss Laxenburg, Austria, they have supported the project and provided us with an excellent work environment, the splendid Habsburg summer palace, where most of this book was written up in the Belvedère, the tower room on top of the palace.

At IIASA we were further supported by the friendly efficiency of Edith Gruber and Barbara Hauser. We are prepared to embark on a new such venture, just not to lose contact with these colleagues, and with Emma of the Laxenburger Hof (Powidltascherln mit Zuckermohn und zerlassener Butter), and the constantly cheerful Johanna Sampl of the IIASA Guest House. It was a pleasure to work with the competent, friendly, and insightful Publications Department of IIASA. We are grateful to Eryl Maedel, John Ormiston, and Lieselotte Roggenland.

Economically, the project was supported by the Adaptive Dynamics Network and the Knut and Alice Wallenberg Foundation. We express our gratitude.

Erich Mayer and Karin Harding read the first and second chapters and provided many helpful comments, from the English speaker's and the biological reader's viewpoints, respectively, for which we thank them.

While we were writing this book, another monograph on branching processes in biology appeared in print, with Marek Kimmel and David Axelrod as authors. This is not a problem, as the books are different, and there are too few, not too many, books that try to use the power of mathematics on the important problems of biology. We are glad to count Marek Kimmel among our contributors.

Finally, branching processes are probabilistic. We believe that much more of mathematical biology ought to be phrased in terms of the erratic, randomly varying, behavior of individuals (molecules, cells, plants, animals) rather than in terms of deterministic mass flows. Probability theory renders it possible to take modeling a step further, beyond the phenomena that are described suitably by the classic applied mathematics of differential equations. However, up to now, most mathematical biology has remained deterministic. We include a special section (Section 4.3) that relates the mathematically most advanced deterministic population theory, that of structured population dynamics, to branching processes, a section that may provide heavy reading. It is included to shed light on the relation between two mathematical approaches to largely the same biological problems, and also to provide background for Section 7.8 on adaptive biological dynamics, in which stochastic and deterministic modeling interact in a fascinating way to model how species come about.

Parts of the text may be hard to digest for the reader with a background in the biological sciences. Some of these we have placed in "boxes" to warn the reader of the mathematical content. Others may be skipped on your own initiative. Even though understanding is important, proof in its technical detail is not, from a wider scientific perspective.

We wish you an informative read!

Laxenburg, October 2003
Patsy Haccou
Peter Jagers
Vladimir Vatutin

Notational Standards

Throughout we adhere to the following notational standards:

∞	Infinity
$\leq$	Less than or equal to
$\geq$	Larger than or equal to
$>$	Larger than
$<$	Less than
$\gg$	Much larger than
$\neq$	Not equal to
$\in$	Element of
$\notin$	Not element of
$\sum$	Summation
$\prod$	Product
$\cup$	Union
$\cap$	Intersection
$\int$	Integral
$|x|$	Absolute value of x
$x!$	x factorial
$1_{[0,T)}$	Indicator function: equals 1 in the interval $[0, T)$ and zero otherwise
$1_D(s)$	Indicator function: equals 1 if $s \in D$ and zero otherwise
$\Gamma(x)$	Gamma function
$\sup_{z \geq 0} r(z)$	Supremum of $r(z)$ over the domain $z \geq 0$
$\mathbb{P}[x]$	Probability of x
$\mathbb{P}[x|y]$	Conditional probability of x given y
$\mathbb{E}[z]$	Expectation of z
$\mathbb{E}[z|y]$	Conditional expectation of z given y
$\mathbb{E}[z; y]$	Expectation of z over the simultaneous distribution of z and y
$\mathrm{Var}[x]$	Variance of x
$\mathrm{Var}[x|y]$	Conditional variance of x given y
CV	Coefficient of variation
$f(s)$	Generating function
M	Mean matrix
$v = (v_1, \dots, v_d)^T$	Left eigenvector of M for the dominant eigenvalue
$u = (u_1, \dots, u_d)^T$	Right eigenvector of M for the dominant eigenvalue
α	Malthusian parameter
β	Mean age at childbearing
T	Life span
ℓ	$= \mathbb{E}[e^{-\alpha T}]$
$L(a)$	Life span distribution
$\mu(a)$	Reproduction function
$\xi(a)$	Reproduction point process
$\mu(s, A \times B)$	Reproduction kernel

1

Generalities

Why should a biologist read a book about branching processes in biology, and why should a mathematician?

This book is aimed primarily at biologists, so let us start with the mathematicians. *You* should read this book because it places a beautiful mathematical theory in a proper context. This is not to say that branching processes cannot be viewed in contexts other than those of population biology. On the contrary, branching processes occur in particle physics, in chemistry, and in computer science. However, mathematics can lose its direction in the jungle of problems that are syntactically well formed and mathematically intriguing, but that have no clear bearing on the outside world. Too many mathematicians, in our view, work on intellectual riddles, while important scientific problems escape their attention. You should read this book to see that branching processes are not only a fascinating mathematical structure, but also can help us to understand fundamental questions of nature.

This applies whatever your field of mathematics happens to be. If your expertise is in parts of mathematics other than probability or statistics, such as traditional applied mathematics oriented on differential equations or physics, there is a further point: this book suggests an alternative, largely discrete approach to population dynamics. It also emphasizes the need to model the complete spectrum from the behavior of individuals up to population phenomena. This is characteristic of modern stochastics and brings modeling a step forward from classic (deterministic) applied mathematics, in which equations are typically derived by intuitive, non-rigorous arguments, and then analyzed in a strict mathematical manner.

And now to those whose attention we really want to catch: biologists. *You* should read this book because many phenomena in areas like population biology, cell kinetics, bacterial growth, or DNA replication are general; they follow logically from the fundamental properties of populations. They are mathematical (i.e., logically inherent) properties of sets of individuals that change because the members generate new individuals. Why is extinction (of families, local populations, or species) so frequent? Is it a consequence of catastrophes or environmental changes external to populations? Or is it rather (or also) an intrinsic property of populations that they tend to die out? If so, how can frequent extinction be compatible with the famous Malthusian law of exponential increase, until resources become scarce? Is there a dichotomy between exponential growth and extinction, or are there other, slower forms of population growth? Can population size stabilize? What happens to the composition of populations if their multiplication persists for a long time? Does it stabilize? And what can be said about the history of surviving or extinct

1

populations? How gradual or abrupt is natural extinction? What can be said about mutational history? Or about the time between now and the most recent common ancestors?

Such broad population dynamics questions are addressed in the second part of this book. Before that we introduce models of varying generality. These sharpen our vague intuitive notions about population development into mathematical concepts that build up strict theories. With the help of these we can understand what must or may happen. However, even if a problem eschews our efforts to provide a mathematical solution, its formulation in mathematical terms makes it possible to simulate the model, and thus learn about reality from what has aptly been termed "experiments in the model." A mathematical model makes it possible to calculate explicitly the values of important population parameters, such as the doubling time or the growth rate, from parameters that describe individual behavior. Finally, it renders it possible to use population data to estimate parameters (e.g., the expected offspring size per individual) and test hypotheses about them, using the appropriate statistical distributions involved.

We emphasize here that branching processes have a role in general population dynamics. However, specific phenomena can (and should) also be analyzed through specific, tailor-made models. Such models are presented in the last part of the book, on topics that range across the spectrum from the smallest living entities to ecosystems and the evolution of life on earth.

Finally, it is one thing to wish to address biologists, but quite another to what extent biologists are prepared to receive our message. We certainly feel that biology is ripe for mathematical analysis, and the increasing role of mathematical formulation in all of biology, from algorithms for DNA sequencing and gene search to modeling of evolution and ecological systems, clearly bears witness to this. Unfortunately, many biologists may feel that they do not have the mathematical prerequisites for a text like ours.

We believe that our book can be read, in full, by an interested biologist with a basic command of calculus, linear algebra, and probability theory, and we dare hope that many others can capture the important ideas, and may even be intrigued enough to pursue a more thorough reading, with an elementary text book at their side. For your benefit we have collected text-book style facts of basic probability into a mathematical Appendix.

1.1 The Role of Models

The most important function of models is to order our thoughts. With models we formulate what we know (or think we know) about the world, and we perform thought experiments through "what if" scenarios. Every scientist makes models of the system that he or she studies. In many instances, initial models are verbal rather than mathematical. There is nothing wrong with verbal models, and they may suffice. As the scenarios become more complicated, however, it becomes increasingly difficult to keep track of verbal arguments and to check their consistency. Verbal models therefore involve the risk that they may overlook important factors and/or introduce logical inconsistencies. Here, mathematics provides a powerful language that forces us to be logically consistent and helps provide an explicitness about assumptions. Although a set of equations may seem daunting and complex, in most instances it is much easier to check the logic of an argument from such a list than when it is formulated in ordinary language. (The latter also takes up much more space!)

Another important use of models is their function as an idealization of the world. Whenever we formulate a model, we are forced to make a choice concerning which aspects of "the real world" we include in our description and which we choose to ignore (for the moment). This is true for verbal and mathematical models alike, but is more easily noted in the latter: mathematical formulations reveal contradictions and implicit assumptions. (This lack of gullibility may be one of the reasons for the unwarranted aversion toward formulation in mathematical terms.)

We cannot simply put everything we know about a system into a model because it soon becomes intractable. Computers help in this respect. They can accommodate many factors, but it is often difficult to determine which of these affect the predictions of the model and in what way. Modeling always involves a compromise between realism and the tractability of mathematics (or verbal arguments), and the inherent conflict between the two becomes more pronounced the more complex the system under study is.

The choice of what is put into a model and what is left out depends not only on the perceived importance of various factors, but also on the purpose of the model: is it to gain insight into a specific question or to address general issues and detect general patterns? Or is it to control a process, as in many engineering applications, rather than to understand it?

When we seek the answer to a specific question in a specific system, we might put in more detailed knowledge about the system. Such models have been called "tactical," since the results often have limited relevance. If we are interested in general patterns and conclusions, we need more of a "strategic" model, alternatively termed "conceptual." In such a model, we want to make assumptions that apply in a large class of systems and we want to draw very general conclusions (e.g., what were the most important factors in the evolution of sexual reproduction?). A strategic model cannot readily be used for any specific system, but it might indicate which general pattern to expect. Finally, if the purpose is prediction, management, control, or even purely descriptive, "black box" models, with a simplistic structure, but with many parameters so that they can be fitted to many data sets, have proved useful, even though their explanatory scientific value seems limited. Examples of such models are time series, such as autoregressive processes or moving averages, or artificial neural networks. Here we do not consider such descriptive models, even though they have been used in population biology, for instance to describe periodicity in the famous Hudson Bay Company data set on lynx and hare [cf. Diggle (1990), which also gives further references].

The models we consider are individual-based. They start from descriptions of, or assumptions about, individual life and reproduction, and deduce the behavior of populations. Such models are sometimes called mechanistic, and the whole approach reductionist, since properties of populations are brought back to the underlying mechanisms of individual life. Population models can also be based directly on phenomena that appear at the population level; these are called phenomenological. For instance, the effects of population density may be hard to describe at the individual level but established much more easily at the population level; an example is the well known phenomenon of "logistic growth."

Still, a word of warning is required here. Phenomenological population models carry tacit assumptions about the individual level. These should be pondered and made explicit, so that they are first of all not self-contradictory, but also so that they do not imply assumptions we are not willing to make.

As an illustration of this, many simple classic formulations of population dynamics, in terms of differential equations, can be shown to imply that individuals have exponentially distributed life spans. Other models, often otherwise quite sophisticated, have a basic Markov structure, which again means the same. However, a property that characterizes exponential distributions is that the conditional distribution of surviving another time period, given that you have survived up to a certain age, is independent of the latter. The biological meaning of this is that individuals do not age. This may actually be acceptable in models, say, of populations of small birds for which the hardships of life mean individuals do not die of old age itself (though the risk of dying is often higher for chicks than for adults). But it certainly matters in demography, or in cell kinetics, in which cells have to perform various tasks, like doubling their DNA, before splitting. For such cases, we demonstrate that one can do without the dubious and often simply false assumption of exponentially distributed life spans.

Finally, a word about the very concept of a *model*. It derives its meaning from scale models, simplified but yet replicas of larger structures, such as buildings or ships. Among the connotations of the concept is therefore a structural similarity between the model and the original, albeit simplified, sometimes even to the extent of caricature, but still there. In phenomenological modeling there is less of this similarity, and in what we have called "black box" modeling above it is almost always absent. Affinity between model and reality, as in the type of models considered here, certainly makes us more confident about our conclusions than mere curve fitting, which may cease to work around the next corner.

1.2 The Role of Randomness

The fate of individuals is stochastic. There is nothing mysterious in such an assertion, simply a recognition that it would be impossible to record all the conditions that have repercussions on the life career of an individual, even if this were possible in principle (a matter over which honorable people can disagree). Thus, life spans are influenced by predation risks, food access, weather, and other processes that can only be described as stochastic. As a result, life span is a random variable and different individuals usually have different life spans.

It is often argued that stochasticity at the individual level can be ignored in the study of large groups of individuals, and so deterministic models can be used to study their fate. However, such models are always approximations, since (obviously) if individuals evolve in a stochastic manner, so do finite populations. The best approach is always to perform this approximation explicitly, that is to use the fully fledged stochastic model and show that by some law-of-large-numbers effect it is well approximated by a deterministic simplification. Thus we may have a chance to estimate the magnitude of the errors involved, and also to discover aspects for which the approximation is unfeasible. We discuss such approximations further in this book, and often come back to the possible existence of stationary – i.e., possibly fluctuating locally, but in the long run stable – populations, for which disregard of stochasticity leads to radically different phenomena. Indeed, the place of extinction in population dynamics, and evolution even, and concepts like viability cannot be understood within the framework of stationary or deterministic population theory.

Population randomness through individual variability is called "demographic stochasticity." Another source of randomness is environmental stochasticity, caused by spatial and/or temporal variation in environmental factors, that affects the population as a whole or their members individually. The environment in its turn can be influenced by the population (e.g., by its size). Thus, it may be interesting to study feedback loops like environment → individual reproductive behavior → population size → environment.

Whereas the impact of demographic stochasticity can diminish for large population sizes, the effects of environmental stochasticity remain important for larger populations and should be included in model formulations, if relevant to the real biological system.

Additional sources of randomness to be incorporated may be the effects of measurement errors or factors not explicitly included in a model, but lumped together into an unspecified random effect (incomplete description).

Mainly, we are concerned here with modeling demographic stochasticity, sometimes in combination with environmental stochasticity. Branching process models were developed originally to account for inter-individual variation in offspring numbers and life spans (i.e., forms of demographic stochasticity). More recently, environmental stochasticity was added to these models. The latter may drastically change predictions about population size or structure, and also what is sometimes called "post-diction," attempts to reconstruct history from data about the present (e.g., the time back to the last common ancestor of those presently alive in the branching process). Other recent developments in branching processes of interest to biologists are the introduction of processes whereby individuals may interact, compete or collaborate, and the above-mentioned feedback scheme, whereby population size or other properties of a population, say mirroring the availability of resources, may influence individual reproduction.

There are several reasons why demographic stochasticity may remain important in populations. Populations may be prevented from growing to relatively large sizes (e.g., because of resource limitations or repeated environmental disturbances). Furthermore, even for large populations care should be taken with deterministic approximations. For instance, when the effects of environmental factors are non-linear, the effects on expected population size are not identical to the expected effects on population size. Therefore, in general it is advisable to formulate an individual-based stochastic model first and then, if possible, derive stochastic or deterministic approximations, rather than to start at higher levels. We illustrate this point in Section 1.4. Other contexts in which demographic stochasticity is important involve the fate of (initially) small subgroups of populations, as in models of evolutionary processes.

Finally, a point that may seem less important: in reality, populations are usually measured by counting their numbers, although there are exceptions when the relevant entity may be something like total body mass. Deterministic theories treat these discrete numbers as continuous and even differentiable functions. This results in assertions such as those in the Ricker model of Section 1.4 (see Figure 1.1), for which the population number might even be irrational. Depending upon ideology, you may feel this to be a mere nuisance or a major epistemological problem, as when certain models claim that an epidemic goes on forever in a cyclic manner, but actually the minimum of each epidemic cycle, though strictly positive, is much less than the one carrier needed to prolong the epidemic. Anyhow, it is certainly an advantage in population modeling that branching processes are integer valued.

1.3   Branching Processes: Some First Words

As pointed out, branching processes are individual-based models for the growth of populations. This property they share with the more advanced deterministic models, in particular those of structured population dynamics (cf. Metz and Diekmann 1986; Diekmann *et al.* 1998, 2001), in which the course of individual lives is described by differential equations, at least in the most prominent cases. In branching processes we content ourselves with probabilistic descriptions of life careers: the basic purpose is thus to deduce properties of the processes (i.e., of populations) from the probability laws of individual childbearing and life spans.

Many methods and techniques used in population biology have a branching process background or interpretation. For example, the Leslie matrices of classic demography are nothing but descriptions of the expected reproductive behavior of individuals of various age groups. Populations in discrete time for which age matters for reproduction are but a special form of multi-type branching processes, which are discussed in Sections 2.3 and 2.4.

An individual is understood in a broad sense: it might be an animal or a plant, but also a cell or an elementary particle – the defining property is that it gives birth, splits into, or somehow generates new individuals. It could even be a whole population that gives rise to new populations through mutation (speciation), or if it lives in one patch by colonization of other patches. The name "branching process" alludes to the family trees thus arising. However, more often the name refers to the simpler stochastic process that records the population sizes at various times only, or records the sizes of subsequent generations, rather than to the complete population tree. Historically, the first question tackled with the help of branching processes was that of population extinction: what is the probability that a population dies out? Other classic topics are the possible stabilization of population sizes, growth rates, and age distributions or other aspects of population composition.

The theory started with (Bienaymé–)Galton–Watson processes (discrete-time models) and Markov branching processes (their continuous-time equivalents). For overviews of its fascinating history, see Kendall (1966), Jagers (1975), and Heyde and Seneta (1977).

The Galton–Watson processes just count generation sizes. In real life, generations and physical time can have quite diverse relationships. Whatever these are, the problem of ultimate extinction is unaffected: if there is a time when the population is extinct, there must also be an empty generation, and vice versa – even if generations can overlap and be shifted drastically in time.

In the simplest case, Galton–Watson processes can also be viewed as real-time processes, provided all the individuals can be assumed to have the same life span of length one (year or season). In such cases, we assume that individuals are born at the beginning of the season.

Somewhat more general discrete-time processes allow individuals to live over several discrete seasons, during which they may give birth repeatedly. Still more general processes allow arbitrary life spans, and arbitrary reproduction during life.

In all traditional models, however, independence in reproduction and survival among different individuals is assumed. The rationale for this is that the models are meant for small populations, in which resource limitations, for example, can be assumed not to play an important role. Obviously, the latter is not true for many biological systems. As pointed out, recent research has concerned these restrictions and analyzed setups in which, for example, reproduction may depend on resources, population size, or density. We report some findings in Sections 2.6, 5.8, and 6.5.

1.4 Stochastic and Deterministic Modeling: An Illustration

Many deterministic models are simply expectation versions of branching processes. These we shall meet in the discussion of the corresponding branching processes. However, where there are non-linear dependences, difficulties may arise. We illustrate this with a well-known deterministic approach to population-size dependence, the so-called Ricker equation. It has the form

$$z_{n+1} = m z_n e^{-b z_n} , \qquad (1.1)$$

which relates population size at time $n + 1$, z_{n+1}, to size during the preceding period, z_n. What happens if modeling is refined down to an individual stochastic level?

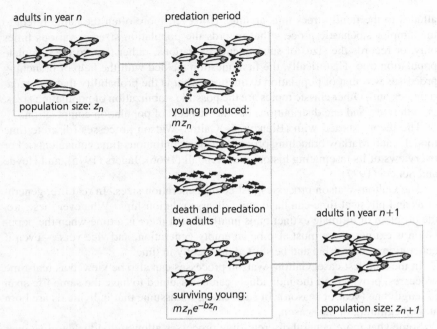

Figure 1.1 The Ricker model: adults predate on juveniles. Those young that survive the vulnerable period form the next generation.

The Ricker equation purports to model predation on juveniles by adults, as happens in many fish species. The rationale behind it is (cf. also Figure 1.1): adults produce m juveniles, on average, so the juvenile population consists of $m \times z_n$ individuals. There follows a period during which predation by adults occurs. The predation rate is assumed proportional to population density z_n/A, where A is the area occupied by the population. By a differential equations argument, the proportion of juveniles that survive the predation period has the form claimed, e^{-bz_n}. Indeed, a conventional argument is as follows: let $j(t)$ denote the number of juveniles alive after a portion t of the predation period has passed, $0 \le t \le 1$. Then, $j(0) = mz_n$, $j'(t) = -bz_n j(t)$ for some constant $b > 0$, and hence

$$z_{n+1} = j(1) = mz_n e^{-bz_n} , \qquad (1.2)$$

in this deterministic framework.

A stochastic version of this is to say that each of the Z_n individuals in the nth generation generates a random number of offspring, with expectation m. In total there are j_{n+1} juveniles, and the expected offspring size in the $(n+1)$th period, given the number of adults, Z_n, is

$$\mathbb{E}[j_{n+1}|Z_n] = mZ_n . \qquad (1.3)$$

Each of the young fish is then subjected to a predation risk $1 - e^{-bZ_n}$ independently. As a consequence, the next generation, Z_{n+1}, is distributed binomially with parameters j_{n+1} and e^{-bZ_n}, given Z_n and j_{n+1} (see the Appendix).

Conditional distributions and expectations, as above, play a crucial role throughout this book. Typically, it is easier to see what the conditional expectation (e.g., given some crucial information) will be, than it is to determine the overall, unconditional expectation directly. However, the latter equals the expectation of the former (see the Appendix).

In the present case, properties of the binomial distribution imply that the expected population size in the next period, given both the number of juveniles, j_{n+1}, and the population size, Z_n, in the preceding period, is

$$\mathbb{E}[Z_{n+1} | j_{n+1}, Z_n] = j_{n+1} e^{-bZ_n} . \tag{1.4}$$

The expectation of this over all possible values of j_{n+1}, given only Z_n, is

$$\mathbb{E}[Z_{n+1} | Z_n] = m Z_n e^{-bZ_n} , \tag{1.5}$$

as given by the binomial distribution. Hence, unconditionally,

$$\mathbb{E}[Z_{n+1}] = m \mathbb{E}[Z_n e^{-bZ_n}] . \tag{1.6}$$

This illustrates well what happens to non-linear relationships, in which a deterministic derivation is replaced by expectations in a stochastic model. In the present terminology, the deterministic model we started from only considers $m_n = \mathbb{E}[Z_n]$; it cannot distinguish between the real population size and its expectation. Therefore, it claims that

$$m_{n+1} = m m_n e^{-bm_n} . \tag{1.7}$$

A more refined analysis, however, results in the recursion (1.6), and certainly

$$m \mathbb{E}[Z_n e^{-bZ_n}] \neq m \mathbb{E}[Z_n] e^{-b\mathbb{E}[Z_n]} = m m_n e^{-bm_n} , \tag{1.8}$$

except in degenerate cases. Actually, Jensen's inequality (see the Appendix) tells us that the expectation of a convex (concave) function is larger (smaller) than the function of the expectation. Since the function Ze^{-bZ} is concave for small z and convex for large, the deterministic model tends to fluctuate less than does the expectation of the stochastic model. In particular, if the population starts from a high level, deterministic theory underestimates its expected size and, conversely if it starts small, size is overestimated.

As is shown in Section 5.2, stochastic Ricker populations inevitably die out. Their so-called quasi-stationary behavior (i.e., what the population development looks like, given that the population is not yet extinct) was discussed for small b by Högnäs (1997). He showed that short-term development follows the deterministic model, at least in a crude sense. We return to this later (Section 6.9).

The Ricker model is just one in a whole family of deterministic discrete-time models that have the mathematical form

$$x_{n+1} = f(x_n) , \tag{1.9}$$

where x_n either equals the population size Z_n or else the population density, $x_n = Z_n/K$; K is the maximal population size in the deterministic model (or a typically approximate maximal level in stochastic cases) and is often referred to as the *carrying capacity*. Thus, in the Ricker approach, $f(x) = xe^{-bKx}$ in terms of density. Another classic, but somewhat ad hoc population model, Verhulst's so-called logistic model from 1845, has $f(x) = x(1 - x), 0 \le x \le 1$, where x stands for density again. In the logistic case, f is concave along its whole interval of definition, so that by Jensen's inequality the deterministic relationship always overestimates the expected population density. The reason for this is simply that the model is not defined for population sizes larger than the carrying capacity.

In biomathematical literature the recursion above is usually formulated in terms of the individual reproduction function R, so that

$$x_{n+1} = x_n R(x_n) \,. \tag{1.10}$$

Mathematically, the two are, of course, equivalent, but the first is slightly more general and, from the point of interpretation, the second form has great advantages.

1.5   Structure of the Book

Chapters 2 and 3 give an overview of branching processes and their main characteristics, with an emphasis on biological interpretation. Chapter 4 is the most technical chapter of the book. It deals with relations between branching processes and other models of population dynamics, in particular diffusion processes and deterministic models. Chapters 5 and 6 describe how important characteristics of population dynamics can be studied with branching processes. Finally, Chapter 7 gives examples of recent applications of branching processes in various fields of biology.

2

Discrete-Time Branching Processes

Both this chapter and the next survey basic branching processes relevant to biological applications. We describe process structure, and give some examples. Underlying assumptions are exhibited and it is indicated when and where a specific model might be applied (which areas, under what conditions). Note that even the simplest models have proved their usefulness to biology in the past, and continue to find intriguing new applications (e.g., in molecular biology, cf. Chapter 7).

The purpose is to give an impression of what is available in the literature, so as to gain an overview of the theory as well as starting points for modeling. Furthermore, we give references in which more details, proofs, and results on these models, and generalizations of them, can be found. When possible we rely on verbal formulations and illustrations, rather than extensive equations, and we do not give the main mathematical results yet. However, we do introduce basic concepts and notation, and also some fundamental facts used in later chapters, in which the results are stated and their relevance to basic biological issues made clear.

This chapter treats discrete-time models. In these, time is represented by integers that indicate reproduction periods. Thus, it is assumed that reproduction can only occur during separate non-overlapping periods. In the simplest forms (described in Sections 2.1 to 2.3) generations cannot overlap either (think of annual plants).

An argument in favor of discrete-time models is that they are mathematically much easier to handle than those in continuous time (see Chapter 3). Furthermore, many species do reproduce only during fixed periods of the year. In other situations, like polymerase chain reactions (PCR, see Section 7.5), the experimental setup creates well-defined discrete reproduction periods: during heating DNA strands separate, and during the subsequent cooling the single strands form counterparts, and thus the next generation of DNA molecules.

On the other hand, "real time" is certainly perceived as continuous, and in many cases discrete processes may seem artificial. Which approach to use depends on the situation or purpose and is also a matter of taste (cf. Chapter 1). Discrete-time models with small time steps can also be viewed as approximations of continuous-time processes. Sometimes, the time steps need not even be that small, as is the case with the yearly counts traditional in demography.

Below we give some general approaches, based on the main characteristics of the life history of the species under consideration. In these, we distinguish between three main types of life histories, according to the duration of the maturation period (from birth to reproductive age) and how births occur.

1. In the simplest case the separate reproduction periods are short compared to the maturation period. Thus individuals born during one reproduction period do not reproduce during that period. There are many examples of such life histories: in temperate zones many bird, fish, and mammal species only reproduce during spring and/or summer and their young do not start to reproduce until the following year. Insects, like butterflies and moths, deposit eggs during spring or summer. Their larvae grow during summer and overwinter as pupae to emerge as adults next spring. Even in tropical regions examples can be found. For instance, large mammals such as lions reproduce during the rainy season when food is abundant. Discrete-time processes can be used readily in such cases. Repeated reproduction of individuals during one period does not create difficulties, since differences in offspring age are negligible as compared to the length of the maturation period. Thus, we can simply count the numbers of surviving offspring at the beginning of the next reproduction period.
2. In the second type of life histories, there are also distinct reproduction periods, but maturation time is much shorter than the length of one period. In such cases the young can start to reproduce during the period in which they were born. Examples are many insect species in temperate zones, like flies, aphids, or parasitic wasps, which have seasonal reproduction, but produce several generations each season and overwinter in diapause. Such systems can be described by distinguishing two time scales. A continuous-time model can be used to model the process within each reproduction period to yield the relation between the population at the start and end of a reproduction period. For population development on a larger time scale, a discrete-time model can then be used to describe the relation between population sizes just before each reproduction period in successive years.
3. Finally, there are species with life histories for which no fixed reproduction periods can be distinguished. Examples are humans, many (semi-) domesticated animals (like house mice, guinea pigs, or goldfish), and many tropical insect species. For such life histories continuous-time models are usually more appropriate. The same applies to cells and prokaryotes, which divide after life cycles of highly variable length.

Throughout this book we use the convention for discrete-time models that the population starts from time zero and that population numbers are counted just before each reproduction period. This is sometimes referred to as performing "prebreeding censuses" (Caswell 2001, p. 25).

It happens that the population thus obtained just before the $(n + 1)$th reproduction period is referred to as the "nth generation." Strictly speaking, this is correct only in the simplest case in which all individuals live exactly one time unit. This is evident in continuous time, in which those alive at any given time belong to many different generations. In addition, whatever the time structure is, we can always count the successive generation sizes, and thus obtain what is often called

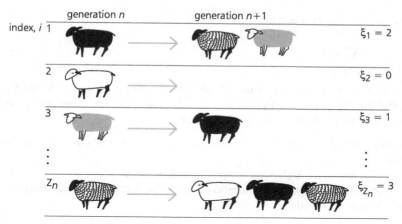

Figure 2.1 Schematic representation of the Galton–Watson branching process.

the *embedded generation process*: the zeroth generation initiates the population, the first contains the immediate offspring (children) of the former, the second generation consists of the grandchildren, and so on.

Even though generations in many real populations tend to disperse over large time spans, some problems (notably extinction) are easier to study in the generation process: it should be clear that a population dies out if and only if its embedded generation process turns to zero (see Section 3.1, in which the embedded process is defined explicitly).

2.1 The Basic Process

The oldest and simplest discrete-time branching processes are the basic Galton–Watson single-type branching processes or, historically more correct, the Bienaymé–Galton–Watson processes (see Heyde and Seneta 1977), sometimes also referred to as *simple branching processes*. These are discrete-time processes with non-overlapping generations, and only one type of individual. Alternatively, these can be viewed as counting successive generations in populations with a more complicated time structure, such as individuals that do not necessarily have the same life spans or that give birth at different ages.

The way these proceed is illustrated in Figure 2.1. We denote the population size in the nth generation by Z_n. The process starts with an initial population of Z_0 individuals just before the first reproduction period. Each individual produces a random number of offspring and then dies. The population size in the next generation, Z_1, can be determined by summing the numbers of offspring. These individuals then reproduce, which gives the next generation, and so on.

Thus, if we number the individuals in the nth generation in an arbitrary way and denote the number of offspring of the ith individual by ξ_i, the population size in the $(n + 1)$th generation is

$$Z_{n+1} = \sum_{i=1}^{Z_n} \xi_i \qquad (2.1)$$

with the ξ_i independent and identically distributed. The complete specification of a model involves a choice of the probability distribution of the offspring numbers.

The scheme has three basic characteristics:

1. All reproductive individuals are of a single type with identical offspring distribution.
2. They do not affect each other's reproduction in any way.
3. The distribution of offspring numbers is assumed to remain the same during the whole period considered.

We discuss each of these basic characteristics and their consequences here, and consider several generalizations of each in the following sections of this chapter.

The first property, that there is no distinction at all between reproductive individuals with respect to distribution of numbers and types of offspring, implies that there is no condition or location variable that affects reproduction and neither is there a distinction between, for example, different genotypes. This is suitable for homogeneous systems with clonal reproduction, such as cell populations or bacteria.

Models with this property might, however, also be used for some systems with sexual reproduction. For instance, in hermaphrodites, such as many plants (so-called *monoecious* plants), only one type of individual needs to be considered. Furthermore, since the assumption only concerns reproductive individuals, even bisexual (or, in plants, *dioecious*) species may be modeled thus, as there is only one reproductive type, namely females (see Figure 2.2). In such cases, however, additional requirements have to be met to satisfy the assumption of independence between individuals, at least approximately. Thus, the supply of males must be sufficient, otherwise the chance that a female reproduces depends on population size. Moreover, since sexual reproduction also introduces genetic variation in the absence of mutation and it is reasonable to assume that genotype affects reproduction, such models are not normally suitable for populations with sexual reproduction, except in special situations, as with the initial spread of mutants. For such a spread, the resident population is assumed to be homozygous and, as long as mutants are rare, the probability that they mate with each other can be neglected. Therefore, initially the population of mutants only consists of one reproductive type: heterozygous individuals with the new allele.

The second assumption, independence between reproductions, implies that the approach is unsuitable when population limitation occurs in any way. It can be used for the initial growth of small populations, when resources are still abundant. It can also be used for invading mutants in evolving systems, as long as the resident population can be assumed to set a fixed background. This implies, for example, that the resource supply is controlled by the resident population and that mutant

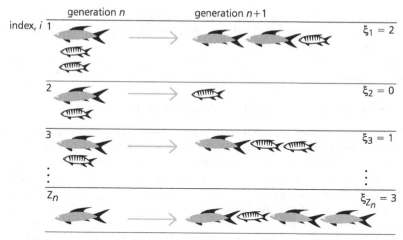

Figure 2.2 A process with two types of individuals: females (big fish) and males (small fish). Only the females are counted, since they are the reproductive type. (See text for further explanation.)

numbers are still so low that they do not affect this. Furthermore, it implies that the resident population is keeping the amount of available resources at a fixed level.

The third property means that there are no essential differences between reproduction events in different periods. Thus, environmental factors that affect reproduction are assumed to be constant. Furthermore, the number of reproducing individuals in different periods must be equivalent. This can be implemented in two different ways. For the moment, we only consider one of these possible implementations, namely to assume that generations do not overlap (i.e., the parent population dies after reproduction). The other implementation does allow generation overlap, and we return to this in Section 2.4.

Although this is a simple scheme, it is useful for the study of various aspects of populations, such as the survival chances of small populations and initial population growth. It should also be considered as an approximation that gives a global insight by focusing on the bare essentials of the dynamics of small populations.

In the next sections we first consider two particularly simple benchmark processes, and then models in which one or more of the three assumptions are relaxed in various ways. Thus, reproductive individuals may have different types (Section 2.3) or states (Section 2.5) that may affect the numbers and/or types (and states) of offspring they produce. Generation overlap and aging are treated in Section 2.4. Furthermore, we discuss the effects of interactions between individuals, which introduce dependences, in Sections 2.6 to 2.8; Section 2.8 is devoted to the complications induced by sexual reproduction. Section 2.9 deals with models with temporal variation in factors that affect offspring distributions. In each section it is assumed that only one of the characteristics of the basic model is generalized, whereas the other two remain the same. Therefore, remarks made in this section concerning these characteristics continue to hold. At the end of some sections, however, we do consider models with combinations of previously described generalizations.

Finally, Section 2.10 is devoted to yet another generalization, namely branching models with immigration from other populations.

2.2   Basic Properties and Two Benchmark Processes

As benchmark processes we use two particularly simple cases: binary splitting and geometrical offspring distributions. Whereas the former has a biological background in cell division or bacterial growth, the latter seems more of a mathematical artifact. However, with a slight modification, it has been used in classic demography (Lotka 1931c), and it avails itself to a relatively elementary and direct mathematical analysis. Also, the two cases complement each other: one has bounded reproduction (never more than two children for binary splitting), and in the other reproduction has no given bound, k children can occur for any number $k = 0, 1, 2, 3, 4, \ldots$ with the probabilities

$$p_k = qp^k \, . \tag{2.2}$$

In this, $q = 1 - p$ and p is some number between 0 and 1. For binary splitting the probability of no children is q and the probability of two is p. No other numbers of offspring can occur.

In general, the mechanism that underlies geometric distributions is that of repeated trials, each with a success probability p until the first failure, which occurs with probability q and thereby ends the process. A biological, albeit somewhat strained, analogy would be that, during one reproduction period, the mother makes repeated independent tries to reproduce, each with success probability p, like laying one egg after the other. Once she fails, she produces no more children during that season.

The slight generalization mentioned is into *zero modified* geometric distributions, also called fractional linear, after the form of the generating function (see Section 5.3 and Appendix): the probability of producing no children does not necessarily equal q, but is, say, $p_0 = r$, $0 < r < 1$, and the other probabilities are duly modified into

$$p_k = (1 - r)qp^{k-1}, k = 1, 2, 3, \ldots \, . \tag{2.3}$$

Indeed, Lotka found that probabilities defined through $r = 0.4825$ and $p = 0.5893$ fitted the American 1920 census data on white males well.

The expected number of children of an individual with reproduction distribution $\{p_k, k = 0, 1, 2, \ldots\}$ is denoted by

$$m = \sum_{k=1}^{\infty} kp_k \, , \tag{2.4}$$

so that

$$m = 2p, \tag{2.5}$$

in the case of binary splitting, and

$$m = \sum_{k=1}^{\infty} kqp^k = p/q, \tag{2.6}$$

for geometric reproduction.

In biological applications that have no bound on the number of children, the Poisson reproduction distribution is often used rather than the geometric one. Then, the probability of having k children is

$$p_k = e^{-m} m^k / k!, k = 0, 1, 2, \ldots, \tag{2.7}$$

where m is some positive number. The reason for choosing the letter m is that the expectation (in other words, mean) of a Poisson distributed random variable ξ equals its parameter:

$$\mathbb{E}[\xi] = m, \text{ if } \mathbb{P}(\xi = k) = e^{-m} m^k / k!, \ k = 0, 1, 2, \ldots . \tag{2.8}$$

An often-used shorthand for a random variable that has the Poisson distribution with parameter m is to say that it is Poisson(m), or even Poi(m). Similarly, one might say that a variable is Geometric(p), or Geo(p), and Binomial(n, p), or Bin(n, p), for the geometric or binomial distributions.

The assumption of a Poisson distribution is sometimes not that well founded, though often there are good reasons for it: if a large number of eggs are produced, and only a few of them grow to maturity, then the number of mature children will follow a Poisson distribution (approximately). This is an instance of what classic probability literature used to call the "Law of Small Numbers". Its historical debut was possibly as applied to the yearly number of Prussian soldiers killed by horse kicks (Von Bortkiewicz 1898). Since then it has been applied successfully to a panoply of situations in which a large number of independent tries are independently scanned and only rarely let through, notably in telecommunications, such as the number of incoming calls at a switchboard. Arguments for it can be found in any textbook, from the elementary to advanced monographs on point processes and random measures. A classic reference is Feller's *Introduction to Probability Theory* (1957).

The mean numbers of offspring in the Poisson and the geometric situations coincide provided $m = p/q$, that is, $p = m/(m+1)$. For large m the two distributions take quite different shapes; whereas geometric probabilities remain decreasing, the Poisson probabilities increase to their maximum at the largest integer that does not exceed the mean, and then start to decrease.

For small m the difference between the two distributions is minute (see Figure 2.3), and so mathematical expediency (though never a good argument) might favor use of the geometric law; we discuss the malleability of geometric distributions in Section 2.3.

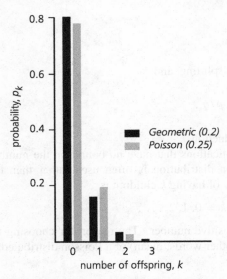

Figure 2.3 Similarity of the geometric and Poisson distributions.

The probability of many children is larger in the geometric case. Also, the variance of the number of children is larger, p/q^2 in case of a Geo(p) distribution as opposed to p/q in the corresponding Poisson (p/q) case.

Now, how do simple Galton–Watson processes evolve? This is the subject matter of Chapters 5 and 6, but we preview expected population sizes here, both because of the neat result and because of the method used to deduce it, whereby the typical conditioning argument of branching processes is introduced gently.

The expected size of the nth generation satisfies

$$\mathbb{E}[Z_n] = \mathbb{E}[\mathbb{E}[Z_n|Z_{n-1}]] . \tag{2.9}$$

However, given the size of the preceding generation (assume that Z_{n-1} is known and fixed), the present generation is the sum of the number of offspring of each of the Z_{n-1} potential mothers. In symbols,

$$Z_n = \xi_1 + \cdots + \xi_{Z_{n-1}} , \tag{2.10}$$

where ξ_i denotes the number of children of the ith member of generation $n - 1$. These all have expectation m, independent of Z_{n-1}, and since the expectation of a sum equals the sum of the expectations (see the Appendix)

$$\begin{aligned} \mathbb{E}[Z_n|Z_{n-1}] &= \mathbb{E}[\xi_1 + \cdots + \xi_{Z_{n-1}}|Z_{n-1}] \\ &= \mathbb{E}[\xi_1|Z_{n-1}] + \cdots + \mathbb{E}[\xi_{Z_{n-1}}|Z_{n-1}] = mZ_{n-1} . \end{aligned} \tag{2.11}$$

Hence, $\mathbb{E}[Z_n] = m\mathbb{E}[Z_{n-1}]$, and repetition yields

$$\mathbb{E}[Z_n] = m\mathbb{E}[Z_{n-1}] = \cdots = m^n\mathbb{E}[Z_0] . \tag{2.12}$$

Of course, if the initial number Z_0 is fixed and known, we need not use its expectation.

Populations with $m < 1, m = 1$, or $m > 1$ are called *subcritical, critical,* and *supercritical*, respectively. From Equation (2.12), we see that as $n \to \infty$ the average size of a subcritical population goes to zero and that of a critical population is stable, whereas the average size of a supercritical population grows unboundedly. Indeed, both these developments occur at the famous exponential rate: there is decay at the rate m^n if $m < 1$, and growth at the rate m^n if $m > 1$.

Exactly critical populations, however, are of less interest for applications, since it is not reasonable to assume that the average number of direct descendants per individual is *exactly* one during many generations. And the slightest deviation results in either of two radically different types of behavior: exponential growth or exponential decay. The situation is quite different for sub- and supercritical populations in which small changes of the mean reproduction affect the rate, but not the character, of the development.

In "near-critical" population models, in which the deviation of m from one is "very small" and varies with, say, time or population size, the situation becomes different. If such a population does not die out, its expected size often grows almost linearly. This topic is discussed in Sections 2.6 and 6.5.3. A related situation arises if we wish to describe the survival chances of populations for which m does not vary with time, but in itself is close to one. It has a bearing on the establishment probabilities of slightly advantageous genes, and is taken up in Section 5.6.

Subcritical processes may be used to investigate populations that are dying out more or less rapidly. Young populations that develop under favorable conditions (enough food and space, and a generally benign environment) tend to be supercritical. This therefore applies to situations such as in vitro – or generally early – bacterial or cell growth, invasion by mutants (i.e., the initial growth of the number of mutants), colonization of new habitats, and many others.

In the deduction of the mean behavior that leads to Equation (2.12) we used only the first and third basic properties of Galton–Watson processes, namely that the reproduction distribution is the same for all individuals at all times. The independence between population members was not called for, but to deduce further properties this is, however, important; independence is required in the calculation of variances. As we shall see in Chapter 5, it plays a crucial role in the determination of extinction risks. Indeed, let

$$\sigma^2 = \mathrm{Var}[\xi] = \sum_{k=0}^{\infty} (k - m)^2 p_k \tag{2.13}$$

denote the *reproduction variance*, and repeat the sequence of arguments that lead to Equation (2.12). Then, of course, independence (or at least correlation zero) is needed for the additivity of the variance,

$$\mathrm{Var}[Z_n | Z_{n-1}] = \mathrm{Var}[\xi_1 + \cdots + \xi_{Z_{n-1}} | Z_{n-1}]$$
$$= \mathrm{Var}[\xi_1 | Z_{n-1}] + \cdots + \mathrm{Var}[\xi_{Z_{n-1}} | Z_{n-1}] = \sigma^2 Z_{n-1} . \tag{2.14}$$

Once this is obtained, we can use the general (and quite useful) relation that variance can always be viewed as comprising two parts: expected variation around a random level ($\mathbb{E}[X|Y]$) and the variance of the level itself,

$$\text{Var}[X] = \mathbb{E}[\text{Var}[X|Y]] + \text{Var}[\mathbb{E}[X|Y]] . \tag{2.15}$$

This is derived in the Appendix, and yields

$$\begin{aligned}
\text{Var}[Z_n] &= \sigma^2 \mathbb{E}[Z_{n-1}] + \text{Var}[mZ_{n-1}] \\
&= \sigma^2 m^{n-1} \mathbb{E}[Z_0] + m^2 \text{Var}[Z_{n-1}] \\
&\ \ \vdots \\
&= \sigma^2 (m^{n-1} + m^n + \cdots + m^{2n-2}) \mathbb{E}[Z_0] + m^{2n} \text{Var}[Z_0] \\
&= \begin{cases} \sigma^2 \frac{m^n(m^n-1)}{m(m-1)} \mathbb{E}[Z_0] + m^{2n} \text{Var}[Z_0] , & \text{if } m \neq 1 \\ \sigma^2 n \mathbb{E}[Z_0] + \text{Var}[Z_0], & \text{if } m = 1 . \end{cases}
\end{aligned} \tag{2.16}$$

In the subcritical case both mean and variance of population size thus decrease to zero, and extinction should be unavoidable. Some of the peculiarities of the critical case can be seen from the joint appearance of a stable mean size and an ever-growing variance, which at first sight seems to indicate that such populations either die out or become extremely large. The latter, however, does not happen; critical populations die out with certainty.

During supercritical growth, the coefficient of variation (i.e., the population standard deviation divided by its expected size) stabilizes quickly. If we denote it by CV, indeed

$$(CV)^2 = \begin{cases} \frac{\sigma^2}{m(m-1)}(1-m^{-n})/\mathbb{E}[Z_0] + \text{Var}[Z_0]/\mathbb{E}[Z_0], & \text{if } m \neq 1 \\ \sigma^2 n/\mathbb{E}[Z_0] + \text{Var}[Z_0]/\mathbb{E}[Z_0], & \text{if } m = 1 . \end{cases} \tag{2.17}$$

For all practical purposes we should thus expect established (i.e., n is large) supercritical populations with the same reproduction distribution and one ancestor, $Z_0 = 1$, to have the coefficient of variation

$$CV = \frac{\sigma}{\sqrt{m(m-1)}} = cv\sqrt{m/(m-1)} , \tag{2.18}$$

where cv is the coefficient of variation of individual reproduction. Thus, $CV \approx cv$, provided m is not too small.

Example 2.1 Observations of the type considered above have been used in studies by Azevedo *et al.* (2000) and Azevedo and Leroi (2001) to explain what they call a power law that relates the variation of cell numbers between individual organs to the average of these numbers. Their investigations started from the widely held belief that there is a so-called *eutely*, meaning absence of variation in cell number, in certain nematodes (like *Caenorhabditis elegans*) claimed to have a deterministic number of cells. Linear regression studies of the logarithm of variance (V) versus the logarithm of mean size (M) rather pointed at a relation of the form

$$\ln V = a + b \ln M , \tag{2.19}$$

with $b = 2$ more often than not. Thus, the coefficient of variation of organ size usually equals a species-specific constant.

They also explained their results in terms of branching processes. If the cell growth of some species occurs according to a Galton–Watson process with mean $m > 1$ and variance σ^2, then certainly we would expect organs of that species to display roughly the same coefficient of variation between individuals, according to the argument above,

$$V/M^2 = (CV)^2 \approx \sigma^2/m(m-1) = (cv)^2 m/(m-1) . \tag{2.20}$$

In the linear regression [Equation (2.19)], a and b thus ought to equal $\ln[\sigma^2/m(m-1)]$ and 2, respectively.

We return to this example in Section 3.3.1, where we discuss branching processes in which individuals have variable life spans.

2.3 Several Types

One generalization of the previous setup is to allow several types of individuals. The type of an individual is defined as an attribute (or a set of attributes) that remains fixed throughout an individual's lifetime. For instance, it may be the individual's genotype, or its size at birth.

The type may affect the distribution of the numbers of offspring. Individuals of the same type are assumed to have identically distributed numbers of offspring, but individuals of different types may differ. The type of a parent can also affect the probability that its offspring has a specific type. The population in generation n is characterized by a vector $Z_n = (Z_{n1}, \dots, Z_{nd})^T$, where d is the number of different types and Z_{nj} denotes the number of individuals of type j in the nth generation. [Here, and in the following, vectors are viewed primarily as columns and the superscript T is used to denote transpose (i.e., the operation of turning a row into a column or vice versa).] The total population size in generation n, $|Z_n|$, is thus the sum of the vector components

$$|Z_n| = Z_{n1} + \cdots + Z_{nd} . \tag{2.21}$$

The types are numbered arbitrarily, so that, for instance, types 1 and 2 are not necessarily closer to each other in any real sense than are types 1 and 3. The offspring of individual i is also denoted by a vector, $\xi_i = (\xi_{i1}, \dots, \xi_{id})^T$, where ξ_{ij} is the number of its offspring of type j. In this notation the analog of Equation (2.1) for multi-type processes becomes

$$Z_{n+1} = \sum_{i=1}^{|Z_n|} \xi_i , \tag{2.22}$$

with ξ_i independently distributed vectors. This, however, conceals the difficulties, since the vectors are no longer identically distributed; the distribution of ξ_i depends on the type of the parent individual i. A complete model involves specification of the different distributions of the offspring vectors for each of the types distinguished, and if we wish to reach a formulation in terms of independent and

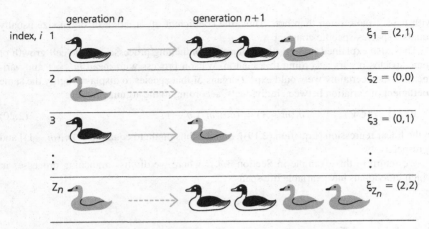

Figure 2.4 Example of a multi-type branching process.

identically distributed offspring vectors, $\xi_i^{(j)}$, where j now indicates the mother's type and i a numbering of the individuals of that type, we have to accept the ominous expression

$$Z_{n+1} = \sum_{j=1}^{d} \sum_{i=1}^{Z_{nj}} \xi_i^{(j)}. \tag{2.23}$$

These branching processes are called *multi-type* Galton–Watson processes. An illustration of how they evolve is given in Figure 2.4.

An example of type that we have already encountered in Section 2.2 is the sex of individuals. We can model a situation with non-overlapping generations by a process with two types: males, who have no offspring, and females, who have identically distributed numbers of offspring that can be of either type (as in Figure 2.2). As mentioned in Section 2.2, a single-type process, in which only females are counted, could also be used, but then we could not keep track of both males and females. Nevertheless, the model remains asexual, since mating is still not required. This may play less of a role in large populations with stable sex ratios, but in small populations the risk that there are no males cannot be disregarded. Also, mating patterns may play a role that is not mirrored in a model in which females can do completely without the interference of males. We return to sexual reproduction in Section 2.8.

Another (and maybe better) example of types is that of genotypes: the genotype of a parent may affect the numbers of its offspring and it obviously affects the genotypes of its offspring. In a model with clonal reproduction and mutation, parents of each genotype can be assumed to produce a stochastic number of offspring according to a certain probability distribution. Each of the offspring can then be assumed to be of the parent's type with a probability close to one and of a mutant type with a very small probability. Different genotypes may, for instance, differ in the probability with which they produce different mutants and/or the distribution of their numbers of offspring. Selection on fertility or survival chances can be modeled by differences in forms of offspring distributions of the different genotypes.

In diploid organisms with sexual reproduction, interactions between individuals generally have to be taken into account, since both parents influence their children's genotypes. Therefore, such systems should initially be viewed along the lines described in Section 2.8.

However, there are cases in which mating can be disregarded. One of particular interest, which we have come across already in Section 2.1, is the situation in which the initial spread of a mutant allele in a resident population is studied. Since, initially, the chance of mating between individuals that carry the mutant allele is negligible, mates of mutants are from a fixed background population. Therefore, the probabilities of combinations with the different resident genotypes are given and the chances for each of the different offspring types are determined by the genotype of the "mutant parent." This holds even if mate choice depends on type, so-called assortative mating. It remains true regardless of the numbers of alleles or loci that are considered, that is, the model can be used for the so-called multi-allele, multi-locus invasion problem (see any textbook on evolutionary genetics, e.g., Roughgarden 1979). Below, we give an illustration for the one-locus, two-allele case.

More generally, in large populations in which the simplifying assumption can be made that individuals choose mates independently from the population, possibly according to a distribution dependent upon their type, the mating aspect can be disregarded, and the type distribution of children determined by the type of the individual considered. The invasion case is then the special situation in which the partner is chosen from a large surrounding background environment.

Example 2.2 Consider a population of bacteria in which a gene has two possible alleles, A and B. We assume that bacteria with allele A produce zero offspring with probability 0.2 or they divide and thus produce two offspring with probability 0.8. The chance that, because of a mutation, exactly one of a bacterium's offspring will have allele B instead of A equals 0.0002. Bacteria with allele B have a higher chance of producing viable offspring, namely 0.9. If a bacterium with allele B divides, the chance that one of its offspring will have allele B and the other one allele A instead of B equals 0.001. We represent the offspring by a vector in which the first position gives the number of offspring with allele A and the second the number with allele B. The setup is illustrated in Figure 2.5 and summarized in Table 2.1.

As we saw in Section 2.2, much of population development is determined by the mean reproduction. In the multi-type case this will be a matrix, with entry m_{hj} giving a type h-individual's expected number of children of type j. We call this

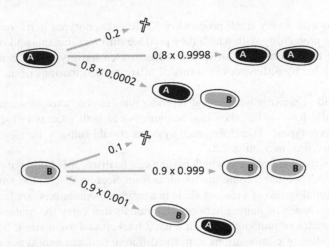

Figure 2.5 Reproduction of bacteria with two types of alleles, A or B. The allele affects mortality risk.

Table 2.1 Probability of offspring vectors for a population of bacteria in which a gene has two possible alleles, A and B.

Offspring vectors	When the parent is an A bacterium	When the parent is a B bacterium
$(0,0)^T$	0.2	0.1
$(2,0)^T$	0.8×0.9998	0
$(0,2)^T$	0	0.9×0.999
$(1,1)^T$	0.8×0.0002	0.9×0.001

the *mean matrix* and denote it by

$$M = \left(m_{hj}\right)_{h,j=1}^{d} . \tag{2.24}$$

For instance, if we call A bacteria type 1, m_{11} is the expected number of A bacteria produced by one A bacterium. From the entries in Table 2.1 it can be seen that

$$m_{11} = 2 \times (0.8 \times 0.9998) + 1 \times (0.8 \times 0.0002) , \tag{2.25}$$

since an A bacterium can either produce two offspring of its own type, or one A and one B offspring, or none.

The expected number of type j individuals in generation n satisfies

$$\mathbb{E}[Z_{nj}] = \mathbb{E}[\mathbb{E}[Z_{nj}|Z_{n-1}]] = \mathbb{E}[\sum_{h=1}^{d} Z_{n-1,h}m_{hj}] = \sum_{h=1}^{d} \mathbb{E}[Z_{n-1,h}]m_{hj} . \tag{2.26}$$

If we further introduce the notational convention that the expectation of a random vector is the vector of component expectations,

$$\mathbb{E}[Z_n] = (\mathbb{E}[Z_{n1}], \ldots, \mathbb{E}[Z_{nd}])^T, \tag{2.27}$$

we can summarize these relations for $h = 1, 2, \ldots, d$ in vector-matrix form,

$$\mathbb{E}[Z_n]^T = \mathbb{E}[Z_{n-1}]^T M, \tag{2.28}$$

according to the established conventions of linear algebra.

Iteration then leads on to

$$\mathbb{E}[Z_n]^T = \mathbb{E}[Z_0]^T M^n. \tag{2.29}$$

(In such relations, and generally, we allow ourselves sometimes not to write the superscript T for transpose.) We can conclude that we obtain subcriticality and extinction if $M^n \to 0$, and supercriticality if $M^n \to \infty$. The question that arises is whether criticality can be described in terms of one scalar parameter ρ.

Obviously, the mean matrix M is non-negative, in the sense that all its elements must be non-negative. Such matrices have interesting properties, best exposed in a book by Seneta (1981). For an extensive, gentle introduction, from a population dynamic perspective, see Caswell (2001). Here, we just mention some indispensable concepts and facts.

If there are non-null vectors $u = (u_1, \ldots, u_d)^T$ and $v = (v_1, \ldots, v_d)^T$ with the property that

$$\rho u_h = \sum_{j=1}^{d} u_j m_{jh}, \ \sum_{j=1}^{d} m_{hj} v_j = \rho v_h, \tag{2.30}$$

or, in matrix form,

$$u^T M = \rho u^T, \ Mv = \rho v, \tag{2.31}$$

for some number ρ, these are called left and right *eigenvectors* of M, and ρ is the corresponding *eigenvalue*. "Eigen" is German and means "own" or "proper." Thus, along the direction of an eigenvector the matrix only changes the scale, according to the corresponding eigenvalue.

Matrices can have many eigenvectors; in particular, multiplying a vector by a constant does not destroy its eigenvector property. We consider the situation in which it is possible to choose left eigenvectors u so that their components are non-negative and sum to one. Then the corresponding right eigenvectors v also have non-negative entries and can be chosen so that $u^T v = 1$. In explicit terms,

$$u^T v = \sum_{j=1}^{d} u_j v_j = 1, \ v_j \geq 0, u_j \geq 0, \sum_{j=1}^{d} u_j = 1. \tag{2.32}$$

We now use this normalization of the eigenvectors. If $\mathbb{E}[Z_0]$ is proportional to a left eigenvector u and c a proportionality constant,

$$\mathbb{E}[Z_n]^T = \mathbb{E}[Z_0]^T M^n = cu^T M^n = cu^T \rho^n \qquad (2.33)$$

by Relation (2.30). Thus, the eigenvalue ρ is the required indicator of criticality in the direction of the eigenvector: the process is subcritical if $\rho < 1$, critical if $\rho = 1$, and supercritical if $\rho > 1$. A correspondingly simple characterization continues to hold, provided any starting configuration can lead to any other composition over types, and so it is not astonishing that the largest eigenvalue (in absolute value) takes over and decides the asymptotics of M^n, as $n \to \infty$. This number is usually called the Perron root of the matrix, and from here on we reserve the symbol ρ for that largest eigenvalue.

In an *indecomposable* branching process, each type of individual eventually may have progeny of any other type. More formally, for every combination of types j and h there is an $r \geq 1$ such that the probability that type j has progeny of type h after r generations is positive. (In other words, for every j and h there must be a positive integer r such that the entry at position jh of M^r is positive.) Clearly, Example 2.2 concerns such a process, since both types of bacteria can have both types of offspring. For a process to be indecomposable it is not necessary that every type can produce all types of offspring immediately.

Indecomposability is thus the property that any initial configuration can lead to any other type composition. Hence, it is natural [but not easy to prove, see Mode (1971)] that in indecomposable populations all types grow at the same rate, ρ. Extinction probabilities may, however, differ depending upon the type of the initial individual: the more fertile the ancestral type, the greater the survival chances.

In *decomposable* processes, distinct groups of types that do not produce types of other groups can be distinguished. Decomposable systems can display great heterogeneity. One type may become extinct whereas another thrives. For cases in which none dies out, different types may grow at different rates. Nevertheless, uniform growth can also occur in some decomposable cases, such as when one type has the largest rate of growth and can produce progeny of all other types. An example of this is a two-type population in which females produce a large number of offspring, female or male, each with probability 1/2, whereas males – as usual – do not reproduce. Other examples occur in cell biology, such as stem cells that produce cells of various kinds (see also Section 2.5). Since the two types of processes behave so differently, we consider them separately in the following subsections.

In this section we restrict ourselves to types as fixed attributes and non-overlapping generations. However, multi-type processes are also used to model age and state dependence (Sections 2.4 and 2.5), so at least mathematically the distinction between this section and those is not strictly necessary. A careful classification of types appears in Sevastyanov's book (1971) on branching processes, unfortunately not available in English [see also the books by Mode (1971), and by

Athreya and Ney (1972)]. The classic paper on decomposable branching processes is Kesten and Stigum (1966a).

In some early approaches to bacterial growth, cell mass was assumed to be the factor crucial for division. Thus, size at birth would enter naturally as a type. Mass being continuous, this would not be a process with a given finite number of types. Such models, and even setups with both continuous time and general infinite type spaces, are mathematically intriguing and can be analyzed. However, the mathematics needed is largely beyond the scope of this book. For our purposes – and most practical modeling, we believe – more complicated type spaces can be approximated by finite ones. Still, such general state spaces occur later in this book. In particular, size-dependent cell models are sketched in Section 3.3.

2.3.1 Indecomposable processes

When considering indecomposable processes, we have to make a formal distinction between periodic and non-periodic processes. However, since any periodic process can be transformed into a non-periodic one by means of a simple transformation of the time scale (see Note below), it suffices to discuss non-periodic processes.

Note. An indecomposable Galton–Watson process is called *periodic* with period $t (\geq 2)$ if the largest common divisor of all n such that all the diagonal elements of the nth power of the mean matrix, $m_{ii}^{(n)}$, are positive is equal to t. An example of a non-periodic process was given above. A process with the following mean matrix:

$$M = \begin{pmatrix} 0 & 1 \\ 1 & 0 \end{pmatrix} ,$$ (2.34)

has period 2, since

$$M^2 = \begin{pmatrix} 1 & 0 \\ 0 & 1 \end{pmatrix} = M^4 = \cdots = M^{2n}, n = 0, 1, \ldots ,$$ (2.35)

whereas

$$M = \begin{pmatrix} 0 & 1 \\ 1 & 0 \end{pmatrix} = M^3 = \cdots = M^{2n+1}, n = 0, 1, \ldots ,$$ (2.36)

and, therefore, for all n, $m_{ii}^{(2n)} > 0$, $m_{ii}^{(2n+1)} = 0$ ($i = 1, 2$).

Any periodic process can be transformed into a non-periodic process. For example, if we consider the population in the previous example only at generation numbers 2, 4, etc. (i.e., we consider the transformed process with mean matrix $M' = M^2$), we obtain a non-periodic process. This is why only such processes are considered here.

$$\diamond \diamond \diamond$$

The celebrated Perron–Frobenius theorem (see the Appendix) shows that for an indecomposable process the powers of M have the property that

$$M^n = \rho^n A + o(\rho^n) ,$$ (2.37)

where A is a $d \times d$ matrix with elements $a_{hj} = v_h u_j$, all > 0, and $o(\rho^n)$ denotes an expression such that $o(\rho^n)/\rho^n \to 0$ as $n \to \infty$. Therefore, for large n, M^n

is approximately equal to $\rho^n A$, which justifies the definition of criticality in terms of ρ.

Various asymptotic properties of the process can be derived from the Perron–Frobenius theorem; the basic one is that

$$\mathbb{E}\,[Z_n]^T \sim \mathbb{E}\,[Z_0]^T \,\rho^n A\,, \qquad (2.38)$$

as $n \to \infty$. (We use the notation $\sim$ somewhat liberally to mean that the ratios of components of the two sides tend to one.) In the long run, thus, the average numbers of individuals of each type are proportional to ρ^n, but the relative frequencies of the different types depend on A and therefore on the eigenvectors u and v.

In particular, if the population initially consisted of only one individual of type h, the expected number of type j individuals at time n is

$$\mathbb{E}[Z_{nj}] \sim v_h u_j \rho^n\,. \qquad (2.39)$$

Hence, in the long run the expected number of individuals of type j divided by the expected total population is

$$\lim_{n \to \infty} \frac{\mathbb{E}[Z_{nj}]}{\mathbb{E}[|Z_n|]} = \frac{u_j}{u_1 + \cdots + u_d} = u_j\,, \qquad (2.40)$$

since the components of u were chosen to sum to one [see Equation (2.32)].

2.3.2   Decomposable processes

Consider a population initiated at time $n = 0$ by one individual and described by a single-type Galton–Watson branching process. To calculate the total number of individuals that existed in the population up to moment n, we define an artificial second type of individuals. When an individual of type 1 dies it produces not only a random number of individuals of type 1 (its offspring in the original process), but also exactly one individual of type 2 (i.e., a dead individual). Thus, the mean matrix becomes

$$M = \begin{pmatrix} m & 1 \\ 0 & 1 \end{pmatrix}. \qquad (2.41)$$

This is a simple example of a decomposable process, and Z_{n2} is the total number of individuals born in the population up to generation $n - 1$. A situation in which such a trick could prove useful might be to study the spread of an infective disease. Then Z_{n1} is the number of individuals in the population that are infected during period n and Z_{n2} is the accumulated number of infections up to time $n - 1$. (It is assumed that an individual only remains infective during one period, after which it dies or becomes non-infective.)

Another interesting possibility is to calculate the total number of individuals that have at least K direct descendants – the number of "hero mothers" or, in cases of epidemic spread (like HIV/AIDS), the number of notorious perpetrators. In this case we define a process in which an individual of the basic type 1 produces a random number of direct descendants ξ, and if $\xi \geq K$ it also produces exactly one

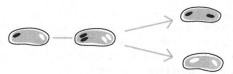

Figure 2.6 A bacterium with two different types of plasmids. Before division the plasmids are copied. The resultant plasmids are divided between the daughter cells. Each daughter receives two plasmids, but not necessarily one copy of each original.

individual of type 2. Similarly, the total number of individuals with at most K or exactly K direct descendants can be studied.

More generally, consider a mean matrix of the form

$$M = \begin{pmatrix} m_{11} & m_{12} \\ 0 & m_{22} \end{pmatrix} . \tag{2.42}$$

It is not difficult to show that

$$M^n = \begin{pmatrix} m_{11}^n & m_{12}^{(n)} \\ 0 & m_{22}^n \end{pmatrix} , \tag{2.43}$$

where

$$m_{12}^{(n)} = \begin{cases} n m_{12} m_{11}^{n-1}, & \text{if } m_{11} = m_{22}, \\[2mm] m_{12} \frac{m_{11}^n - m_{22}^n}{m_{11} - m_{22}}, & \text{if } m_{11} \neq m_{22} . \end{cases} \tag{2.44}$$

For instance, if M is as defined in (2.41)

$$M^n = \begin{pmatrix} m^n & m_{12}^{(n)} \\ 0 & 1 \end{pmatrix} , \tag{2.45}$$

where $m_{12}^{(n)}$ is the expected accumulated number of individuals produced in the initial process up to moment n,

$$m_{12}^{(n)} = \begin{cases} n, & \text{if } m = 1 , \\[2mm] \frac{m^n - 1}{m - 1}, & \text{if } m \neq 1 . \end{cases} \tag{2.46}$$

Thus, if an infected individual infects 1.5 others ($m = 1.5$) on average, and there was one infective individual to begin with $[Z_0 = (1, 0)^T]$, then the expectation of the population state after 10 infective periods is

$$\mathbb{E}\left[Z_{10}^T \right] = Z_0^T M^{10} = \left(1.5^{10}, \frac{1.5^{10} - 1}{0.5} \right) \approx (57.7, 113.3) . \tag{2.47}$$

After 10 generations, we thus expect around 58 individuals to be infected and more than 113 people to have been hit since the outbreak of the disease.

Example 2.3 Most bacteria contain plasmids, pieces of extra-chromosomal DNA that can cause resistance against antibiotics. Before cell division, plasmids are doubled, so that there are two copies of the original set. We consider the case in which both daughter cells contain the same number of plasmids at division, but not necessarily the same plasmids. (Alternatively, it may happen that the plasmids are not equally shared between daughters, so that one can have more than the other.) As a result, some offspring lines may have lost plasmid types their ancestors used to have (see Figure 2.6).

Assume the plasmids to be of two different kinds, A coding for antibiotic resistance and B for non-resistance. Let each bacterium contain two plasmids that can be of either kind. Thus, there are, in principle, three types of bacteria: AB, AA, and BB, which we call types 1, 2, and 3. A bacterium of type 1 (AB) can have daughters of all three types, but bacteria of types 2 and 3 produce only bacteria of their own sort. The four plasmids are shared equally, but at random, by the daughter cells, so that if a bacterium of type AB divides, it produces two bacteria of type AB with probability 2/3, whereas with probability 1/3 the two bacteria are of types AA and BB. Here is an argument: first, a daughter cell receives one plasmid, be it A or B, then for the second one there are two candidates of the other sort, and only one of the former. We assume that the risk for a bacterium to die without division is 0.1.

The alternative to death is division into two daughter cells, so that the expected number of type 1 bacteria produced by a type 1 bacterium equals $0.9 \times 2 \times \frac{2}{3} = 1.2$. In this manner, we find the mean matrix

$$M = \begin{pmatrix} 1.2 & 0.3 & 0.3 \\ 0 & 1.8 & 0 \\ 0 & 0 & 1.8 \end{pmatrix}. \tag{2.48}$$

The top-left 2 by 2 submatrix

$$\begin{pmatrix} 1.2 & 0.3 \\ 0 & 1.8 \end{pmatrix}$$

determines the corresponding part of the matrix M^n (as can be checked straightforwardly by multiplying M by itself). It is a matrix of the form in (2.42), and thus the corresponding submatrix of M^n is given by (2.44), to yield

$$\begin{pmatrix} 1.2^n & \frac{1}{2}(1.8^n - 1.2^n) \\ 0 & 1.8^n \end{pmatrix}.$$

By the same method the remaining elements can be found. (The elements m_{11}^n, m_{31}^n, m_{13}^n, and m_{33}^n are also determined solely by the corresponding submatrix of M.) Hence,

$$M^n = \begin{pmatrix} 1.2^n & \frac{1}{2}(1.8^n - 1.2^n) & \frac{1}{2}(1.8^n - 1.2^n) \\ 0 & 1.8^n & 0 \\ 0 & 0 & 1.8^n \end{pmatrix}. \tag{2.49}$$

So, from a start with a single bacterium that contains both plasmid types, the expected numbers of different types of bacteria after n generations is

$$\mathbb{E}\left[Z_n^T\right] = (1, 0, 0)M^n = [1.2^n, (1.8^n - 1.2^n)/2, (1.8^n - 1.2^n)/2]. \tag{2.50}$$

Thus, the proportion of the population that has antibiotic resistance after n generations is

$$\frac{(1.8^n + 1.2^n)/2}{1.8^n} = \frac{1}{2}\left[1 + \left(\frac{1.2}{1.8}\right)^n\right], \tag{2.51}$$

and after many generations we can expect about half of the population to remain resistant.

$$\diamond \diamond \diamond$$

Example 2.4 Taneyhill *et al.* (1999) give several examples of the application of branching processes in parasitology. One of these concerns the relationship between microsporidian parasites, *Octosporea effeminans*, and several species of the genus *Nosema*, and their host, *Gammarus duebeni*, an amphipod that lives in brackish water along the coasts of Europe and Northern America. The parasites are transmitted vertically, from mother to offspring, and their effect is to turn genetic males into phenotypic females. To be transmitted to a following generation of hosts, the parasite must find its way to the host's so-called germline tissue (which forms the gametes). Dunn *et al.* (1998) used a multi-type process to examine which mode of transmission the parasite most likely uses.

In the model, host cell-type is determined by the number of parasites a cell contains. In each time unit, the host cells divide into two. Beginning with a host zygote that contains h parasites, the parasites are sorted into daughter cells during cleavage of the embryo, and they may replicate within the host cells by binary fission. In each generation of host cells, parasites then split independently into two daughters with a probability p. Furthermore, there is a probability q of being transmitted to a given daughter of the host cell. For instance, if q equals 1/2, both daughter cells are equally likely to receive a given parasite. If $q \neq 0.5$, one cell line has a higher probability of receiving parasites than the others. Thus, a host cell of type a produces two daughter cells, which may have types 0 to $2a$. The probability that one of its daughter cells has type j is

$$p_j = \sum_{k=a}^{2a} \binom{a}{k-a} p^{k-a} (1-p)^{2a-k} \binom{k}{j} q^j (1-q)^{k-j} . \qquad (2.52)$$

Starting with a zygote that contains h parasites, an embryo of size 2^n (i.e., after n divisions of the host cells) can contain up to $h2^n$ parasites. Thus, conditional on a positive number of parasites, there are $h2^n$ different possible cell types.

Dunn *et al.* (1998) applied this model to study the distribution of parasites in embryos of *Gammarus duebeni* that consisted of $64 = 2^6$ cells, so $n = 6$. They calculated the reproduction mean matrix, M, for different values of q that corresponded to an unbiased transmission of the parasite to daughter cells ($q = 0.5$), and to a biased transmission ($q \neq 0.5$). It was estimated that zygotes contain approximately 200 parasites. Assuming this to be the initial zygote type, they calculated expected distributions of cell types conditional on containing a positive number of parasites after n generations and compared them with observed distributions. The observed distributions fitted best to a model of biased transmission. They concluded that there is a cell line that receives a high proportion of the parasites.

2.4 Generation Overlap

So far we have either counted successive generations or, in a real-time interpretation, assumed that individuals live through one season only, so that different generations do not overlap. We now allow overlap. Initially, we stay with single-type Galton–Watson processes, since they can be used to model special cases of generation overlap, namely those without aging.

When individuals can live for longer than one generation, but age does not affect offspring numbers or survival chances, adults and newborns are equivalent with respect to future reproduction. Therefore, we do not have to make a distinction between parents and their offspring in the next generation. Instead, a surviving

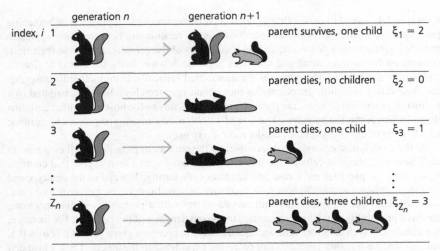

Figure 2.7 Illustration of how a process with overlapping generations can be represented by a Galton–Watson process.

parent is included in the offspring counted by the number ξ_i. Thus, we simply define the "offspring" of a reproducing individual as its children plus the individual itself, if it survives (see Figure 2.7). To a biologist, this may seem awkward, but it is a convenient mathematical device that renders the above approaches applicable to situations with generation overlap. Note, however, that it involves the assumption that each individual has a constant risk of dying, irrespective of age. As a consequence, life spans must be distributed geometrically. For most organisms this is obviously not realistic, but it may be a good approximation if life spans are determined mainly by external random factors, such as large predation or starvation risks, as may be the case for many small bird or mammal species. By using such ideas, we can calculate properties such as the expected population size and extinction probability of a population. However, to capture the population age structure, more elaborate modeling is needed.

Example 2.5 Small mammals, such as squirrels, have one reproduction period per year and a large yearly mortality risk, because of starvation or predation. We therefore assume that each year a female squirrel has a constant probability r of dying, a probability q of surviving without reproduction, and a probability p of producing one offspring. (We count only female individuals.) If these probabilities are assumed to be independent of age, the population can be modeled as a single-type Galton–Watson process, in which one individual produces one offspring with chance q, two offspring with chance p, and no offspring with chance r. The expected number of offspring per individual thus equals $2p + q$, and so the expected population size in the nth season is $(2p + q)^n$. If $2p + q$ exceeds one, the expected population size increases exponentially. Otherwise, the population dies out in the long run (see Chapter 5).

The *parity* of an individual is the total number of offspring she has produced. The (asymptotic) distribution of the parity in a population can be calculated by means of a multi-type branching process; type corresponds to parity in this case. We count to a maximum parity only if a female has had N or more offspring, she has type N. Thus, we distinguish $N + 1$ types of females. If a female who has produced $k (< N)$ offspring does not reproduce, but survives, she "produces" one individual of her own type. This happens with chance q. If she reproduces, she has one daughter of type 0 and produces one individual of type $k + 1$. This happens with chance p. Females of type N produce one female of their own type and one female of type 0 when they reproduce.

The mean matrix of the process has size $(N + 1) \times (N + 1)$ and looks as follows:

$$M = \begin{pmatrix} p+q & p & 0 & 0 & \cdots & \cdots & 0 \\ p & q & p & 0 & \cdots & \cdots & 0 \\ p & 0 & q & p & \cdots & 0 & \cdots \\ \cdots & \cdots & \cdots & \cdots & \cdots & \cdots & \cdots \\ \cdots & 0 & 0 & \cdots & \cdots & p & 0 \\ \cdots & \cdots & \cdots & \cdots & \cdots & q & p \\ p & 0 & 0 & \cdots & \cdots & 0 & p+q \end{pmatrix} . \tag{2.53}$$

Clearly, the process is indecomposable since every female can have daughters of type 0, which in turn can reach any parity. The maximal eigenvalue of the matrix, ρ, is $2p + q$ and the corresponding (normalized) transposed right eigenvector has the form

$$v^T = (1, 1, \ldots, 1) , \tag{2.54}$$

while the left eigenvector (properly normalized and transposed) equals

$$u^T = (u_0, \ldots, u_k, \ldots, u_{N-1}, u_N) = \left(\frac{1}{2}, \ldots, \frac{1}{2^{k+1}}, \ldots, \frac{1}{2^N}, \frac{1}{2^N} \right) . \tag{2.55}$$

By the Perron–Frobenius theorem (see the Appendix),

$$m_{hj}^{(n)} \sim v_h u_j \, (2p + q)^n , \tag{2.56}$$

for large n. If the population is initiated by a single female with no previous offspring,

$$\mathbb{E}\left[Z_n^T\right] = (1, 0, \ldots, 0) \qquad M^n = \left(m_{00}^{(n)}, \ldots, m_{0N}^{(n)} \right) . \tag{2.57}$$

So, after a large number, n, of generations, the expected number of females with parity $h \leq N - 1$ is approximately $(2p + q)^n / 2^{h+1}$. Furthermore, the expected number of females with at least N daughters is $(2p + q)^n / 2^N$.

The "parity distribution" of the population tends to

$$\lim_{n \to \infty} \frac{m_{0k}^{(n)}}{m_{00}^{(n)} + m_{01}^{(n)} + \cdots + m_{0N}^{(n)}} = \frac{u_k}{u_0 + \cdots + u_N} = \frac{1}{2^{k+1}} ,$$
$$\text{for } k = 0, \ldots, N - 1 , \tag{2.58}$$

and

$$\lim_{n \to \infty} \frac{m_{0N}^{(n)}}{m_{00}^{(n)} + m_{01}^{(n)} + \cdots + m_{0N}^{(n)}} = \frac{1}{2^N} . \tag{2.59}$$

Note that this result does not depend on the particular values of p and q. However, it has a simple interpretation only if $2p + q > 1$, since then the process is supercritical and the expected population size increases. Otherwise, the population dies out and we may have to consider concepts of quasi-stationarity (see Chapter 6).

$$\diamond \diamond \diamond$$

Table 2.2 Probabilities of offspring vectors for a strictly biennial plant species.

Transposed offspring vectors	For a 1-year-old parent	For a 2-year-old parent
$(0, 0)$	$1 - p_1$	e^{-m}
$(0, 1)$	p_1	0
$(1, 0)$	0	$me^{-m}/1!$
$(2, 0)$	0	$m^2 e^{-m}/2!$
$\vdots$	$\vdots$	$\vdots$

If reproduction or survival are age-dependent, multi-type processes must be used throughout. If no other factors are involved, type corresponds to age.

Example 2.6 As a further illustration, consider a biennial plant species. If it were strictly biennial, 1-year-old individuals produce no seeds and have a positive chance p_1 of survival, but 2 year olds would always produce seeds and die. Since such plant species usually have enormous amounts of seeds, only a few of which germinate, a Poisson offspring distribution seems natural. Thus, assume that 2 year olds have a Poisson(m) distributed surviving number of offspring that are 1 year old in the next generation. In this case, if individual i is 1 year old, ξ_i is of the form $(0, I_i)^T$, where I_i is 1 year with probability p_1 and 0 otherwise. If individual i is 2 years old, ξ_i is of the form $(\xi, 0)^T$, where ξ is a Poisson(m) distributed random variable. This is represented in Table 2.2.

In most instances, biennial plant species do not follow this strict rule, but instead there is only a chance that they flower, and then die, when they are 2 years old. If they do not flower, they can survive for another year (up to a certain maximum age), but as soon as they have produced flowers they die. We can model this as follows. Suppose that the maximum age of the plants is 3 years. As before, 1-year-old individuals survive with chance p_1. We assume that 2-year-old individuals flower with chance q_2 and, if they do not, they survive with probability p_2. If they do produce flowers the number of offspring is Poisson(m_2) distributed. Individuals who are 3 years old always produce flowers with the number of offspring Poisson(m_3) distributed. This is illustrated in Figure 2.8. This results in Table 2.3.

The mean reproduction matrix is

$$M = \begin{pmatrix} 0 & p_1 & 0 \\ q_2 m_2 & 0 & (1 - q_2) p_2 \\ m_3 & 0 & 0 \end{pmatrix}. \tag{2.60}$$

We consider the specific values $p_1 = 3/4$, $m_2 = 36/5$, $q_2 = 5/9$, and $p_2 = 3/4$, so that $q_2 m_2 = 4$ and $(1 - q_2) p_2 = 4/9 \times 3/4 = 1/3$, and $m_3 = 8$. With these

$$M = \begin{pmatrix} 0 & 3/4 & 0 \\ 4 & 0 & 1/3 \\ 8 & 0 & 0 \end{pmatrix}. \tag{2.61}$$

The maximal absolute value of the eigenvalues is $\rho = 2$, the process is supercritical, and the expected population grows exponentially. The right and left (transposed) eigenvectors are

$$v^T = \left(\frac{23}{36}, \frac{46}{27}, \frac{23}{9} \right); \quad u^T = \left(\frac{16}{23}, \frac{6}{23}, \frac{1}{23} \right), \tag{2.62}$$

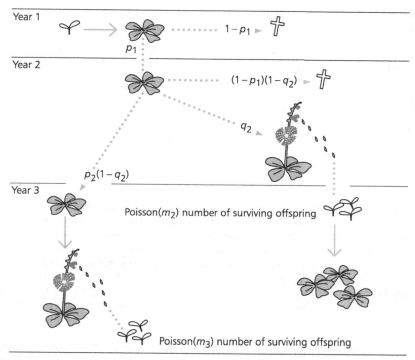

Figure 2.8 Population dynamics of a biennial plant with possible delay in reproduction.

the components normed according to Convention (2.32). Therefore, as $n \to \infty$,

$$\mathbb{E}[Z_n^T] \approx \rho^n \mathbb{E}[Z_0^T] \begin{pmatrix} v_1 u_1 & v_1 u_2 & v_1 u_3 \\ v_2 u_1 & v_2 u_2 & v_2 u_3 \\ v_3 u_1 & v_3 u_2 & v_3 u_3 \end{pmatrix} = 2^n \mathbb{E}[Z_0^T] \begin{pmatrix} \frac{4}{9} & \frac{1}{6} & \frac{1}{36} \\ \frac{32}{27} & \frac{4}{9} & \frac{2}{27} \\ \frac{16}{9} & \frac{2}{3} & \frac{1}{9} \end{pmatrix} . \qquad (2.63)$$

So, if 10 1-year-old plants initiate the population, after n seasons the population contains on average some $\frac{40}{9} \times 2^n$ 1 year olds, $\frac{10}{6} \times 2^n$ 2 year olds, and $\frac{10}{36} \times 2^n$ 3 year olds. In the long run the proportions of these different age classes in the population stabilize at, respectively, $\frac{16}{23}$, $\frac{6}{23}$, and $\frac{1}{23}$.

◇ ◇ ◇

Possibly not only the age, but also the type of an individual may affect its survival and numbers and types of offspring. This can be modeled by a multi-type branching process in which the population state Z_n is characterized by a matrix, rather than a vector. For instance, the rows may represent the different ages and the columns the types. Whereas the age of newborns is always 1, the distribution of types of children may depend on the parental type. Furthermore, since an individual's type is fixed, survival of an individual of age l years involves that it produces one offspring of the same type, with age $l + 1$.

Table 2.3 Probabilities of offspring vectors for a biennial plant species with delay in reproduction.

Transposed offspring vectors	For a 1-year old parent	For a 2-year old parent	For a 3-year old parent
$(0, 0, 0)$	$1 - p_1$	$(1 - q_2)(1 - p_2) + q_2 e^{-m_2}$	e^{-m_3}
$(0, 1, 0)$	p_1	0	0
$(0, 0, 1)$	0	$(1 - q_2)p_2$	0
$(1, 0, 0)$	0	$q_2 m_2 e^{-m_2}$	$m_3 e^{-m_3}$
$(2, 0, 0)$	0	$q_2 m_2^2 e^{-m_2}/2!$	$m_3^2 e^{-m_3}/2!$
$\vdots$	$\vdots$	$\vdots$	$\vdots$

Example 2.7 As mentioned in Section 2.1, two-sex populations can be modeled by disregarding the mating aspect and even the sex of individuals. Either only females are counted (as in the squirrel example) or the sex can be disregarded. When generations do not overlap, the chance that an individual of unknown sex has no offspring, p_0, is the chance that the individual is male plus the chance that it is female and has no offspring. The chance that it produces $k > 0$ offspring, p_k, is the chance that it is female and has k children. Denote the probability that a female has k children by a_k, and suppose that the probability that a child is male is 1/2. Then we obtain a Galton–Watson process with the offspring distribution

$$p_0 = 1/2 + a_0/2, \quad p_k = a_k/2, k = 1, 2, \ldots . \tag{2.64}$$

With overlapping generations, the same reasoning could be applied. If, however, individuals can survive more than one reproduction period, it is more natural to resort to a two-type model. Then, males can have either one or no "children," the former corresponding to survival and the latter to death of the individual concerned. Females can have different numbers of children, depending on whether or not they survive and whether or not they reproduce. If we denote survival chance by s, and other chances are as before, the risk that a female has no offspring is $1 - s$. The chance that she has one female offspring, of age 1, *viz.* herself, is sa_0, and the chance of having one daughter (or one son) aged 0 plus one female "offspring" of her own age plus 1 is $sa_1/2$. The chance that she has exactly two female offspring of age 0 plus one female "offspring" of her own age plus 1 is $sa_2(1/2)^2$ (and equals the chance that she has two male offspring of age 0 plus one female "offspring" of her own age plus 1). Generally, the chance of a female surviving and having n children of which k are female and $n - k$ male is

$$sa_n \binom{n}{k} 2^{-n}, k = 0, 1, \ldots, n . \tag{2.65}$$

The risk that a male has no offspring (meaning that he dies) is $1 - s$ and the chance that he has one male "offspring" of his own age plus 1 is s.

$$\diamond \diamond \diamond$$

2.5 State Dependence

In Section 2.3 we define the type of an individual as an attribute that remains fixed throughout its life. Here, more general (sets of) attributes are considered that may change in the course of time. Such a set of attributes is referred to as the state of an individual. More formally, we define an individual's *state* as a set of properties such that if the current values of these are known, no further information about the individual's past is needed to determine the probabilities of its future reproductive events. Such descriptions are called *Markovian*, and in much literature the use of the word "state" implies Markovianness. Components of (or partial information about) a state are called *state variables*. (The reason for this is that we usually think of states as vectors of variables.) For instance, an individual's type may be one of the state variables. An individual's state can include part of its history (e.g., past reproduction). In demography, the *parity* of mothers (i.e., how many children they have had) is often used in this way. However, especially with long-lived species, such models can become quite complicated and difficult to analyze. One should strive for a state description that is as economic as possible (e.g., the complete life history of an individual is always a state, but not a very parsimonious one).

The state of an individual may affect the state and number of its offspring, its survival chance, and its own future states. Age is a special kind of state variable, since the ages of newborns and of surviving individuals are determined by the laws of time. Thus, age is an example of a deterministically changing state. At the other extreme, the future state can be completely independent of the current state. For instance, individuals of very mobile species may travel large distances between periods, so their locations in successive periods may be virtually independent. (Of course, this also depends on the spatial scale that is considered.) There is also a range of intermediate possibilities, in which the current state affects the probability distribution of future states, of both the individual and its progeny.

When generations do not overlap, there is only one census point in each individual's lifetime, namely just before reproduction. Thus, even if the state of an individual at reproduction is different from its state at birth, we need not model this explicitly. Instead, we can define a relation between the state of a parent just before reproduction and the distribution of states of its offspring just before they reproduce. With this formulation, the models of Section 2.3 can be applied directly to state-dependent processes with non-overlapping generations, and states can be viewed as types.

If age, or any other state variable, is thought of as continuous, new modeling problems appear. This is not so if there are only a finite number of states, in which case a full specification involves the distributions of offspring numbers and states, the state-dependent survival chances, and the transition probabilities to other states, for each possible state. For mathematical simplicity it may then be useful to think of a surviving individual as her own offspring, as in Section 2.4.

We illustrate this by an example with two state variables, one of which is age (Example 2.8). It is based on a continuous-time model of brain-cell differentiation, formulated by Yakovlev *et al.* (1998). We have made several simplifications for the sake of the illustration.

Example 2.8 During the ontogeny (i.e., embryonic development) of vertebrates, specific cell types first appear at precisely regulated moments. The timing of such processes is strikingly similar in different individuals. One of the major challenges in developmental biology is to study the principles that underlie such appropriately timed cell generation. During the development of embryos, cells change from dividing "precursor cells" into differentiated (and often non-dividing) progeny cells. Yakovlev *et al.* (1998) formulated a model for this process, based on in vitro observations of precursor cells of the central nervous system. These so-called oligodendrocyte type-2 astrocyte (O-2A) progenitor cells divide and generate oligodendrocytes.

As mentioned, we assume that the state of a cell consists of two variables. One is its age; cells may belong to one out of four age classes, where classes 1, 2, and 3 mean ages 1, 2, and 3 (time units), respectively, whereas class 4 means age 4 or older. The other state variable either has the value (0) for a progenitor cell or (1) for a differentiated cell (i.e., an oligodendrocyte). The latter do not change, but a progenitor can become differentiated.

Each progenitor cell first completes a mitotic cycle of random length. It is assumed that the probability that a cell of age class 1 to 3 completes its mitotic cycle at the next time step is q, whereas this chance is r for cells of age class 4. After the end of the mitotic cycle the cell either divides into two daughter progenitor cells, with probability p, or it differentiates into an oligodendrocyte, with probability $1 - p$. Cell death is not included in the model, since in this case it is very rare during the considered period of observation.

Since state is two-dimensional, the device of modeling state variable changes by letting an individual be her own "offspring" yields a reproduction vector with two parts; the first four elements represent the numbers of produced progenitor cells of age classes 1 to 4, and the fifth element the number of oligodendrocytic offspring.

For instance, a progenitor cell of age class 1 has the offspring vector $(0, 1, 0, 0; 0)$ with chance q, which signifies its own survival into the next period without any proper offspring. It has the vectors $(0, 0, 0, 0; 1)$ with chance $(1 - q)(1 - p)$ and $(2, 0, 0, 0; 0)$ with chance $(1 - q)p$. Progenitor cells of age class 4 have the reproduction vector $(0, 0, 0, 1; 0)$ with probability r. The chances for the other two reproduction vectors are as above, with r replacing q.

Note that a differentiated cell can only produce one "offspring" of its own kind, so that the process is decomposable, but of a type for which the asymptotics are easily describable, since there is really only one reproducing type.

Using a more complicated version of this model, Yakovlev *et al.* (1998) examined whether cell differentiation depended solely on cell age (the intrinsic biological-clock model) or was also affected by an external hormonal signal. By comparison of model predictions to data they showed that thyroid hormone reduced the mean duration of the mitotic cycle of progenitor cells and increased the probability of cell differentiation, which provided evidence in favor of the latter hypothesis.

2.6 Dependence on the Population Itself

Up to this point, we have assumed that individuals reproduce independently from one another. In real life, however, individuals usually affect each other's offspring numbers as well as types or states in various ways. Whereas classic branching processes with independent individuals have a long history, only recently has it become possible to treat various types of interaction mathematically.

We start with some generalizations of the simple branching processes of Section 2.1 and either assume that generations are non-overlapping or that there is no aging. First, consider situations in which offspring numbers depend upon population size. Of course, this does not necessarily mean that population size must influence the reproduction mechanism physically. Rather, the situation might be that population size mirrors some aspect of the environment that in its turn affects offspring numbers. Thus, a large population often means that resources are being exhausted. However, there are also more direct effects of population size, as in the Ricker model in which adults eat juveniles (see Sections 1.4 and 6.9).

Galton–Watson processes dependent upon population size were, essentially, explored by Klebaner in the 1980s (see Klebaner 1989c). In connection with continuous-time Markov branching (see Chapter 3) similar processes had been studied by Boiko (1982). The framework is Galton–Watson processes as defined by requirements 1, 2, and 3 (see Section 2.1) being valid only conditionally upon population size in each generation. Thus, given Z_n, reproduction in the nth generation remains independent and identically distributed. However, different values of Z_n imply different offspring number distributions. This is a very general scheme, and results are also general, relating the growth rate to the growth rate of a sort of limiting classic branching process (i.e., which is independent of population size).

The idea behind the scheme is that unhampered populations, even with dependences, either die out or become large, indeed grow beyond any limits, as explored in Section 5.2. In the latter circumstances, the effects on individual reproduction converge to what would have been the case in an extremely large (potentially infinite) surrounding population. Often the result is to establish exponential growth, but in a *near critical case* (i.e., a case in which reproduction approaches one child per individual), as Z_n grows to infinity, new phenomena can appear, such as linear rather than geometric growth.

Usually, population size dependence is more or less tacitly assumed to be negative in the sense that the mean reproduction decreases as the population increases. Even though this is biologically natural, it is not mathematically necessary.

Example 2.9 In the Ricker model of Section 1.4, assume that each adult initially produces a Poisson(m) distributed number ξ of offspring. If, furthermore, each of these young independently runs a predation risk of $1 - e^{-bZ_n}$, given the adult population size Z_n, by the Poisson thinning property (derived in the Appendix), the numbers of young that survive are again conditionally Poisson, but with the parameter me^{-bZ_n}. The expected number of surviving children per individual equals the expected offspring number times the survival chance, and the offspring numbers of different individuals are independent, given the generation size, since the original childbearing was independent, as was predation.

The expected size of the population remains bounded,

$$\mathbb{E}[Z_n] = \mathbb{E}[\mathbb{E}[Z_n|Z_{n-1}]]$$
$$= \mathbb{E}[Z_{n-1}me^{-bZ_{n-1}}] \leq \sup_{z \geq 0} zme^{-bz} = me^{-1}/b . \tag{2.66}$$

As is shown later, it follows from the general arguments in Section 5.2 that such a population must die out. This is natural, since the mean reproduction $m(z) = me^{-bz}$ would become very little, should the population size z happen to become large. Ultimate exponential growth would occur if the reproduction distribution were, say, Poisson with some other parameter $m(z)$, where m is a decreasing function of z with $\lim_{z\to\infty} m(z) > 1$. Linear growth would appear in the case where $z(m(z) - 1)$ approaches a positive limit, as $z \to \infty$. We return to these matters in Section 6.5.

$$\diamond \diamond \diamond$$

Another situation in which population size may affect reproduction is when there is competition for space. As an illustration, consider a plant population that occupies an area with s suitable sites. In summer, each plant produces a certain number of seeds that disperse within the area. The parent population dies in winter and the next spring seeds germinate. However, since each site can only harbor one adult plant, seeds that have landed on the same site compete and only one can survive. Therefore, the chance that seedlings survive depends on the total number of seeds produced by the previous generation.

For such a process, the notation used until now is no longer practical. Instead, we denote the number of seeds produced by the ith individual by γ_i, so that the total number of seeds produced in the nth generation is

$$X_n = \sum_{i=1}^{Z_n} \gamma_i \,, \tag{2.67}$$

where γ_i are independent and identically distributed random variables. The population size Z_{n+1} of the next generation then equals the number of sites that have been occupied.

If seeds are uniformly and independently distributed over the sites (think of the feathered seeds of dandelions), determination of the number of occupied sites is a classic probabilistic problem called the *occupancy problem* [see Feller (1957), Chistyakov *et al.* (1978), or, for a textbook exposition, Durrett (1995)]. Indeed, as shown in these references, the exact probability that exactly k of the sites remain empty if $X_n = x$ is

$$p_k(x, s) = \binom{s}{k} \sum_{j=0}^{s-k} (-1)^j \binom{s-k}{j} \left(1 - \frac{k+j}{s}\right)^x . \tag{2.68}$$

This follows from the so-called inclusion–exclusion formula of elementary probability (see the Appendix), in conjunction with that the probability of j given sites not being occupied must be

$$\frac{(s-j)^x}{s^x} = (1 - j/s)^x \,, \tag{2.69}$$

for which see any of the references above.

If there are fewer seeds than sites, the number of occupied sites is close to the seed number. If, on the contrary, there are far more seeds, few sites remain empty, and $Z_{n+1} \approx s$. The intermediate case is less obvious.

We examine these three cases in terms of expected population size z in a season, given that there are x seeds. This can be computed in the following manner. Number the sites somehow from 1 to s and let $\eta_i = 1$ if site i is empty and zero otherwise. Then, there are

$$s - z = \eta_1 + \cdots + \eta_s \tag{2.70}$$

empty sites, but all η_i have the same probability of equaling one,

$$\mathbb{P}(\eta_i = 1) = \left(1 - \frac{1}{s}\right)^x \text{ and } \mathbb{E}[\eta_i] = \mathbb{P}(\eta_i = 1). \tag{2.71}$$

(For simplicity we do not spell out the conditioning on the number of seeds.) Hence

$$\mathbb{E}[s - z] = \mathbb{E}[\eta_1] + \cdots + \mathbb{E}[\eta_s] = s \left(1 - \frac{1}{s}\right)^x. \tag{2.72}$$

However, if s is large (and x much smaller than s^2),

$$s \left(1 - \frac{1}{s}\right)^x \approx s e^{-\frac{x}{s}}, \tag{2.73}$$

and

$$\mathbb{E}[z] \approx s(1 - e^{-\frac{x}{s}}). \tag{2.74}$$

If seeds are much fewer than sites, $x/s \approx 0$,

$$s(1 - e^{-\frac{x}{s}}) \approx x, \tag{2.75}$$

and, as we concluded above, the (expected) population given the number of seeds has a size similar to the latter. If there are many more seeds than sites, $x \gg s$, $se^{-x/s} \approx 0$, which yields $\mathbb{E}[z] \approx s$. However, if s and x are both large, and $se^{-x/s} \approx c$, rather straightforward calculations (see the cited textbooks) show that the number of empty sites have an approximate Poisson(c) distribution. Thus, if in one season there are x seeds, the population has size z, where $s - z$ is (approximately) Poisson distributed. In particular, the expected population size is $s(1 - e^{-x/s})$.

When the number of young that can be produced depends on a limited population resource that cannot be mirrored through population size, we may have a more complicated situation. For example, past population sizes may affect current resources. However, sometimes there is dependence upon the *accumulated population size*,

$$Y_n = Z_0 + Z_1 + \cdots + Z_n, \tag{2.76}$$

somewhat inadvertently referred to as the total population size in branching process literature. This case is actually treated more easily than dependence upon the present population size, since the latter can increase and decrease, whereas accumulated populations do not decrease. Furthermore, it is natural to assume that the mean number of children m is a non-increasing function of either the present or accumulated population. Hence, it follows directly that $m(Y_n)$ must tend to a limit, as $n \to \infty$, and thus mean reproduction stabilizes with time. But since Z_n need not be non-decreasing, convergence cannot be concluded for $m(Z_n)$. Indeed, the latter can fluctuate wildly as $Z_n \to \infty$, even if the mean is a monotonous function of actual population size. This matter is pursued further in Section 6.5.2.

Such observations make it tempting to replace (accumulated) population size as an influencing factor by some general resource variable, χ_n, during the nth season. This would be a random entity influenced by all preceding resources and population sizes, $\chi_0, \ldots, \chi_{n-1}, Z_0, \ldots, Z_n$, but not necessarily fully determined by them. On one extreme, χ_n could be a function of Z_n, as above, and on the other there could be independent, identically distributed environments – in which χ_n is completely unaffected by the population or its history. Such models are discussed again in Section 2.9. Unfortunately, frameworks that encompass the whole range of resource models, including as different situations as environment and population dependence, seem so general that little substantial can be deduced about them. However, specific examples are certainly of interest, provided they are both realistic and amenable to analysis; an example is the near-critical model discussed by Jagers and Klebaner (2004).

2.7   Interaction Between Individuals

Population size dependence is one form of *global* dependence: it is the population as a whole that has repercussions on individual behavior. Another type of dependence is *local* interaction between a limited number of siblings or close relatives, or between neighbors in a geographically structured population. A third form of interaction is that between different populations or subpopulations, as in competition or sexual reproduction, where males and females interact.

Interaction between geographic neighbors has been a favorite area of probability modeling during past decades. It is an intriguing subject for mathematicians, because conceptually simple models result in rich mathematics [see the books by Durrett (1988), Grimmett (1999), and Liggett (1999)]. From a biological viewpoint, the models can, however, only be viewed as idealized benchmark processes.

We focus on the mathematically much simpler problem of sibling dependence. For biological literature on populations situated on lattices, see Dieckmann *et al.* (2000), in particular the chapters by Iwasa (2000), and Sato and Iwasa (2000).

Galton–Watson processes with sibling dependence (or even more general interactions within the same generation) can be defined exactly as in Section 2.1, except that the ξ variables in the definition of the next generation are no longer independent. However, this does not enter into the expectation calculations of Section 2.2 and it follows that expected growth is not affected by dependence between siblings. Much more generally, dependence between the offspring numbers in the same generation does not change the expected population size. When we return to the dynamic behavior of branching populations, we shall see that this conclusion remains valid for growth rates and composition generally. In contrast, extinction risks can be influenced strongly (see Olofsson 1997).

Actually, there is a very simple way to remove sibling dependence: just consider the whole sibship as a *macro-individual* that collectively gives birth to the children

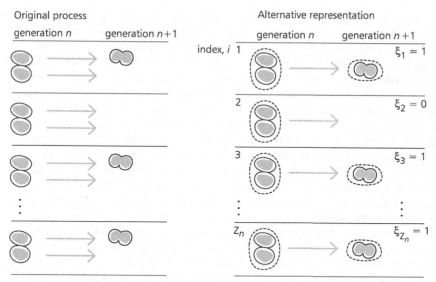

Original process

generation n generation $n+1$

Alternative representation

generation n generation $n+1$

index, i 1 $\xi_1 = 1$

2      $\xi_2 = 0$

3      $\xi_3 = 1$

Z_n $\xi_{Z_n} = 1$

Figure 2.9 Alternative representations of a process with sibling dependence.

of its members. The process of these macro-individuals has independently reproducing elements, as different sibships are independent of each other. The numbers of "macro children" need, however, not be distributed identically (e.g., we might expect a large sibship to give birth to more children than a small one). Thus the macro-process is a multi-type branching process, and in the Galton–Watson case simply has sibship size as its type.

Example 2.10 Imagine a cell population in which if one cell divides, her sister cell cannot, but dies without trace. Then each individual (= cell) can produce zero or two children, but not independently. The two possibilities are that one sister cell divides, say with probability p, or that none does. The population of sibling pairs is again a single-type process, which produces one child (= pair) with probability p, and none with probability $1 - p$. If there are no dependences beyond that between sisters, the process of sister pairs is a classic independently reproducing branching process. This is illustrated in Figure 2.9.

2.8   Sexual Reproduction

G. Alsmeyer

Up to now we have ignored that in species with sexual reproduction changes in population size depend on the formation of couples. Sometimes this may be justified. For instance, if mating is polygynous (one male may have many females), and the number of males is sufficiently large, the total number of females in a generation determines future population size. Another example is the number of initially rare mutants in an evolutionary process, for which, provided the residence population is large enough, pair formation is not a limiting factor. In many instances, however, mating is an important factor that cannot be dismissed. Bisexual Galton–Watson processes take mating into account explicitly. Unfortunately, this complicates life considerably, and few mathematical results are available.

So-called *bisexual* branching processes – a somewhat unfortunate name for processes with mating – were introduced by Daley (1968). In them it is convenient to represent population size by the *number of couples* rather than the number of individuals. Denote the number of couples in the nth generation by Z_n. The process starts with an initial number of couples, Z_0, just before the first reproduction period. Each couple has random numbers of female and male offspring. Thus, the offspring of a couple is represented by a pair of numbers, which are independently distributed for different couples, with an identical distribution that is constant over generations. In the simplest case it is assumed that, conditional on the total number of offspring, the offspring sex is determined at random (like flipping a, possibly biased, coin). Another model of sex determination assumes that the numbers of male and female offspring are independent with a possibly different distribution. The two mechanisms are equivalent for Poisson-distributed numbers of male and female offspring, since the sum of two independent Poisson variables is Poisson again. As in the Galton–Watson process, generations are non-overlapping.

The most realistic scenario is to form pairs according to a stochastic process. In existing theory, however, pair formation is assumed to be deterministic. Thus, the formation of couples is determined by a *mating function* $\zeta(x, y)$, which specifies the total number of couples formed if there are x females and y males. If mating is strictly monogamous and if we assume that the population is well-mixed, so that all possible pairs are formed, $\zeta(x, y)$ equals the minimum of x and y. In a polygynous mating system, in which each male may inseminate up to d females, $\zeta(x, y)$ is equal to the minimum of x and dy. Another, less common example, is unilateral promiscuous mating, where $\zeta(x, y)$ equals x times $\min(1, y)$. All these examples belong to the class of so-called *common sense mating functions*, which satisfy three natural conditions, i.e., a void generation cannot produce offspring, the number of couples increases with the numbers of males and females, and if the numbers of males and females tend to infinity so does the number of couples. Furthermore, these functions are *super-additive* (Hull 1982), which means that for all $x_1, x_2, y_1, y_2 = 0, 1, 2, \ldots$

$$\zeta(x_1 + x_2, y_1 + y_2) \geq \zeta(x_1, y_1) + \zeta(x_2, y_2) .   \tag{2.77}$$

Super-additivity of the mating function implies that, if the population is divided into two subgroups that each form couples, the total number of couples is at most equal to that of the undivided population.

Common sense mating functions do not always make biological sense. As an example, consider the function $\zeta(x, y) = xy$, which could be called bilateral promiscuous mating. This function is unrealistic whenever there are more than just a few males: it implies that the total expected number of offspring per female increases linearly with the number of males in the population, since each female mates with every male and the numbers of offspring produced per couple (i.e., mating) are independent.

We number the couples in the nth generation in an arbitrary way and let X_k and Y_k denote the numbers of female and male offspring, respectively, of the kth couple. Conditionally, on the number of couples in the $(n-1)$th generation, Z_{n-1}, the number of couples in the next generation equals

$$Z_n = \zeta \left(\sum_{k=1}^{Z_{n-1}} X_k, \sum_{k=1}^{Z_{n-1}} Y_k \right). \tag{2.78}$$

A process is extinct when Z_n equals zero. Bruss (1984) pointed out that the growth of a population is determined by the "average unit reproduction means"

$$m_j = \frac{1}{j} \mathbb{E} \left[Z_n \mid Z_{n-1} = j \right], \; j \geq 1. \tag{2.79}$$

(Note that, in the standard Galton–Watson process, the expectation equals j times m, so m_j equals m.) Daley *et al.* (1986) showed that for common sense, super-additive mating functions, as j tends to infinity, m_j converges to a limit m_∞ with $m_j \leq m_\infty$ for all j, and that extinction is certain (irrespective of the initial population size) if this number is less than or equal to one. If $m_\infty > 1$, there is a positive survival probability provided the initial population size is large enough. If, however, the mating function and reproduction functions are such that, for all i,

$$\mathbb{P} \left(Z_n > i \mid Z_{n-1} = i \right) > 0, \tag{2.80}$$

then $m_\infty > 1$ implies a positive survival probability for any initial positive population size.

From the above we can conclude

$$\mathbb{E} \left[Z_n \mid Z_{n-1} \right] = m_{Z_{n-1}} Z_{n-1} \leq m_\infty Z_{n-1}, \tag{2.81}$$

and then

$$\mathbb{E} \left[Z_n \right] \leq m_\infty \mathbb{E} \left[Z_{n-1} \right] \leq \cdots \leq m_\infty^n \mathbb{E} \left[Z_0 \right]. \tag{2.82}$$

Thus, the expected population size does not necessarily change exponentially, as in the asexual branching process models already considered. Instead, decline may be more rapid if $m_\infty \leq 1$, and growth may be slower in the case of $m_\infty > 1$. The growth rate depends on characteristics of the mating function and offspring distributions. In Section 6.6 we examine population growth of these processes in more detail.

Example 2.11 Consider a small population of swans. Since these birds are strictly monogamous, we can use a model with mating function $\zeta(x, y) = \min(x, y)$. We assume that each couple has a chance $1 - q$ of producing zero offspring and a chance q of producing three surviving offspring. Given that a couple has offspring, the number of females is distributed Bin(3, p), with $p = 0.5$. Denote by $\mathbb{P}(x, y \mid j)$ the conditional probability that, given the adult population size equals j, the total number of female offspring equals x and the total

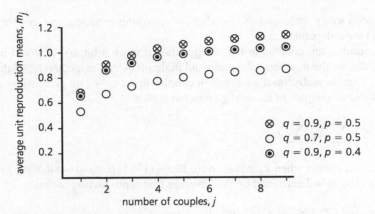

Figure 2.10 Values of m_j for the model given in Example 2.11, with different parameter values.

number of males equals y. Then

$$\mathbb{E}\left[\min\left(X,Y\right)\mid j\right]=\sum_{k=1}^{3j}\mathbb{P}\left(\min\left(X,Y\right)\geq k\mid j\right)=\sum_{k=1}^{3j}\sum_{x=k}^{3j}\sum_{y=k}^{3j}\mathbb{P}\left(x,y\mid j\right),\qquad(2.83)$$

since $\mathbb{E}\left[X\right]=\sum_{k\geq1}\mathbb{P}\left(X\geq k\right)$ (see the Appendix). The probabilities $\mathbb{P}(x,y|j)$ can best be derived using probability generating functions (see the Appendix and Section 5.3). We used Mathematica to do these rather tedious calculations. Figure 2.10 shows m_j as a function of j for two different values of q. When q equals 0.7, m_∞ is about 0.85, so the population certainly becomes extinct. The expected number of offspring per couple is 2.1 in this case, so the expected number of offspring per individual equals 1.05 and an asexual branching process with this value of m would have a positive survival probability. When q equals 0.9, m_∞ equals approximately 1.1 and, accordingly, there is a positive chance that the population will persist. The survival probability of the population also depends on the proportion of females that is produced. When the chance of female offspring is not 1/2, the value of m_∞ drops, since the number of pairs that can be formed becomes lower. As shown in Figure 2.10, when $q = 0.9$ the process becomes nearly critical when the chance of having daughters, p, is 0.4 rather than 0.5. This illustrates that in sexually reproducing species, not only are the expected numbers of individuals important, but also their sex ratio is a critical factor in determining extinction probability. Obviously, this also depends on the mating system. When there is (slight) polygyny, for instance, we can expect that female-biased sex ratios are less risky than male-biased ones. The magnitude of such effects may be studied with bisexual branching process models. Extinction risks of processes with sexual reproduction are further discussed in Section 5.9.

2.9 Varying Environments

The simple Galton–Watson processes, discussed in Section 2.1, give a general impression of what kind of phenomena branching models can elucidate. However, as we have emphasized, they lack many important properties of real-world populations. One such property is that the distribution of offspring numbers may vary over seasons, because of factors such as food supply or weather conditions. Such phenomena can be described by branching processes in *varying environments*. Environment is interpreted in a broad sense, but it should be basically exogenous, even though population size can rightly be viewed as part of the environment for individual reproduction, as discussed in Section 2.6. Like the processes there, those considered here do not satisfy the third basic characteristic in Section 2.1. The other assumptions, that there is merely one type of individual and that these reproduce independently, remain in force.

We consider both deterministic variation in offspring distributions, for an environment that varies in a predictable way, and random variation.

2.9.1 Deterministically varying environments

The arguments of this section differ only slightly from those of non-varying distributions of offspring numbers (Section 2.2, in particular). As there, the expected size of the nth generation satisfies

$$\mathbb{E}[Z_n] = \mathbb{E}[\mathbb{E}[Z_n | Z_{n-1}]] , \tag{2.84}$$

and given the size of the preceding generation, Z_{n-1}, the present population size is the sum of the numbers of offspring of each of the potential mothers. The expectation of these numbers now varies from generation to generation and is denoted by $m(n-1)$. Since the expectation of a sum is the sum of expectations, $\mathbb{E}[Z_n | Z_{n-1}] = m(n-1)Z_{n-1}$. Hence, $\mathbb{E}[Z_n] = m(n-1)\mathbb{E}[Z_{n-1}]$, and repetition yields

$$\mathbb{E}[Z_n] = m(n-1)\mathbb{E}[Z_{n-1}] = \cdots = \prod_{k=0}^{n-1} m(k)\mathbb{E}[Z_0] . \tag{2.85}$$

As before, if the initial population size is fixed and known, $\mathbb{E}[Z_0]$ can be replaced by Z_0. And, like for simple Galton–Watson processes, independence between the numbers of offspring of different individuals is not needed to calculate expected growth. For more delicate matters it is, however, essential.

The different ways in which $m(k)$ might vary are endless, so it is impossible to give an exhaustive description of branching processes with varying environments. For illustrations here, we consider only two types of variation.

The first is *constantly improving* or *deteriorating* environments (these may be caused, for instance, by human interference, such as environmental measures or pollution, or by long-range temperature changes). Assume that the influence of the environment is such that the sequence of values of $m(n)$ increases monotonically and $\lim_{n\to\infty} m(n) = m > 1$. It is intuitively clear that the number of individuals in the population then grows exponentially, so the process may be viewed as supercritical (which is, indeed, the case almost without additional assumptions; see Section 5.10.1). If, on the contrary, $m(n)$ decreases and $\lim_{n\to\infty} m(n) = m < 1$, the process dies out sooner or later and may be treated as subcritical. Note that

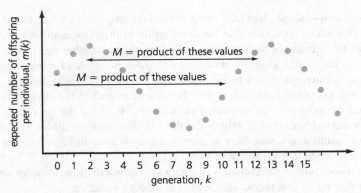

Figure 2.11 A periodic environment: expected numbers of offspring per individual in successive generations. M is the product of these values over one period.

for any specific case the expected population size after n generations is calculated easily. For instance, suppose that the environment deteriorates in such a way that $m(k) = 1/(k+1)$. Then, using Equation (2.85), we find that the expected population size in generation two is half the original size.

The second type of variation concerns *periodic* environments. Assume that the distributions of offspring numbers vary periodically, that is, there is an integer $T \geq 2$ such that for any $k = 0, 1, \ldots$, the offspring distribution of the individuals of the kth generation coincides with that of the individuals of the $(k+T)$th generation (see Figure 2.11). If we let

$$M = m(0)m(1)\cdots m(T-1),$$ (2.86)

then, clearly,

$$m(k-1)m(k-2)\cdots m(k-T) = M, k = T, T+1, \ldots.$$ (2.87)

Therefore, for any positive integer r

$$\begin{aligned}
\mathbb{E}[Z_{rT}] &= m(rT-1)\mathbb{E}[Z_{rT-1}] = \cdots \\
&= m(rT-1)m(rT-2)\cdots m((r-1)T))\mathbb{E}[Z_{(r-1)T}] \\
&= M\mathbb{E}[Z_{(r-1)T)}] = \cdots = M^r\mathbb{E}[Z_0].
\end{aligned}$$ (2.88)

Hence, we see that if $M < 1$ the expected size of a population in a periodic environment goes to zero, if $M > 1$ the expected size grows unboundedly, and if $M = 1$ the average size is stable. Equation (2.88) indicates that an auxiliary process Z_{rT} can be derived from the periodic process Z_n. The former is a standard (non-periodic) Galton–Watson process, and can be studied by standard theory. The expected offspring numbers in that process are equal to M. This corresponds to the expected numbers of offspring that an individual in the original process has left T generations later.

Figure 2.12 General appearance of *Bicyclus* butterflies. Left, wet-season form; right, dry-season form.

Example 2.12 The genus *Bicyclus* of butterflies contains around 80 species in sub-Saharan Africa. Many of its members produce different morphs in the wet and dry seasons (see Figure 2.12). Brakefield and Larsen (1984) argued that the differences between the forms are adaptations to the different types of environments. The wet-season forms generally have large eyespots, which are thought to deflect predators, who are likely to aim at the eyespots rather than at the body. Whereas wet-season forms are rather active, dry-season forms remain mostly inactive and do not reproduce until the start of the wet season. They are more uniformly brown and probably rely on crypsis (hiding between dry brown leaves) to avoid predation.

We model this process as a time-varying branching process with period $T = 2$. Consider a population at the start of a wet season. Each of the (female) individuals in this population has a random number of descendants by the end of that season (i.e., the beginning of the following dry season). We denote the probability of having j live descendants at the end of the wet season by $p_j(w)$ and the expected number of descendants per individual by m_w. During the subsequent dry season, each of these descendants produces a random number of progeny. The probability that one of these butterflies has j descendants at the end of that season is denoted by $p_j(d)$ and her expected number of offspring by m_d.

If we count population sizes only at the beginning of wet seasons, we obtain a non-periodic, classic Galton–Watson process. The probability that a butterfly at the beginning of a wet season has k living descendants at the beginning of the subsequent wet season is

$$\mathbb{P}\,(\xi = k) = \sum_{j=1}^{\infty} p_j\,(w) \sum_{k_1+\cdots+k_j=k} p_{k_1}\,(d)\dots p_{k_j}\,(d)\ . \tag{2.89}$$

The expected number of offspring that she has at the beginning of the next wet season is $M = m_w m_d$. Thus, if, for instance, $m_w = 1.1$ and $m_d = 0.9$, then $M = 1.1 \times 0.9 = 0.99 < 1$ and the population dies out sooner or later, while if $m_w = 1.2$ and $m_d = 0.9$, then $M = 1.2 \times 0.9 = 1.08 > 1$ and the population is supercritical. (Observing the population only at the beginning of dry seasons yields exactly the same result.)

2.9.2 Random environments

Approaches along the lines of Subsection 2.9.1 can be used if the sequences of offspring distributions in different reproduction periods are fixed. In many cases, however, environmental factors that affect reproduction and mortality, such as weather, may have strong stochastic components.

For example, suppose that there are favorable (F) and unfavorable (U) years, which appear independently with probabilities p_F and $p_U = 1 - p_F$, respectively. In a favorable year the offspring mean is m_F and in an unfavorable one, m_U. Clearly, this situation is more complex than that described before: we do not have a deterministic alternation of offspring distributions, but rather a variety of possible scenarios.

Models with randomly varying offspring distributions are called *branching processes in random environments*. In a sense, branching processes in deterministically varying environments are particular cases of branching processes in random environments, for which a specific sequence of offspring distributions occurs with probability one.

In the example given above, let $m(k), k = 1, 2, \ldots, n$ denote the average number of offspring per individual in the kth year. Then, clearly, $m(k)$ is a random variable, equal to m_F with probability p_F and to m_U with probability p_U. Thus, the average number of offspring equals

$$\mathbb{E}[m(k)] = p_F m_F + p_U m_U, k = 1, 2, \ldots, n , \qquad (2.90)$$

and is independent of k. In the following we denote $\mathbb{E}[m(k)]$ by μ.

There are four possible scenarios for two consecutive years

$$FF, \quad FU, \quad UF, \quad UU . \qquad (2.91)$$

The average numbers of offspring per individual in each of the scenarios are, respectively,

$$m_F m_F, \quad m_F m_U, \quad m_U m_F, \quad m_U m_U . \qquad (2.92)$$

The expected number of individuals (from one mother) at the end of the second year is the expectation over all possible scenarios

$$\mathbb{E}[m(1)m(2)] = p_F p_F m_F m_F + p_F p_U m_F m_U + p_U p_F m_U m_F$$
$$+ p_U p_U m_U m_U = \mathbb{E}[m(1)]\mathbb{E}[m(2)] = \mu^2 . \qquad (2.93)$$

We can calculate this directly, since $m(1)$ and $m(2)$ are assumed to be *independent* random variables. The same arguments show that

$$\mathbb{E}[m(1)m(2)\ldots m(n)] = \mathbb{E}[m(1)]\mathbb{E}[m(2)]\ldots\mathbb{E}[m(n)] = \mu^n . \qquad (2.94)$$

It is tempting to believe that the value of μ determines population growth in random environments, as for simple Galton–Watson processes. However, in random environments the value of $\mathbb{E}[\ln m(k)]$ is different from $\ln \mu$, and it is the former that decides the asymptotic growth rate of a population.

To understand this, consider any specific realization of $m(1)$ and $m(2)$. Observe that a population that grows by first $m(1)$ children per individual and then $m(2)$ has the same expected growth as a population for which, during two consecutive periods, the average number of offspring per individual equals $\sqrt{m(1)m(2)}$. Similarly, growth by a sequence of $m(1), m(2), \ldots, m(n)$ is equivalent to growing by the geometric mean $[m(1)m(2)\ldots m(n)]^{\frac{1}{n}}$ during each of n years. Furthermore,

$$[m(1)m(2)\ldots m(n)]^{\frac{1}{n}} = e^{\frac{1}{n}(\ln m(1)+\ln m(2)+\cdots+\ln m(n))} . \tag{2.95}$$

Since $m(1),\ldots,m(n)$ are independent random variables, it follows that so are $\ln m(1),\ldots,\ln m(n)$. According to the *Law of Large Numbers*,

$$\lim_{n\to\infty} \frac{1}{n}(\ln m(1) + \ln m(2) + \cdots + \ln m(n)) = \mathbb{E}[\ln m(k)] \tag{2.96}$$

with probability one (a qualification that we sometimes drop). Hence, the asymptotic population growth rate is

$$\lim_{n\to\infty} [m(1)m(2)\ldots m(n)]^{\frac{1}{n}} = e^{\mathbb{E}[\ln m(k)]} . \tag{2.97}$$

For this reason a process in random environments is called *supercritical* if $\mathbb{E}[\ln m(k)] > 0$, *critical* if $\mathbb{E}[\ln m(k)] = 0$, and *subcritical* if $\mathbb{E}[\ln m(k)] < 0$. This terminology is analogous to that for classic Galton–Watson processes, so subcritical and critical processes in random environments die out sooner or later, whereas supercritical populations can survive forever.

Note that according to Jensen's inequality (see the Appendix)

$$\mathbb{E}[m(k)] = \mathbb{E}[e^{\ln m(k)}] \geq e^{\mathbb{E}[\ln m(k)]} . \tag{2.98}$$

Hence, there are subcritical processes in random environments such that $\mathbb{E}[m(k)] > 1$. For example, assume that favorable and unfavorable years each occur with probability $1/2$. Suppose, further, that if the environment is favorable the expected number of offspring of an individual, m_F, is 3, while in unfavorable environments the expected number of offspring, m_U, is 1/4. Under these conditions

$$\mathbb{E}[m(k)] = \frac{1}{2} \times 3 + \frac{1}{2} \times \frac{1}{4} = 1\frac{5}{8} > 1 , \tag{2.99}$$

whereas

$$\mathbb{E}[\ln m(k)] = \frac{1}{2} \times \ln 3 + \frac{1}{2} \times \ln \frac{1}{4} = \frac{1}{2}\ln\frac{3}{4} < 0 . \tag{2.100}$$

Thus, even though the expected number of offspring per individual exceeds one, the process is subcritical, and therefore the population becomes extinct.

That $\mathbb{E}[\ln m(k)]$ rather than $\mathbb{E}[m(k)]$ is crucial in random environments has been noted by many authors in the biological literature (e.g., Cohen 1966; Lewontin and Cohen 1969). Random, independent, and identically distributed environments were introduced into branching processes by Smith and Wilkinson (1969). Athreya and Karlin (1971a, 1971b) generalized this to ergodic, stationary environments (see Athreya and Ney 1972).

A second-order Taylor approximation of $\mathbb{E}[\ln m(k)]$ around $m(k) = \mu$ gives

$$\mathbb{E}[\ln m(k)] \approx \ln\mu - \frac{1}{2}\left(\frac{\text{Var}[m(k)]}{\mu^2}\right) . \tag{2.101}$$

Equation (2.101) demonstrates that in random environments both the expectation and the variance of the expected offspring numbers $m(k)$ have a large effect on the growth rate. In the context of behavioral ecology, this led to the discovery that mixed strategies, so-called "bet-hedging strategies," are often optimal in time-varying environments, because they reduce such variance (e.g., Schaffer 1974; Seger and Brockmann 1988; Philippi and Seger 1989).

Our examples assumed the quality of seasons to vary independently. A natural generalization is to consider stationary sequences of not necessarily independent variables. A first step is then Markovian dependence, the environment this year being influenced by the preceding year's, but not further back. If the character of the dependence remains the same (so that the Markov chain of environments is *time-homogeneous*), the above classification remains valid with $\mathbb{E}[\ln m(k)]$ still in the crucial role. A further generalization leads to the above-mentioned ergodic environmental sequences. These allow more general dependences (e.g., weather conditions one year may continue to have effects for some time). It is assumed, however, that dependences fade away, so that the processes are *ergodic*, which implies that a form of the Law of Large Numbers remains in force.

A very general scheme, suggested by Jagers and Lu (2002), is that of branching processes in *random, deteriorating environments*. Here, the very point is in the dependence between successive environments, which become worse and worse in the sense that $m(k + 1) \leq m(k)$. By monotonicity, the (random) limiting mean reproduction $m(\infty) = \lim_{k \to \infty} m(k)$ exists, from which it is clear that whether this is above or below one is important for the development of the process. The case $m(\infty) = 1$ turns out to be quite subtle. Such more complicated situations are discussed in Chapters 5 and 6.

2.10   Migration

In the preceding section we have considered isolated or closed populations that evolve from a given number of ancestors. In this section, migration is incorporated into the setup. First, processes with immigration are introduced, and later emigration is discussed.

In a *branching process with immigration*, a random number of immigrants may arrive during each reproduction period. Assume that at time $n = 0$ the population consists of Z_0 individuals. Let η_n, $n = 1, 2, \ldots$, be the number of immigrants during period n. If the local population before immigration is of size Z_{n-1}, the population size at time n, Z_n, is determined by the relation [which corresponds to the basic Equation (2.1) for populations without immigration]

$$Z_n = \xi_1 + \cdots + \xi_{Z_{n-1}} + \eta_n \tag{2.102}$$

where, as in Section 2.1, ξ_i denotes the number of offspring of the ith population member in reproduction period $n - 1$. Thus, the nth generation comprises the offspring of the individuals of the $(n - 1)$th generation plus η_n invaders. (Although this is formally not correct, we use the term generations here.) We assume that all individuals are of the same type (i.e., offspring distributions are the same for all individuals, including immigrants).

Suppose that the average number of immigrants per generation is constant, $\lambda = \mathbb{E}[\eta_n]$. As before, the average number of offspring of each of the Z_{n-1} potential mothers of the $(n-1)$th generation is denoted by m. Then, for fixed Z_0,

$$
\begin{aligned}
\mathbb{E}[Z_n] &= \mathbb{E}[\mathbb{E}[Z_n|Z_{n-1}]] = \mathbb{E}[\mathbb{E}[\xi_1 + \cdots + \xi_{Z_{n-1}} + \eta_n]|Z_{n-1}]] \\
&= \mathbb{E}[\mathbb{E}[\xi_1|Z_{n-1}]] + \cdots + \mathbb{E}[\mathbb{E}[\xi_{Z_{n-1}}|Z_{n-1}]] + \mathbb{E}[\mathbb{E}[\eta_n|Z_{n-1}]] \\
&= \mathbb{E}[mZ_{n-1}] + \lambda = m\mathbb{E}[Z_{n-1}] + \lambda = m(m\mathbb{E}[Z_{n-2}] + \lambda) + \lambda \cdots \\
&= m^n\mathbb{E}[Z_0] + \lambda m^{n-1} + \lambda m^{n-2} + \cdots + \lambda m + \lambda \\
&= Z_0 m^n + \lambda m^{n-1} + \lambda m^{n-2} + \cdots + \lambda m + \lambda .
\end{aligned}
\tag{2.103}
$$

Thus, if $m \neq 1$, then

$$
\mathbb{E}[Z_n] = Z_0 m^n + \lambda \frac{m^n - 1}{m - 1} ,
\tag{2.104}
$$

and if $m = 1$, then

$$
\mathbb{E}[Z_n] = Z_0 + \lambda n .
\tag{2.105}
$$

Hence, if the branching process involved is subcritical, $m < 1$, then

$$
\lim_{n \to \infty} \mathbb{E}[Z_n] = \lim_{n \to \infty} \left(Z_0 m^n + \lambda \frac{1 - m^n}{1 - m} \right) = \frac{\lambda}{1 - m} ,
\tag{2.106}
$$

and the expected number of individuals remains bounded, while in the critical and supercritical cases the average size of the population is proportional to n and to m^n, respectively. See Box 2.1 for an alternative derivation.

Since, in the subcritical case, the average size of the population becomes constant, it could be that the distribution of the population size stabilizes also, provided the whole immigration process is stationary. In Section 6.7 this is shown to be the case: if the numbers of immigrants are independent and identically distributed, the population size approaches its stationary distribution rather quickly. And since immigration does not cease, the population cannot ultimately die out, even in subcritical cases, $m < 1$.

Indeed, after a temporary extinction, the population site is re-colonized. Thus, periods when the site is occupied, *life periods*, alternate with periods during which it is empty. The distribution of periods of emptiness is determined exclusively by the arrival chance of immigrants per period. For instance, if the chance of arrival of at least one immigrant is constant over time, as we now assume, such periods have a geometric distribution. It is not an easy mathematical task to investigate properties of life periods. In Section 6.7, however, we show that the expected length of such periods can be determined.

We now turn to processes with *emigration*, and distinguish three different modes. The first is independent emigration, in which, just after birth, each individual may choose to leave according to a given probability and independently of

Box 2.1 Some technical remarks

From a mathematical point of view it is sometimes convenient to treat branching processes with immigration as decomposable two-type processes, as described in Section 2.3.2. Type 1 individuals are introduced as an artificial device to handle immigration. The local population is constituted by type 2 individuals. Assume that at time $n = 0$ the population consists of one individual of type 1 ($Z_{01} = 1$) and Z_{02} individuals of type 2. At the end of each reproduction period, the individual of type 1 dies after having produced exactly one individual of type 1 and a random number of individuals of type 2 (these are the immigrants) and each type 2 individual gives birth to a random number of children, with expected value m (these are the local offspring). Thus, the mean matrix M of the process has the form

$$M = \begin{pmatrix} 1 & \lambda \\ 0 & m \end{pmatrix}. \tag{a}$$

Therefore (recall Section 2.3.2),

$$M^n = \begin{pmatrix} 1 & m_{12}^{(n)} \\ 0 & m^n \end{pmatrix}, \tag{b}$$

with

$$m_{12}^{(n)} = \begin{cases} \lambda n, & \text{if} \quad m = 1, \\ \lambda \dfrac{1 - m^n}{1 - m}, & \text{if} \quad m \neq 1. \end{cases} \tag{c}$$

Hence, we have

$$\mathbb{E}[(Z_{n1}, Z_{n2})] = (1, Z_{02}) M^n = (1, Z_{02}) \begin{pmatrix} 1 & m_{12}^{(n)} \\ 0 & m^n \end{pmatrix}$$

$$= (1, Z_{02} m^n + m_{12}^{(n)}), \tag{d}$$

and we arrive at Equations (2.104) and (2.105) again.

other individuals. Mathematically, this does not introduce any novelties. Indeed, we may consider emigration as an immediate death and not count emigrating individuals as offspring. If q is the probability of emigration and p_k the probability of having k children, the probability of having k non-emigrating children is

$$\sum_{j \geq k} \binom{j}{k} (1 - q)^k q^{j-k} p_j. \tag{2.107}$$

If m denotes the mean reproduction, as usual, then the mean number of non-emigrating children is $m^* = m(1-q)$. Thus, this type of emigration can transform a supercritical process into a subcritical one, if $m^* < 1 < m$. However, it remains a classic branching process, and can be analyzed as such.

A second possibility is that emigration chance depends on the total number of newborn individuals or the size of the adult population. If all individuals are assumed to make the choice whether or not to emigrate independently of each other, the population dynamics can be studied with branching processes that are population-size dependent (see Sections 2.6 and 6.3).

More difficult, but probably also more interesting, are cases in which emigration decisions of different individuals within one generation are not independent. Unfortunately, as yet there are no results on the dynamics of such processes.

3

Branching in Continuous Time

Chapter 2 introduces discrete-time branching processes. Mathematically, these are much simpler objects than branching processes in continuous time. We have also seen that they occur naturally in many situations, such as generation counting and populations with seasonal regularity in reproduction, and in models for demographic changes recorded annually. Furthermore, it can be argued that data are never recorded continuously, but rather at regular or irregular (albeit sometimes short) intervals. Thus, models in continuous time are not necessarily needed.

The need is more on the conceptual or possibly perceptional side. We certainly conceive of time as a continuous flow, and if mathematical models are to mimic such firsthand conceptions of reality, they should be formulated in continuous time. Similarly, 19th century scientists thought of matter, such as fluids or metals, as self-evidently continuous in the same way as we perceive time. This view has been changed drastically by modern particle- and quantum-based discrete physics.

However, to what extent our perception of time is a cultural, psycho-biological, or physical phenomenon lies outside the scope of this book. We content ourselves with the observation that a continuous-time development of discrete populations is closest to our spontaneous perception of population growth in the flow of time, and that there are good classic mathematical tools for analyzing such situations.

The price to be paid for continuous-time modeling is that the foundation (the rigorous construction of probability spaces and processes) requires more advanced mathematics. We try to conceal this by avoiding explicit construction of the stochastic processes involved. For that, we refer to the mathematical literature.

What we gain for this price is a modeling system that is much closer to many real populations, such as animals or humans in stable conditions: individuals can have arbitrary life spans and give birth repeatedly, according to arbitrary distributions, and life span and reproduction need not be independent of each other. However, the basic (and not always realistic) independence and stability conditions of Section 2.1 remain in force even in general branching processes, though we return to models that allow further dependence in Section 3.5.

3.1 Generations in Real Time

In discrete-time branching processes with non-overlapping generations, a population only contains individuals of one generation at a time. In continuous time this is usually not so. As in human life, aunts can be younger than their sisters' daughters.

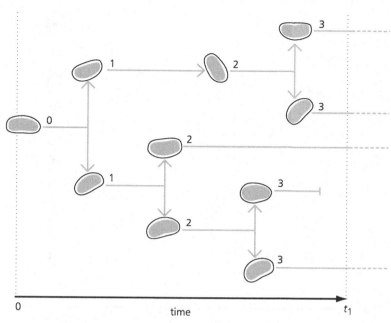

Figure 3.1 Family tree of a continuous-time branching process in which individuals die immediately after a single reproduction event. The life span of each individual is denoted by a branch. Originally, there is one individual who belongs to generation 0. The branches are labeled with their generation number.

Even in the simplest form of continuous-time processes, in which mothers reproduce only once and then die, individuals of different generations can be alive at the same time, because of differences in survival times. An illustration is given in Figure 3.1, in which the population consists of one individual of generation 2 and three of generation 3 at time t_1. Figure 3.2 illustrates the case in which mothers can give birth repeatedly.

The *embedded generation process* counts the numbers of individuals of the various generations. These numbers are random variables (except, usually, for generation 0), and are denoted by ζ_0 (the number of ancestors), ζ_1 the number of children of ancestors, ζ_2 (the number of ancestors' grandchildren), etc. In Figure 3.1, $\zeta_0 = 1$, $\zeta_1 = 2$, and $\zeta_2 = 3$. The embedded process is a discrete-time branching process, like those discussed in Chapter 2. Indeed, if ξ_1, ξ_2, and so on, are the total offspring numbers of the various individuals in the nth generation during all their lives, we retrieve the basic updating relation

$$\zeta_{n+1} = \sum_{i=1}^{\zeta_n} \xi_i \,. \tag{3.1}$$

Therefore, if individuals reproduce independently of each other and the reproduction distribution does not change, the embedded process is precisely a Galton–

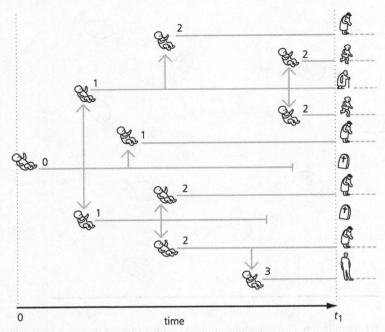

Figure 3.2 Family tree of a general process in which individuals can give birth repeatedly. Originally, there is one individual who belongs to generation 0. The branches are labeled with their generation number.

Watson process. However, it contains only parts of the information about how the population develops. We cannot see its pace, family structure, size, or composition over ages at a given time.

Nevertheless, some fundamental results for discrete-time processes carry over to the continuous-time case via embedded processes. A little afterthought should convince the reader that extinction of the original process occurs if and only if the embedded process dies out. Therefore, the extinction probability of a general branching process is determined by the much simpler embedded process, though the time of extinction, of course, remains undecided, as does the rate of increase in the case of non-extinction – see later sections.

Single-type processes are characterized by the life-span distribution and the so-called reproduction process, which gives the birth-event ages and numbers of offspring. In the simplest cases there is only one reproduction event (at the end of an individual's life), and reproduction is independent of the life span. Such processes are considered in Sections 3.2.1 and 3.2.2. In Section 3.2.3 we generalize and allow the reproduction distribution to depend upon the life span. In Section 3.3, we meet the really general processes, in which individuals may reproduce several times during their lives. Finally, we allow multiple types, in which case the embedded process, of course, also becomes multi-type, counting the numbers of individuals of the various types in generation after generation.

3.2 Reproducing Only Once

Processes in which each individual lives for a random time and then produces a random number of offspring are called *splitting* processes, since individuals are replaced by their offspring. A schematic representation of how such processes proceed is given in Figure 3.1.

In Sections 3.2.1 and 3.2.2, we consider processes in which the numbers of offspring are independent of the mother's life span and individuals reproduce independently of one another. It is easy to see that for such processes the expected total life-time reproduction, m, is equal to the expectation of the offspring distribution. For the simplest process, discussed in Section 3.2.1, it is even possible to find an explicit expression of the expected population size. Unfortunately, this is not achieved so easily for other continuous-time processes, unlike the discrete-time case. In Section 3.2.3, we consider a more general type of process in which the reproduction may depend on life span. In this case, the expected life-time reproduction is also affected by the life-span distribution, but is still relatively easy to calculate.

Note that formally it is not necessary to make the assumption that individuals die immediately after reproduction in the splitting case, as long as they only reproduce at most once during their life. All the models in this section can be reformulated easily to account for life after childbearing. What is referred to below as "life span" thus represents the age at reproduction, and it is the size of the reproductive population only that is measured, rather than the total population size. In most cases, however, it is more realistic to use the more general processes of Section 3.3 for "non-splitting" situations, since these also allow repeated reproductive events (see Figure 3.2).

3.2.1 Markov branching and birth-and-death processes

In the Galton–Watson branching processes of Section 2.1 the future development of the process, given the number Z_n of individuals alive at time period n, is independent of the past. This means that we cannot obtain better information about the destiny of the population with a more detailed description of the present or from the history of the process. This is the defining property of *Markov* processes; the reader may recall the discussion of states in Section 2.5.

The Galton–Watson processes of Section 2.1 thus have the Markov property. They can also be viewed as splitting, since individuals live for one season only, whereupon they are replaced by their offspring. Markov branching processes are the counterparts of Galton–Watson processes in continuous time; that is, they are splitting Markov processes with the three characteristic properties: just one single type of reproductive individual, independence between individuals, and a reproduction distribution that is the same all the time.

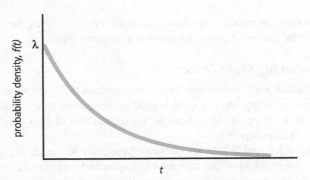

Figure 3.3 Example of an exponential probability density function.

Their structure is as follows: a newborn individual lives for an exponentially distributed life span, and then it splits into a number ξ of new individuals. Life lengths and offspring numbers of different individuals are independent and identically distributed. Furthermore, offspring distributions are independent of the age of the parent. The stochastic process Z_t, which gives the population size at any time $t \geq 0$, is determined completely by this structure and a starting condition, that Z_0 should equal some number of initial individuals or ancestors.

The exponential distribution has a probability density function of the form $\lambda e^{-\lambda t}, t \geq 0$ for some $\lambda > 0$ (see Figure 3.3). Its characteristic property is that the probability distribution of the remaining lifetime, given that a certain age has been attained, is independent of the latter. In mathematical symbols, a real-valued non-negative random variable T follows an exponential distribution if and only if

$$\mathbb{P}(T > t + u | T > u) = \mathbb{P}(T > t) \tag{3.2}$$

for all $t, u \geq 0$. In other words, individuals with an exponentially distributed life span do not age. It is precisely the non-aging property that is singled out by the Markov property. Indeed, if it were not present, we could obtain further information about the future from knowing the ages of the individuals alive, and the process would not be Markov.

It can be derived easily by integration that here $\mathbb{P}(T > t) = e^{-\lambda t}$ and Equation (3.2) follows straightforwardly from that. The reverse, that this relation implies an exponential distribution, is shown in the Appendix.

Markov branching processes are characterized by two properties: the parameter λ of the exponential lifetime distribution and the distribution of the offspring numbers ξ. The parameter λ is given many names, such as death rate, hazard, or intensity, and measures the chance per time unit of splitting. For exponential life-time distributions, the hazard rate is constant and, as can be shown easily, the expected life span equals $1/\lambda$. In Subsection 3.2.2 we consider processes with non-constant hazard rates. Hazard rates play an important role in many other biological applications, such as survival analysis (see Kalbfleisch and Prentice 1980) and behavioral analysis (e.g., Haccou and Meelis 1992). These books give a more detailed exposure of hazard rate functions.

The non-aging assumption implied by the exponential life-span distribution appears to be rather unbiological. Nevertheless, Markov models remain popular, probably because they are the simplest continuous-time branching processes, and Markov modeling has an appeal of its own. It can also be claimed that, in the hardship of wildlife, individuals of many species, such as small mammals or insects, do not become old enough to experience physical aging. They may, however, reproduce several times during their life rather than once. The model may be adjusted to accommodate for this by the same trick as used in Section 2.4 to model overlapping generations by Galton–Watson processes. We return to this in Section 3.3.

As we did with Galton–Watson processes, at this juncture we have a preliminary look at expected population sizes and their evolution. From the Markov property, the expected size at time $t + u$ satisfies

$$\mathbb{E}[Z_{t+u}] = \mathbb{E}[\mathbb{E}[Z_{t+u}|Z_t]] \,. \tag{3.3}$$

Given the population size at time t, Z_t, the population size at $t + u$ is the sum of the numbers of living descendants of each of the Z_t individuals produced during the time interval $[t, t + u]$. However, these progenies constitute Markov branching processes themselves, each evolving as the original one. Hence,

$$\mathbb{E}[Z_{t+u}|Z_t] = Z_t\mathbb{E}[Z_u] \,, \tag{3.4}$$

and taking expectations on both sides gives

$$\mathbb{E}[Z_{t+u}] = \mathbb{E}[Z_t]\mathbb{E}[Z_u] \,. \tag{3.5}$$

If u is small, we can disregard the risk that more than one death occurs in each line. With this approximation, Z_u is either ξ or 1, dependent upon whether the ancestor of the process has died by u (and then obtained ξ children) or not. If, as before, $m = \mathbb{E}[\xi]$, it follows that

$$\mathbb{E}[Z_u] \approx \mathbb{P}(T > u) + m\mathbb{P}(T \le u) = e^{-\lambda u} + m(1 - e^{-\lambda u}) \,. \tag{3.6}$$

Insert this into Equation (3.5) to obtain

$$\mathbb{E}[Z_{t+u}] - \mathbb{E}[Z_t] \approx (m - 1)\mathbb{E}[Z_t](1 - e^{-\lambda u}) \,. \tag{3.7}$$

Then divide by u and let u tend to 0, which gives that $\mathbb{E}[Z_t]$ must satisfy a differential equation. Indeed, if we simply write that $M(t) = \mathbb{E}[Z_t]$, then

$$M'(t) = \lambda(m - 1)M(t) \,, \tag{3.8}$$

since the derivative of $e^{-\lambda u}$ at 0 is λ,

$$\lim_{u\downarrow 0} \frac{1 - e^{-\lambda u}}{u} = \lambda \,. \tag{3.9}$$

This differential equation has the unique solution

$$M(t) = M(0)e^{\lambda(m-1)t} \,, \tag{3.10}$$

if $M(0)$ was the expected population size at time $t = 0$. As for Galton–Watson processes, we see that we have an exponential increase or decrease if $m > 1$ or < 1, respectively.

Example 3.1 Microorganisms can reproduce very quickly when the circumstances, such as temperature, acidity, and humidity, are favorable. For example, the bacterium *Listeria monocytogenes*, which occurs in contaminated food and causes fever, muscle aches, nausea, and diarrhea, has a doubling time of 41 minutes at a temperature of 35°C. If we assume that $m = 2$ and further that each bacterium has a constant chance per time unit λ of splitting – a common though not biologically well-founded assumption – then

$$\lambda = \frac{\ln 2}{41} = 0.017 \text{min}^{-1}, \tag{3.11}$$

at this temperature. At 0°C, the doubling time is 7.5 days, so by keeping infected food at a low temperature, λ is reduced to 0.092 per day. The bacteria can be killed by heating. At a temperature of 60°C reproduction stops, $m = 0$, and populations are reduced by 90 percent after 2.85 min. This implies that λ equals the mortality risk and is about 0.037min^{-1}. (*Source*: www-seafood.ucdavis.edu; page now deleted from website.)

◇ ◇ ◇

Markov branching processes are related to a special class of frequently used continuous-time models for population dynamics, the so-called *birth-and-death* processes. These are commonly defined as integer-valued Markov processes X_t with the property that the intensity of jumping one step upward from the population size $X_t = j$ is jb, whereas the intensity of decrease by one step is jd. The (positive) numbers b, d are usually referred to as the birth and death rates, respectively. The argument is that if $X_t = j$, there are j individuals present, each with a birth rate b. Thus, the whole population should have an intensity of increase jb, and similarly the downward rate is jd, from the j individuals at risk. On the individual level this birth-and-death process is a binary Markov branching process. Life spans are distributed exponentially with the parameter $b + d$, so that the mean life span is $1/(b + d)$, and the offspring distribution is

$$\mathbb{P}(\xi = 0) = d/(b + d) \text{ and } \mathbb{P}(\xi = 2) = b/(b + d) . \tag{3.12}$$

Thus, the model implies the assumption that an individual has a chance per time unit of dying without offspring equal to $d/(b + d)$ and a chance per time unit of splitting into two of $b/(b + d)$. For most biological systems this is not a realistic model, which illustrates the risk of modeling at a population level. It is, therefore, often a good idea to test a proposed model presented at the population level by clarifying the requirements implied at the individual level.

This comment also applies to deterministic models, which (like Markov stochastic models) are usually formulated at the population level. As an illustration, consider the most famous differential growth equation of all:

$$y' = \alpha y . \tag{3.13}$$

In itself, this does not seem to imply many assumptions. However, it has probably been derived by an argument that says that change is proportional to population size, since individuals multiply independently and are also subject to independent death hazards. It is also probably assumed (tacitly) that the equation is valid for any initial age distribution of the population. Otherwise the number $y(0)$ would not suffice as the initial condition. Combining these interpretative comments turns $y(t)$ into the expected size of a Markov branching process, $M(t)$, with

$$\alpha = \lambda(m - 1) . \tag{3.14}$$

The birth-and-death processes introduced above are sometimes referred to as *linear*, as opposed to Markov, processes with a growth rate not of $j\mu$ if $X_t = j$, but more generally of μ_j. This corresponds to branching dependent on population size in continuous time, to be introduced in Section 3.5.

3.2.2 Bellman–Harris processes

If the exponential distribution of life spans is unbiological, why not remove that assumption? *Bellman–Harris processes* are branching processes in which individuals have an arbitrary life-span distribution and produce a random number, ξ, of offspring at death, so the three basic characteristics of Section 2.1 remain in force. As in Markov branching processes, the distribution of ξ is independent of the life span. Curiously enough, in the mathematical literature such processes are called *age-dependent branching processes*.

The reason is probably that, in terms of the hazard rate, the assumptions mean that the chance per time unit of splitting may depend on age. We thus have a hazard rate $\lambda(a)$, where a denotes the age of the individual. It is related directly to the life-span distribution function. Indeed, if the latter (the probability that a life span does not exceed a) is denoted by $L(a)$, and it is assumed that it has a density, $g = L'$, then the hazard rate at an age a such that $L(a) < 1$ is

$$\lambda(a) = \frac{g(a)}{1 - L(a)} . \tag{3.15}$$

Conversely, a hazard $\lambda(a)$ implies that the life-span distribution is

$$L(a) = 1 - e^{-\int_0^a \lambda(u)\, du} . \tag{3.16}$$

Specification of a model involves choosing a form of either $L(a)$ or $\lambda(a)$, and an offspring distribution.

Example 3.2 One of the most interesting special cases is provided by binary splitting, in which ξ can take only the values 0 or 2. This yields natural models for cell kinetics and bacterial growth. To elaborate somewhat, proliferating cells are supposed to pass through four stages (see Figure 3.4). The first, usually denoted by G_1, the letter G for "gap," is viewed as a preparation for the synthesis stage S, when the DNA content is doubled. After another gap period, G_2, the cell proceeds to mitosis M, and ends with a completed cell division. That a certain mission has to be completed before cell division indicates that the Markov branching processes of the previous subsection would yield unsuitable models. A

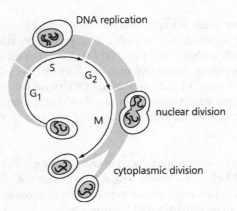

Figure 3.4 The cell cycle. Letters indicate the different stages (see text).

first try might be to model the gap lengths by independent exponentially distributed random variables, whereas the durations of the S and M phases are probably more-or-less normally distributed. However, in reality the picture is much more complex. We return to these matters in Section 3.2.3 and, later, in Sections 6.1 and 7.3.

Remark. The hazard rate does not affect the expected lifetime reproduction m, so for the criticality of the process it does not matter whether a Markov branching process or a Bellman–Harris process is used, as long as the same distribution of offspring numbers is used. The expected population size at any given time is, however, not the same. For Bellman–Harris processes it does not have the simple explicit expression of Equation (3.10). Also, population age structure as a function of time differs in the two types of processes. More is given about this in Chapter 6.

3.2.3   Sevastyanov processes

The idea that not only the hazard rate, but also the distribution of offspring numbers ξ could depend upon age was introduced by B.A. Sevastyanov (1971). It is a fairly natural generalization, and also occurs in simple biological contexts. For example, in Example 3.2 of Subsection 3.2.2, an early cell death might well indicate that something has gone wrong in the cell cycle, so that we would obtain no, rather than two, new cells. Thus, it would be a mistake to model life span and number of daughter cells as mutually independent.

Sevastyanov processes are characterized by a hazard rate $\lambda(a)$ [or, equivalently, a life-span distribution $L(a)$] and offspring distributions that are conditional on the life span T. For these processes, the calculation of expected lifetime reproduction is slightly more complicated than in the previous two cases, since now the life span, which in these models equals the age at reproduction, has to be taken into account. If the life-span distribution has probability density g, and the expected number of offspring, given that the life span is u, equals $m(u)$, the expected life-time reproduction is given by

$$m = \int_0^\infty g(u)m(u)\,du\ . \tag{3.17}$$

As an illustration, consider a model in which the hazard rate is constant (admittedly not very realistic, but nevertheless illustrative), so the life-span distribution is exponential and, conditionally on $T = u$, the numbers of offspring are Poisson distributed with expectation $m(u) = be^{-cu}$, where b and c are positive constants. This implies that at very low ages the expected number of offspring is close to b, whereas at high ages it is close to 0. The rate at which it approaches 0 depends on c. The expected lifetime reproduction for individuals in this process is given by

$$m = \int_0^\infty \lambda e^{-\lambda u} be^{-cu}\,du = \frac{\lambda b}{\lambda + c}\ . \tag{3.18}$$

Thus, the process is supercritical if

$$\frac{\lambda b}{\lambda + c} > 1 \text{ or } b > 1 + \frac{c}{\lambda}\ . \tag{3.19}$$

Now the hazard rate does matter.

Example 3.3 More about age-dependent cell growth: we now generalize Example 3.2 and allow the possibility of quiescence (i.e., newborn cells possibly do not enter the cell cycle, but just remain there without dividing), and also the risk of cell death, which interrupts the cell cycle or quiescence.

The cell cycle has a distribution function F, with the probability density function f. A suitable model would be, for example, a normal distribution, with a small variance as compared to the mean. Newborn cells become quiescent with a probability $1 - p$. Any cell can die (disintegrate), and then not divide. If the death rate can be taken as age independent, say δ, a quiescent cell survives age t with probability $e^{-\delta t}$, whereas a cycling cell does so with probability $(1 - F(t))e^{-\delta t}$. The overall survival probability is thus

$$\mathbb{P}(T > t) = (1 - p)e^{-\delta t} + p(1 - F(t))e^{-\delta t}\ , \tag{3.20}$$

and the life-span distribution of cells turns into

$$L(t) = \mathbb{P}(T \le t) = pF(t)e^{-\delta t} + 1 - e^{-\delta t}\ . \tag{3.21}$$

The probability density function of the life span is found by differentiation:

$$g(t) = pf(t)e^{-\delta t} + \delta e^{-\delta t}(1 - pF(t))\ . \tag{3.22}$$

The number of offspring is either zero or two, and the probability of the latter being the case, if the event occurs at age a, is

$$\frac{pf(a)e^{-\delta a}}{pf(a)e^{-\delta a} + (1 - pF(a))\delta e^{-\delta a}} = \frac{pf(a)}{pf(a) + \delta(1 - pF(a))}\ . \tag{3.23}$$

$m(a)$ equals twice this probability and thus, in this case, the expected lifetime reproduction equals

$$\int_0^\infty g(u)m(u)\, du = 2p \int_0^\infty f(u)e^{-\delta u}\, du \,. \tag{3.24}$$

$$\diamond \diamond \diamond$$

In biology, it is only the simplest populations that grow by splitting. Therefore, it is worth pointing out that the trick used (in Section 2.4) to model overlapping generations by Galton–Watson processes also works here. View individuals as having exponentially distributed life spans and as giving birth at events that occur as a Poisson process during their lives. Each birth event results in a random, independently, and identically distributed number of children. This is a very special form of the general processes that we are going to consider, but as inter-event times in a Poisson process are distributed exponentially, the result is a Markov branching process.

The number of birth times of an individual in this process follows the geometric distribution. Thus, if there is one child per birth event, the offspring number is distributed geometrically, and the embedded Galton–Watson process that counts generation sizes reduces to our geometric benchmark process. The argument that leads to this is a straightforward calculation: up to age t, k births have occurred with probability

$$e^{-ct}\frac{(ct)^k}{k!}, k = 0, 1, 2, \dots \,, \tag{3.25}$$

if the birth process has intensity c. However, an exponentially distributed life span has a density function of the form $\lambda e^{-\lambda t}, t \geq 0$. Integration yields the probability of k offspring as

$$\int_0^\infty e^{-ct}\frac{(ct)^k}{k!}\lambda e^{-\lambda t}\, dt = \frac{\lambda/c}{(1+\lambda/c)^{k+1}} \,, \tag{3.26}$$

which is the familiar geometric distribution with parameter

$$\frac{\lambda/c}{1+\lambda/c} = \frac{\lambda}{\lambda+c} \,. \tag{3.27}$$

3.3 General Branching Processes

The natural, general way to describe reproduction is to say that children can be born at several ages during a potential mother's life. Each birth event can produce single or multiple offspring. The mathematical device that describes such random series of events is a *point process*.

Point processes play an important role in many parts of probability theory. The most well-known are the already mentioned Poisson processes, which can be described as streams of events separated by exponentially distributed waiting times.

These can arise in many ways, the two most important being "thinning" or "super-position." Thinning means random removal of events in a general point process, so that only a few of the original events remain or turn out successful. The more a point process is thinned, the more Poison-like it becomes. Thinning has also given rise to the name The Law of Small Numbers (or Rare Events) for the Poisson distribution. Superposition means addition of many independent point processes, which again results in a Poisson tendency. We refer to the appropriate probabilistic literature for more detail (e.g., Kallenberg 1983; Karr 1986; Daley and Vere-Jones 1988).

Poisson processes can also be viewed as special cases of so-called renewal processes. These are characterized by independent and identically distributed inter-event times. As is evident from their name, they have a background in the study of technical systems: the event could be the replacement of a failed lamp or component of a machine. In population dynamics, the argument could be that the mother needs time to recover after each birth event. In this case, however, the time from a mother's birth to her first birth event cannot be assumed in general to follow the same distribution as the time between successive birth events. As an example, it may take around 2 years to attain fertility, and then litters may follow yearly. Such renewal(-like) processes are called *delayed.*

It is important to stress that the point process chosen for a population model is a matter decided on biological grounds. For cell populations the special case of binary splitting (a point process with just one event!) is evident; animals and plants may have regularly spaced reproduction periods (e.g., each year in spring, like squirrels). Point processes that describe such reproductive patterns could, for instance, have inter-birth times that are deterministic and equal to 1 year, but for a minor normally distributed variation. A more appropriate model might link birth times not to the previous birth, but to calendar time, again with a normal variation some days or weeks up or down.

It can, for example, be assumed that a squirrel has either zero, one, or two off-spring with particular probabilities. Furthermore, its life-span distribution might be exponential if we assume a heavy and constant predation risk. Humans display birth patterns that are more irregular, a very non-exponential distribution of life length, and the reproductive period is surrounded by long periods of infertility.

We refer to the point process that gives the childbearing ages and numbers of children at each birth event as the *reproduction process.* It is a point process on the age interval that starts from 0 and ends at the death of the individual. A general single-type branching process is determined by its life-span distribution and reproduction process. Its structure is given by the population or family tree built from the reproduction processes of all individuals born, the tree that also gave branching processes their name.

In the squirrel illustration we met with reproduction determined (but for random variation) by calendar time, rather than mother's age. The translation to a point process that gives the successive childbearing ages presupposes that we know the mother's birthday (i.e., we have to discern individuals of different types). Thus, *multi-type* general branching processes already arise in very simple settings, if we wish to go beyond modeling in discrete time.

Another simple multi-type process in continuous time is the Bell–Anderson cell model. In this, it is assumed that cell division is not dependent on age, but rather on cell size (or mass): cells divide when they reach a critical mass. Hence, the size of a newborn cell matters for the length of its life cycle.

Again, the need for multi-type models is obvious, as is even the need for (or at least naturalness of) continuous types, such as birth time and birth mass. Thus, we assume that at birth individuals inherit a *type* from their mother. Often we interpret this as a genotype, but (as we have seen here and in Chapter 2) type could also be some simple property of newborn cells that affects their future lives.

3.3.1　Expected growth

Turning back to the single-type case, let us follow the pattern of Galton–Watson and continuous-time Markov processes, and say some words about the development of expected population size, $\mathbb{E}[Z_t]$. For this some notation is needed.

In simple processes individual reproduction is given by one random variable, the number ξ of children, and there is no doubt as to when these were born. In the general case, we not only need the sizes of the various litters, $\nu_1, \nu_2 \ldots$, but also the mother's ages, τ_1, τ_2, etc., at those successive birth events. Alternatively, as described in the preceding section, we can use the *reproduction point process*

$$\xi(a) = \sum_{\tau_i \le a} \nu_i \,, \tag{3.28}$$

which gives the number $\xi(a)$ of children obtained up to each age a (see Figure 3.5).

In branching processes, expected reproduction is described by the *reproduction function*

$$\mu(a) = \mathbb{E}[\xi(a)] \,. \tag{3.29}$$

In demography and related biological population dynamics there is a slightly different tradition in which life span and reproduction are separated. First, a maternity function is introduced to give the reproduction rate, provided the potential mother is alive at the age in question, and the reproduction function is obtained through multiplication by the survival probability. This has advantages in some modeling (as in the example beneath), but in general it only makes notation more cumbersome.

Differential equations have a (too) strong position in such traditional mathematical population dynamics, and it is virtually always assumed that the reproduction function, or corresponding entities, have derivatives. This is not at all necessary for the theory, and, indeed, renders it impossible to formulate a comprehensive theory for both discrete and continuous time. Its advantage, however, is that Riemann integration known from elementary calculus suffices, and no more advanced concepts are needed (not that the so-called abstract Lebesgue

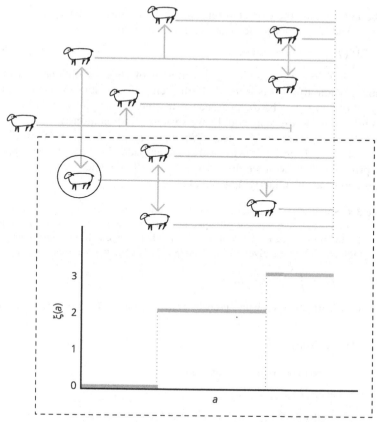

Figure 3.5 Family tree of a general branching process (top) with the reproduction point process of the encircled individual (bottom).

or Lebesgue–Stieltjes integrals are logically more difficult than the Riemann approach, but what is learned at school always seems easiest .... In Section 3.3.2, we however provide some short words about general integrals.)

Thus, whenever convenient for the exposition, we make the unnecessarily strong assumption that the reproduction function has a derivative (i.e., a child-bearing intensity),

$$\mu(a) = \int_0^a \mu'(u) \, du \, . \tag{3.30}$$

It is clear that the mean reproduction of the embedded generation process will satisfy

$$m = \mu(\infty) \left(= \int_0^\infty \mu'(u) \, du\right) , \tag{3.31}$$

since the latter yields the expected total number of children during the mother's entire life. And if T denotes life span, we could alternatively write

$$m = \mathbb{E}[\xi(T)] \tag{3.32}$$

in terms of $\xi(T) = \sum v_i$, which is the number of children obtained during an individual's whole life. (This is unless birth events after death can occur, which is actually an option not to be disregarded, as in epidemic applications where birth event is infection and life span could have various interpretations.)

General processes are *supercritical, critical,* or *subcritical* according to whether $m > 1, = 1$, or < 1, and it is evident that the embedded Galton–Watson process has offspring numbers with the distribution of $\xi = \xi(T)$, and thus the same criticality properties as the underlying general process.

Example 3.4 Suppose that children are born singly in a Poisson process with parameter c until the mother dies, and that her life span is distributed exponentially, with parameter λ. The expected number of children of a living mother of age a is then ca, whereas the expected number of children, given that the mother died at age $u < a$, is cu. Hence,

$$\mu(a) = e^{-\lambda a} ca + \int_0^a cu\lambda e^{-\lambda u} \, du . \tag{3.33}$$

If the life-span distribution is L with a probability density function $L' = g$, the more general relation

$$\mu(a) = [1 - L(a)]ca + \int_0^a cug(u) \, du \tag{3.34}$$

follows. Let $a \to \infty$ to derive the natural relation

$$m = \mu(\infty) = c\mathbb{E}[T] , \tag{3.35}$$

or $m = c/\lambda$ if life spans are exponential with hazard rate λ.

$$\diamond \diamond \diamond$$

For Bellman–Harris and Sevastyanov processes, $\xi(a)$ is the total number of children if $T \le a$ and 0 otherwise. Hence, in the distribution function notation introduced in Example 3.4 (which we adhere to), $L(a) = \mathbb{P}(T \le a)$,

$$\mu(a) = mL(a) \tag{3.36}$$

for Bellman–Harris processes and there is a birth intensity if and only if the life span is a continuous random variable, so that $\mu'(a) = mg(a)$.

Now consider the random variable

$$\sum_i v_i e^{-s\tau_i} ,$$

where the sum is over all birth events, and $s \ge 0$. If $s = 0$, then $e^{-s\tau_i} = 1$ and the sum reduces to the number of children obtained, $\xi(T)$. Hence, its expectation is m. When s is large, $e^{-s\tau_i}$ is small, and the same should be true for its expectation. Since the decrease occurs in a continuous fashion, it should not come as a surprise that, in the supercritical case, there must exist an intermediate choice of s such that the expectation of the sum equals 1. This number, denoted by α, is called the *Malthusian parameter*. It is defined implicitly by

$$\mathbb{E}[\sum_i v_i e^{-\alpha \tau_i}] = 1 . \tag{3.37}$$

For critical processes, of course $\alpha = 0$, whereas in the subcritical case α must be negative, if it exists, as it does in all biologically relevant situations. (Of course, there is a reason for the name of α. It appears shortly!)

Example 3.5 In the case of splitting processes, there is only one birth event age, T, and the defining Equation (3.37) reduces to

$$\mathbb{E}[\xi e^{-\alpha T}] = 1 , \tag{3.38}$$

where $\xi = \sum v_i = \xi(T)$ denotes the number of children. For Bellman–Harris processes, offspring number and life span are independent and Equation (3.37) simplifies further to

$$\mathbb{E}[\xi]\mathbb{E}[e^{-\alpha T}] = m\mathbb{E}[e^{-\alpha T}] = 1 . \tag{3.39}$$

This also indicates how to calculate the Malthusian parameter for this model. If the life-span distribution has the density g, then

$$1/m = \mathbb{E}[e^{-\alpha T}] = \int_0^\infty e^{-\alpha a} g(a) \, da . \tag{3.40}$$

After integration by parts, this takes the form

$$1/m = \mathbb{E}[e^{-\alpha T}] = \int_0^\infty \alpha e^{-\alpha a} L(a) \, da , \tag{3.41}$$

an equality that actually holds even if the density does not exist.

$$\diamond \diamond \diamond$$

In the general case,

$$\mathbb{E}[\sum_i v_i e^{-\alpha \tau_i}] = \int_0^\infty e^{-\alpha a} \mu'(a) \, da = \int_0^\infty \alpha e^{-\alpha a} \mu(a) \, da = 1 , \tag{3.42}$$

in which the first integral presupposes that the derivative μ' exists. Even if such an equation cannot be solved analytically and explicitly, numerical procedures are readily available in programs like Mathematica.

The proper time scale of the development of any population, its generation time, is the *mean age at childbearing*. In the single-type case, this entity, which we denote by β, turns out to be

$$\beta = \mathbb{E}[\sum_i v_i \tau_i e^{-\alpha \tau_i}] \tag{3.43}$$

(see Jagers 1975). Again, if the reproduction function can be differentiated

$$\beta = \int_0^\infty a e^{-\alpha a} \mu'(a) \, da . \tag{3.44}$$

Finally, life spans are important, not only because they (usually) delimit reproduction, but also because they tell how long an individual will be present in the population, and thus counted. As shown below, the life span enters into the asymptotic growth of populations through the equation

$$\ell = \mathbb{E}\left[e^{-\alpha T}\right] . \tag{3.45}$$

Clearly, $\ell < 1$ in the supercritical case and otherwise equal to one or larger. If the life-span distribution has the density g,

$$\ell = \int_0^\infty e^{-\alpha a} g(a)\, da . \tag{3.46}$$

Through integration by parts the following equation can be derived from this

$$\ell = \int_0^\infty \alpha e^{-\alpha a} L(a)\, da , \tag{3.47}$$

which is also valid if the life-span distribution has no density.

Example 3.5 (*continued*) For Bellman–Harris processes, it is always true that $m\ell = 1$, by Equations (3.39) and (3.45). If life spans are exponentially distributed we find, using Equation (3.46)

$$\int_0^\infty e^{-\alpha a} \lambda e^{-\lambda a}\, da = \frac{\lambda}{\alpha + \lambda} = \frac{1}{m} , \tag{3.48}$$

and thus $\alpha = (m - 1)\lambda$. In this case the birth intensity equals

$$\mu'(a) = mg(a) = m\lambda e^{-\lambda a} \tag{3.49}$$

and so, from Equation (3.44)

$$\beta = m\lambda \int_0^\infty a e^{-(\alpha + \lambda)a}\, da = \frac{m\lambda}{(\alpha + \lambda)^2} . \tag{3.50}$$

Substitution of the expression found for α gives $\beta = 1/(\lambda m)$.

$$\diamond \diamond \diamond$$

Remark. Before formulating the exponential growth property of general continuous-time branching processes, we must exclude those that evolve in discrete rather than continuous time. This may seem finical to our biological readers, who might feel that there is nothing like reproductions that can only occur *exactly* at lattice ages, say $\Delta, 2\Delta, 3\Delta$, etc., as opposed to having births that occur, say, yearly, but with a normally distributed variation around the exact year. However, logically this is a possibility that has to be observed and relegated to the discrete case. (For precise formulations and theorems we refer again to the mathematical monographs cited above.) In dealing with continuous time we consider only non-lattice processes from this point.

$$\diamond \diamond \diamond$$

To make the exposition more elementary, we deduce the formula for expected growth under the assumption that there is a childbearing intensity. As pointed out, this is not at all necessary, but renders it possible to argue in terms of elementary Riemann integrals.

At any time t, the population can be divided into the subpopulations that stem from the various children of the ancestor (plus, possibly, the ancestor herself, if she is still alive – for simplicity, we consider a single ancestor, newborn at time 0). If the ancestor gave birth at age a, at time t the process thus started has persisted for $t - a$ time units. Its expected size is $\mathbb{E}[Z_{t-a}]$, and the rate at which the ancestor starts such subpopulations is given by the childbearing intensity, $\mu'(a)$, for $0 \leq a \leq t$. We conclude that

$$\mathbb{E}[Z_t] = \int_0^t \mathbb{E}[Z_{t-a}]\mu'(a)\,da + 1 - L(t)\,. \tag{3.51}$$

Here, $1 - L(t)$ is the probability that the ancestor survives up to time t.

In the critical case, where

$$m = \mu(\infty) = \int_0^\infty \mu'(a)\,da = 1\,, \tag{3.52}$$

this is a classic relationship of the form

$$f(t) = \int_0^t f(t - u)\phi(u)\,du + h(t)\,, \tag{3.53}$$

with $h(t) \geq 0$ and $\int_0^\infty \phi(u)\,du = 1$, known as a *renewal equation*. By well-established theory (see Feller 1966 or later editions), the expected population size must then satisfy

$$\lim_{t\to\infty} \mathbb{E}[Z_t] = \int_0^\infty (1 - L(a))\,da \Big/ \int_0^\infty a\mu'(a)\,da\,. \tag{3.54}$$

The general situation differs from the critical case only in the regard that the integral of μ' no longer has mass 1. However, note that

$$1 = \mathbb{E}\Big[\sum_i v_i e^{-\alpha \tau_i}\Big] = \int_0^\infty \alpha e^{-\alpha t}\mu(t)\,dt = \int_0^\infty e^{-\alpha a}\mu'(a)\,da\,. \tag{3.55}$$

Therefore, if we multiply Equation (3.51) by $e^{-\alpha t}$, the result is a renewal equation,

$$e^{-\alpha t}\mathbb{E}[Z_t] = \int_0^t e^{-\alpha(t-a)}\mathbb{E}[Z_{t-a}]e^{-\alpha a}\mu'(a)\,da + e^{-\alpha t}(1 - L(t))\,, \tag{3.56}$$

since the only lacking property was that we must integrate $\mathbb{E}[Z_{t-a}]$ with respect to a probability density. Indeed, this is a renewal equation of the form above, with

$$f(t) = e^{-\alpha t}\mathbb{E}[Z_t]$$
$$\phi(a) = e^{-\alpha a}\mu'(a)$$
$$h(t) = e^{-\alpha t}(1 - L(t))\,. \tag{3.57}$$

We conclude that

$$\lim_{t \to \infty} e^{-\alpha t} \mathbb{E}[Z_t] = \int_0^\infty e^{-\alpha a}(1-L(a))\,da \Big/ \int_0^\infty ae^{-\alpha a}\mu'(a)\,da = \frac{1-\ell}{\alpha\beta}, \quad (3.58)$$

and thus understand that the Malthusian parameter has inherited its name from the exponential growth postulated by Malthus. This relation is valid for sub- as well as supercritical processes. In the subcritical case, $\alpha < 0$ and $\ell > 1$, so the right-hand side is still positive. It follows that, in this case, the expectation of Z_t declines exponentially.

Example 3.6 For the special case of a Markov branching process, considered in Example 3.5, filling in the values calculated for ℓ, α, and β gives

$$\lim_{t \to \infty} e^{-\alpha t} \mathbb{E}[Z_t] = \frac{1-1/m}{(m-1)\lambda/\lambda m} = 1. \quad (3.59)$$

This result has also been derived in a different way, see Equation (3.10), with $M(0) = 1$.

$$\diamond \diamond \diamond$$

The variance of Z_t can be analyzed by similar arguments as for the expectation, by conditioning upon the reproductive behavior of the ancestor (assuming $z_0 = 1$). In the Bellman–Harris case this becomes rather cumbersome, and the argument does not add many new ideas. We refer to the mathematical literature [Harris (1963) for Bellman–Harris processes, and Jagers (1975) for general processes]. The outcomes are formulas such as

$$\mathrm{Var}[Z_t] \sim e^{2\alpha t} \left(\frac{m-1}{\alpha m\beta}\right)^2 \frac{(\sigma^2+m^2)\int_0^\infty e^{-2\alpha t}g(t)\,dt - 1}{1 - m\int_0^\infty e^{-2\alpha t}g(t)\,dt}, \quad (3.60)$$

as $t \to \infty$, for Bellman–Harris processes with life-span density g, $m > 1$, and σ^2 denoting the reproduction mean and variance, as before.

Example 3.7 We return to the example of variation of organ sizes in Section 2.2. Clearly, the assumption there of a Galton–Watson structure is not realistic, even though, following Azevedo and coauthors (2000, 2001) we allowed reproduction other than merely binary splitting (in the optimistic hope that several binary splittings in close succession could be modeled as a birth of more children than one in a generation). Here we model the organ formation process as a binary splitting Bellman–Harris process. (A binary Sevastyanov model could also be formulated easily.) If p denotes the chance of splitting and $q = 1 - p$ the risk of cell death, $m = 2p$ and $\sigma^2 = 4pq$, and the right-hand side of Equation (3.60) reduces to

$$\sim e^{2\alpha t} \left(\frac{2p-1}{\alpha 2p\beta}\right)^2 \frac{4p\gamma - 1}{1 - 2p\gamma}, \quad (3.61)$$

with

$$\gamma = \int_0^\infty e^{-2\alpha t}g(t)\,dt. \quad (3.62)$$

[The mathematical reader might feel uneasy about such expressions, and wish to check that it is positive, as a variance should be. It is, and Harris (1963) gives an argument for that on p. 146.] We conclude that, as time passes, the coefficient of variation – i.e., the standard deviation divided by the mean – should stabilize around

$$\sqrt{e^{2\alpha t}\left(\frac{2p-1}{\alpha 2p\beta}\right)^2 \frac{4p\gamma-1}{1-2p\gamma}\left(e^{\alpha t}\frac{1-\ell}{\alpha\beta}\right)^{-1}} \tag{3.63}$$

and, because $\ell = 1/m = 1/2p$, this equals

$$\sqrt{\frac{4p\gamma-1}{1-2p\gamma}}\ . \tag{3.64}$$

As an illustration, let life spans be distributed exponentially with a parameter λ. From Example 3.6, we found that in this case $\alpha = \lambda(m-1) = \lambda(2p-1)$. Furthermore, calculation of γ from Equation (3.62) yields

$$\gamma = \int_0^\infty e^{-2\alpha t}\lambda e^{-\lambda t}\,dt = \frac{\lambda}{\lambda+2\alpha}\ , \tag{3.65}$$

so $\gamma = 1/(4p-1)$ and thus the limiting coefficient of variation equals $\sqrt{2p-1}$. Note that the cell cycle length $1/\lambda$ is not involved (though it determines the time scale and thus the rate of convergence).

3.3.2 Multi-type general processes

Eventually, we arrive at the most general branching processes that encompass all imaginable forms of reproduction and heredity patterns. An individual of a general branching population gives births (single or multiple) at random ages, the children can have arbitrary types, and the mother's own type may affect all aspects of her childbearing: ages at birth events, litter sizes, and types of offspring. The type space can be very rich, practically speaking even arbitrary, though certain, very generous mathematical requirements should be fulfilled.

Any finite space will do as a type space, but actually also any space with a countable number of elements. The latter may seem exaggerated and unneeded to a biologist, but in modeling it can be suitable to think of an unlimited number of possible types. The real line, or two- or three-dimensional spaces, or, indeed, spaces of any dimension are included. For a modern biologist it is natural to identify type with genotype, but certainly other types of life conditions can be included, such as birth weight, geographic conditions, or social circumstances, in demographic modeling.

Besides the simple multi-type processes, already discussed in discrete time, types are often real numbers, such as body mass or energy content in cases of physical particles splitting. The rise of molecular biology should open possibilities for more subtle multi-type branching modeling, with huge type spaces equipped with special topologies to reflect distance, in various senses, between genes and gene function.

The probabilities that govern these structures can also be quite general. The basic structure remains that of independent individuals, given their types. However,

since type can be very rich, indeed it could even summarize all relevant information about the environment and population history, this is no severe restriction.
Restrictions arise because we wish to be able to conclude sensible things about the
population, to determine extinction probabilities, growth rates, and the like. Some
property such as the indecomposability discussed in Section 2.3.1 is required.

Under the appropriate conditions, the known array of results can be established
very generally. The reproduction function is replaced by a *reproduction kernel*,
$\mu(s, A \times B)$, which gives the expected number of children with types in the set A
born during the age interval B by a mother herself of type s. Unless they are very
weird, such kernels define a Malthusian parameter α, and the expected growth (or
decrease) of population size can be established through so-called Markov Renewal
Theory. This is mathematically fairly advanced, and the reader is referred to Jagers
(1989) for more information.

It is enough here to point out that if S denotes the whole type space, the process
is subcritical, critical, or supercritical according to the sign of α, so that, as in
Equation (3.58)

$$\mathbb{E}_s[Z_t] \sim v(s)e^{\alpha t}(1 - \ell)/\alpha\beta \, . \tag{3.66}$$

Here, s indicates that there was one ancestor of type s, and v is the eigenfunction

$$v(s) = \int_0^\infty \int_S v(r)e^{-\alpha a}\mu(s, \, dr \times da) \, , \tag{3.67}$$

about which see Box 3.1, and ℓ, β are expectations,

$$\ell = \mathbb{E}_\pi[e^{-\alpha T}], \beta = \int_0^\infty \int_S ae^{-\alpha a}v(s)\mu(\pi, \, ds \times da) \tag{3.68}$$

for an individual whose type is random according to the *stable type distribution* π.
The latter is defined by the requirement that

$$\pi(A) = \int_S \int_0^\infty e^{-\alpha a}\mu(s, A \times da)\pi(ds) \, , \tag{3.69}$$

with normalization chosen such that

$$\pi(S) = 1 \text{ and } \int_S v(s) \, ds = 1 \, , \tag{3.70}$$

and

$$\mu(\pi, A \times B) = \int_S \mu(s, A \times B)\pi(ds) \, . \tag{3.71}$$

The reason for calling π the stable-type distribution is that, as time passes, the
proportion of all individuals born with types in any set A converges to $\pi(A)$. Similarly, if we compare two supercritical processes, one with an ancestor of type s, the
other initiated by a type-r individual at the same time, then their ratio converges to
$v(s)/v(r)$. Therefore, v is called the *reproductive value* of the type. It provides a
sort of relative type-fitness measure.

Box 3.1 Strange integrals

The multi-type general processes equations contain general integrals that could have been introduced naturally much earlier in this book, but that we have avoided and tried to conceal the need for. At this juncture, this is no longer possible. The reader who is unacquainted with these notions should think of them as joint notation for both elementary integrals and sums:

$$\int_S v(s)\pi(ds) \text{ means } \sum_j v(s_j)\pi(s_j) \tag{a}$$

if $S = \{s_1, s_2, \ldots\}$, and it means

$$\int_{-\infty}^{\infty} v(s)\pi'(s)\, ds \tag{b}$$

if S is the real line and π' is the density (derivative) of π. Now, it is easy to understand that situations can arise in which sums and elementary integrals are combined. For example, it could be the case that π is discrete on some interval A and continuous on another, B. Then

$$\int v(s)\pi(s)ds = \sum_{s_j \in A} v(s_j)\pi(s_j) + \int_B v(s)\pi'(s)\, ds \,. \tag{c}$$

Further, the abstract integrals can be still more broadly defined for situations in which π is what is called a *measure*, that is, a generalized probability that need not have mass 1, such as the concept of length on the line, or area in the plane, and S could be virtually any type of space. Notation like $\mu(s,\ dr \times\ da)$ means that for a fixed s we integrate both over type and over age space, with respect to the measure $\mu(s, A \times B)$. This has the interpretation given above.

Our biological readers are advised to try and live with these enigmatic hints. Otherwise, there is much literature on abstract measure and integration, both of a probabilistic and a general mathematical nature. We recommend the books by Billingsley (1979) and Rudin (1987), respectively.

The exponential growth (decline) of expected population sizes was made precise in Equation (3.58) above for the total population. As for discrete-time processes with finitely many types, we can also consider the number of individuals at time t that have types in some set $D \subseteq S$ and conclude that $1 - \ell$ has to be replaced by

$$\mathbb{E}_\pi[1 - e^{-\alpha T} : D]\,, \tag{3.72}$$

where the notation means that the expectation is only to be taken over types in D. Equivalently, this could be expressed through the indicator function $1_D(s) = 1$ if $s \in D$ and 0 otherwise:

$$\mathbb{E}_\pi[1 - e^{-\alpha T} : D] = \mathbb{E}_\pi[(1 - e^{-\alpha T})1_D(X)]\,, \tag{3.73}$$

where X is the ancestral type, distributed according to the stable-type distribution π and T, the ancestor's life span.

Example 3.8 Size-dependent cell division We take a look at cells with size-dependent (as opposed to age-dependent) individual behavior, the classic Bell–Anderson model (see Diekmann *et al.* 1984; Diekmann 1986; Arino and Kimmel 1993). The basic assumption is that there is a splitting intensity $b(x) \geq 0$, where x is the individual cell size. Similarly, there is a death intensity of $\delta(x) \geq 0$, death meaning that the cell disappears without giving birth to daughter cells. When a cell splits, its mass is assumed to be divided equally between the daughters. Individual cell growth is deterministic (i.e., the same for all cells with given birth size), described by the differential equation $x' = g(x)$, with $g(x) > 0$, and $x(0)$ is the size at birth. Assume there are minimal and maximal cell sizes a and $4a$, respectively, so that $0 < a \leq x(0) \leq 2a$, and no cell smaller than $2a$ can divide. Such models are, of course, both simplistic and outdated now that we have entered the era of molecular biology. A modern description of cell growth and division could start from a much greater insight into the cell, its metabolism, and cycle control (see Novak and Tyson 1995; Tyson *et al.* 1997). First, denote the chance *per time unit* of splitting by $\widetilde{b}(t)$ and, similarly, the chance per time unit of death by $\widetilde{\delta}(t)$, so the probability that a cell splits during an infinitesimally small interval $(t, t + dt)$ equals

$$\widetilde{b}(t)e^{-\int_0^t (\widetilde{b}(v)+\widetilde{\delta}(v))\, dv}dt . \tag{3.74}$$

The Bell–Anderson growth equation yields $dt = dx/g(x)$, and therefore it follows that the probability of splitting at a size that lies in the interval $(s, s + ds)$ is

$$b(s)e^{-\int_{x(0)}^s (b(r)+\delta(r))\frac{dr}{g(r)}} \frac{ds}{g(s)} . \tag{3.75}$$

To produce two y-sized daughter cells, the mother must herself attain size $2y$ and the expected number of y-sized daughters becomes

$$2b(2y)e^{-\int_x^{2y} (b(r)+\delta(r))\frac{dr}{g(r)}} \frac{2dy}{g(2y)} , \tag{3.76}$$

where we use the notation $x(0) = x$. Once y has been fixed, the age at division is given by

$$\int_x^{2y} \frac{dr}{g(r)} = h(y|x) . \tag{3.77}$$

Using the notation

$$f(x) := \int_a^x (b(r) + \delta(r))\frac{dr}{g(r)} , \tag{3.78}$$

we can write the reproduction kernel

$$\mu(x, dy \times du) = 4\frac{b(2y)}{g(2y)}e^{-(f(2y)-f(x))}1_{\{h(y|x)\}}(du)dy, a \leq x, y \leq 2a , \tag{3.79}$$

where $1_{\{h(y|x)\}}(du)$ denotes the probability measure that puts all its mass on the point $h(y|x)$.

$$\diamond \diamond \diamond$$

3.4 Age-distribution and Other Composition Matters

Within the framework of general processes many questions can be answered that could not even be formulated, or would have trivial answers, in the simple Galton–Watson case. Thus, the composition of a population of identical individuals is not a very interesting matter, as opposed to the composition of a population in which individuals have different ages, belong to different generations, have different birth orders, or are of different type (see also Section 6.2).

In applications, composition matters may be important, since the fraction of a population with special properties can often be measured. As an example, the mitotic index, which is the proportion of cells in mitosis, is a natural indicator of cell population growth. Alternatively, in a wildlife context, the fraction of females pregnant might yield a corresponding index.

The classic composition matter in population studies is, however, that of age distribution. Consider the number of individuals aged a or less in a single-type general branching process at time t and denote it by Z_t^a. What can we say about its expectation? It satisfies a relation such as Equation (3.51), the only difference being that the ancestor is counted only if not older than a. Thus, in terms of the indicator function $1_{[0,a]}(t) = 1$ or 0, according to whether $0 \le t \le a$ or not, the renewal style equation

$$\mathbb{E}[Z_t] = \int_0^t \mathbb{E}[Z_{t-a}]\mu'(a)\, da + 1_{[0,a]}(t)(1 - L(t)) \tag{3.80}$$

follows. It is analyzed exactly as Equation (3.51), with the outcome that

$$\lim_{t \to \infty} e^{-\alpha t}\mathbb{E}[Z_t^a] = \int_0^a e^{-\alpha u}(1 - L(u))\, du \Big/ \int_0^\infty u e^{-\alpha u}\mu'(u)\, du , \tag{3.81}$$

and therefore

$$\lim_{t \to \infty} \frac{\mathbb{E}[Z_t^a]}{\mathbb{E}[Z_t]} = \frac{\int_0^a e^{-\alpha u}(1 - L(u))\, du}{\int_0^\infty e^{-\alpha u}(1 - L(u))\, du} . \tag{3.82}$$

This is the famous *stable age-distribution* of the process. Note that it depends on the reproduction function $\mu(a)$ only through the Malthusian parameter α.

Here, note that the ratio between expectations is not the same as the expectation of the ratio $\mathbb{E}[Z_t^a/Z_t]$. The latter yields the real expected age distribution. We show later that, as time passes, this also converges to the stable age-distribution of the process (provided it does not die out).

Example 3.9 Many wild moose populations are controlled through predation by wolves and bears, and hunting by humans. For such cases, an exponential life-span distribution is presumably a good assumption. The maximum life span mentioned on the website of the Alaskan Department of Fish and Game (http://www.state.ak.us/adfg/adfghome.htm) is 16 years, so an average life span of, say, $1/\lambda = 5$ years appears to be reasonable,

The stable age-distribution then equals:

$$\frac{\int_0^a e^{-\alpha u}(1 - L(u))\,du}{\int_0^\infty e^{-\alpha u}(1 - L(u))\,du} = \frac{\int_0^a e^{-\alpha u}e^{-\lambda u}\,du}{\int_0^\infty e^{-\alpha u}e^{-\lambda u}\,du} = 1 - e^{-(\alpha+\lambda)a}\,. \tag{3.83}$$

To estimate the rate of change in population size α it thus suffices to have a sample in which two age groups can be discerned, say calves and adult animals. The usual breeding age is about 2 years, so the proportion of calves equals

$$c = 1 - e^{-2(\alpha+\lambda)} \tag{3.84}$$

and thus

$$\alpha = -\frac{\ln(1 - c)}{2} - \lambda = -\frac{\ln(1 - c)}{2} - 0.2\,. \tag{3.85}$$

Thus, a population is supercritical ($\alpha > 0$) if $-\ln(1 - c) > 0.4$, or the proportion c of calves is larger than about one-third; if c is smaller, the population is bound to become extinct. Of course, the value of c is usually not known, but estimated. Equation (3.85) together with the sample size (and sampling method) can be used to derive confidence intervals for the rate of growth and tests of whether or not a population is subcritical (and thus in imminent danger of extinction).

$$\diamond \diamond \diamond$$

Age is an example of a property that is analyzed relatively easily, because an individual's age is determined by events of her own life solely. Such properties are referred to as *individual*, as opposed to those that involve several individuals (e.g., the individual's whole sibship). Thus, another individual property is one's generation, and a non-individual characteristic (of less interest for biological than social science, maybe) would yield the fraction of firstborns. Actually, the generation of individuals alive at time t tends to be distributed normally, with the maximum at t/β, or rather the integer closest to this number (Martin-Löf 1966). Also, the probability of being firstborn converges to

$$\mathbb{E}[e^{-\alpha\tau_1}]\,, \tag{3.86}$$

where τ_1 is the mother's age at her first birth event. Indeed, the underlying renewal equation for the expected number $f(t)$ of firstborns, not counting the ancestor, becomes

$$f(t) = \int_0^t f(t - a)\mu'(a)\,da + \int_0^t (1 - L(t - a))\phi(a)\,da\,, \tag{3.87}$$

if ϕ is the probability density function of τ_1. We do not develop this further, but refer to Jagers (1981, 1991) and Section 6.2.

In the multi-type case, we point out that the composition over types converges in Section 3.3.2. The stable age-distribution appears as before, though L has to be interpreted as the life-span distribution of an individual, whose type is random according to the stable-type distribution. Since we have not given the complicated Markov renewal equations that describe expected population size, or expected subpopulation sizes, when there are several types, we confine ourselves to referring to the other literature cited.

3.5 Interaction, Dependence upon Resources, Varying Environment, and Population

Interaction local in the pedigree, such as dependence between siblings, can be introduced into general processes as in the discrete-time setup. We must specify only the group in which there is interaction, like a sibship, and then give the probability distribution of the reproduction of the whole group.

Population-size dependence turns out to be much harder to define. Previously, we could simply say that in the nth generation reproduction occurs in the classic Galton–Watson fashion, but that the number Z_n must be considered as given. In continuous time, population size changes during an individual's life. We may want to consider cases in which not only are the numbers of children at given birth events affected by population size, but also the birth event ages and life spans.

One case that can be dealt with is that of a *Markovian age structure*. This is the natural, albeit not always realistic, setup whereby an individual's future life and reproduction is determined by the individual's age (and possible external factors, such as population size), but nothing in the individual's past history, besides age, has repercussions on its life to come. In some animal and plant modeling such an assumption might be defended more easily than in demography – humans tend to remember whether they just had children or not! On the other hand, breeding seasons must then be modeled through ages (e.g., from knowing the birth times).

Anyway, the reproduction process of a branching process that is Markovian in its age structure can be described by an age-dependent childbearing intensity, $\beta(a)$, a similarly age-dependent probability distribution for the numbers (and types) of offspring at a birth event at a certain age, and finally a death rate, $\lambda(a)$. These can be made dependent on population size easily; if at time t a certain individual has age a, then the childbearing intensity turns into a number $\beta(a, Z_t)$, Z_t denoting, as before, the number of individuals at time t. The other entities are modified in the same manner. Of course, Z_t can be replaced so that dependence is upon entities other than population size, such as resources or (as a simple measure of resources used) the *accumulated population* of all, living or dead, up to t, Y_t, or, maybe more suitably, the integral $\int_0^t Z_u \, du$.

From this point of view, the general birth-and-death processes, mentioned in Section 3.1, appear not so general, but rather as the special case of birth-and-death intensities independent of age. Nevertheless, much older literature on stochastic population dynamics actually deals with birth-and-death processes, and even with special cases, such as the Ricker (or logistic) or even simpler forms of the birth intensity, and a constant death intensity [see MacArthur and Wilson (1967) and Richter-Dyn and Goel (1972) for classic examples].

4

Large Populations

4.1 Approximations of Branching Processes

T.G. Kurtz

4.1.1 Galton–Watson processes and diffusion

Approximations of stochastic models can be useful for several reasons:

- The approximating model may be simpler than the original model;
- The qualitative behavior of the approximating model may be easier to understand;
- The number of unknown parameters (i.e., parameters that need to be estimated from data) may be smaller for the approximating model; and
- The approximating model may be computationally more tractable.

The types of approximations we have in mind for branching processes are typically justified when the population under consideration is large and the time scale is fast (i.e., we consider the population over many generations). An examination of the Galton–Watson process given by the iteration

$$Z_{n+1} = \sum_{i=1}^{Z_n} \xi_i \qquad (4.1)$$

motivates these approximations.

Recall that ξ_i are independent and identically distributed, so Z_{n+1} is a (random) sum of such variables. The basic limit theorems of probability (the Law of Large Numbers and the Central Limit Theorem, see the Appendix) suggest ways to approximate Equation (4.1). As before, write $\mathbb{E}[\xi_i] = m$ and $\text{Var}[\xi_i] = \sigma^2$. Then, if the population is large, we must have

$$\frac{Z_{n+1}}{Z_n} = \frac{1}{Z_n} \sum_{i=1}^{Z_n} \xi_i \approx m , \qquad (4.2)$$

so that, in a sense, we can approximate Z_n by $Z_0 m^n$, and we see that, in large populations, Malthusian geometric growth captures the general behavior of the branching process.

The implications of the Central Limit Theorem are somewhat more subtle. Suppose that $m = 1$, that is, on average, each individual has one offspring. Then the

Central Limit Theorem suggests that as the population grows the distribution of

$$\frac{1}{\sqrt{Z_n}\sigma} \sum_{i=1}^{Z_n} (\xi_i - 1) \tag{4.3}$$

should converge to the standardized normal law. Therefore:

$$Z_{n+1} = \sum_{i=1}^{Z_n} \xi_i = Z_n + \sum_{i=1}^{Z_n} (\xi_i - 1) \approx Z_n + \sigma\sqrt{Z_n}\zeta_{n+1}\,, \tag{4.4}$$

where ζ_{n+1} is distributed normally with mean zero and variance one. Suppose that the population is of the same order of magnitude as a large number N, and let $\tilde{Z}_n$ denote the *normalized* population size

$$\tilde{Z}_n = \frac{Z_n}{N}\,. \tag{4.5}$$

In one generation the change in the normalized population size is then approximately

$$\tilde{Z}_{n+1} - \tilde{Z}_n \approx \sigma\sqrt{\tilde{Z}_n}\frac{1}{\sqrt{N}}\zeta_{n+1}\,. \tag{4.6}$$

The right side of Equation (4.6) is, of course, small, so over a few generations, there is very little change in the normalized population size. How many generations must pass before a significant change occurs? The ζ_n in the approximate model are independent, and properties of the normal distribution imply that

$$\sum_{n=1}^{N} \frac{1}{\sqrt{N}}\zeta_{n+1} \tag{4.7}$$

is distributed normally with mean zero and variance one. Consequently, the natural time scale for this model is to have N generations per unit time. Define

$$W_N(t) = \sum_{n=1}^{[Nt]} \frac{1}{\sqrt{N}}\zeta_{n+1}\,, \tag{4.8}$$

where $[Nt]$ denotes the largest integer less than or equal to Nt. Then $W_N(t)$ is distributed approximately normally with mean zero and variance $N^{-1}[Nt]$, and $\text{Cov}[W_N(t), W_N(s)] = N^{-1}[N \cdot \min(t, s)]$. If we define $X_N(t) = \tilde{Z}_{[Nt]}$, then since

$$\tilde{Z}_{[Nt]} = \tilde{Z}_0 + \sum_{n=0}^{[Nt]-1} \left(\tilde{Z}_{n+1} - \tilde{Z}_n\right) \approx \tilde{Z}_0 + \sum_{n=0}^{[Nt]-1} \sigma\sqrt{\tilde{Z}_n}\frac{1}{\sqrt{N}}\zeta_{n+1} \tag{4.9}$$

we can write

$$X_N(t) \approx X_N(0) + \int_0^t \sigma\sqrt{X_N(s-)}\,dW_N(s)\,. \tag{4.10}$$

Box 4.1 Stochastic integrals

As with ordinary integrals (the type studied in calculus), stochastic integrals are very useful for describing the dynamics of physical or biological systems. For example, suppose we want to model a quantity V that evolves randomly in time in such a way that for small $\Delta t > 0$,

$$V(t + \Delta t) \approx V(t) + U(t)(Y(t + \Delta t) - Y(t)), \tag{a}$$

where U and Y are stochastic processes. Then simply write

$$V(t) = V(0) + \int_0^t U(s-) \, dY(s) \approx V(0) + \sum_{i=0}^{m-1} U(t_i)(Y(t_i + \Delta t_i) - Y(t_i)), \tag{b}$$

where $0 = t_0 < \cdots < t_m = t$, $\Delta t_i = t_{i+1} - t_i$, and, again, $U(s-)$ denotes the value of U immediately before time s. The use of $U(s-)$ as the integrand reflects that in Equation (a), we take the value of U at the beginning of the time interval $(t, t + \Delta t]$ and multiply it by the increment of Y. To make this description rigorous requires a substantial amount of mathematical machinery. It suffices here simply to warn that the integral is not defined for every pair of processes U and Y and that, in general, replacing $U(t)$ by $U(t + \Delta t)$ in Equation (a) gives a very different definition of the integral.

The "integral" in Equation (4.10) is just the sum in Equation (4.9) and $X_N(s-)$ denotes the value of X_N immediately before time s [which, of course, is the same as $X_N(s)$ unless X_N jumps at time s]. Details about such integrals are provided in Box 4.1.

Intuitively, very little should change if we replace W_N by a stochastic process W with normally distributed values that have mean zero and covariance $\text{Cov}[W(t), W(s)] = \min(t, s)$. Such a process is called a *standard Brownian motion*, and the resultant equation (eliminating N throughout)

$$X(t) = X(0) + \int_0^t \sigma \sqrt{X(s)} \, dW(s) \tag{4.11}$$

is called a *stochastic differential equation*. The solution to this equation is the *Feller diffusion approximation* (Feller 1951) for the critical Galton–Watson process, and it is a valid approximation in the sense that the probability distribution of $X_N(t)$ is close to the probability distribution of $X(t)$, but much more is valid. For example, the distribution of $\sup_{s \leq t} X_N(s)$ is close to the distribution of $\sup_{s \leq t} X(s)$ and the distribution of $\tau_N = \inf\{t : X_N(t) = 0\}$ (the extinction time) is close to the distribution of $\tau = \inf\{t : X(t) = 0\}$. Lindvall (1974) discusses the convergence of these and other functionals of the process.

The diffusion X can also be written as the time change of a Brownian motion, that is,

$$X(t) = X(0) + \tilde{W}\left(\int_0^t \sigma^2 X(s)\, ds\right), \tag{4.12}$$

where $\tilde{W}$ is a standard Brownian motion. It follows that $Z = \sup_{t \leq \tau} X(t) = \sup_{u \leq \gamma}(X(0) + \tilde{W}(u))$, where $\gamma = \inf\{u : X(0) + \tilde{W}(u) = 0\}$, and $P\{Z > b\}$ is the probability that $X(0) + \tilde{W}$ hits b before it hits zero. For $b > X(0)$, a simple calculation gives

$$(P\{Z > b\}) = \frac{X(0)}{b}. \tag{4.13}$$

For example, see Bhattacharya and Waymire (1990), Proposition V.2.5, for a more general result that applies to the more general diffusion discussed next.

We can generalize this approximation by assuming that $\mathbb{E}[\xi_i] = 1 + \beta N^{-1}$, that is, since N is large, the average number of offspring is very close to 1, but not exactly equal to 1. Then Equation (4.6) becomes

$$\tilde{Z}_{n+1} - \tilde{Z}_n \approx \sigma \sqrt{\tilde{Z}_n}\frac{1}{\sqrt{N}}\zeta_{n+1} + \beta\frac{1}{N}Z_n, \tag{4.14}$$

and Equation (4.11) becomes

$$X(t) = X(0) + \int_0^t \sigma\sqrt{X(s)}\, dW(s) + \int_0^t \beta X(s)\, ds. \tag{4.15}$$

Note that this diffusion approximation depends only on the mean and variance of the original offspring distribution; that is, we have replaced a model that depends on the full offspring distribution by one that depends only on two parameters. This shows how one diffusion process can approximate many branching processes. In interpreting the diffusion approximation, one needs to keep in mind that N is both the number of generations per unit time and the order of the population size. Another way to say the same thing, which applies to models in which the generations are not separate, is that the time unit needs to be selected so that the number of births per unit time has the same order as the square of the population size.

The theory of stochastic differential equations is extensive, and many properties of Equation (4.15) can be derived. For example, if $\beta \leq 0$, X (not surprisingly) hits zero in finite time, just as in a critical or subcritical branching process. Similarly, if $\beta > 0$, there is a positive probability of X tending to infinity.

Essentially, we have derived a "Law of Large Numbers" and a "Central Limit Theorem" for a Galton–Watson branching process. The diffusion approximation was introduced by Feller (1951) and made rigorous by Jiřina (1969). Lindvall

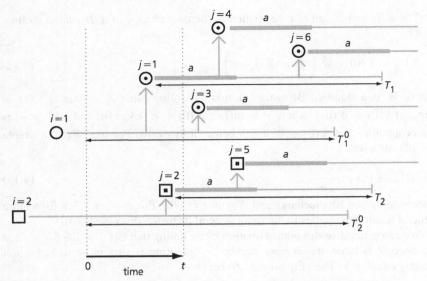

Figure 4.1 Example of a process with two ancestors of different initial ages. Formal definitions of the variables shown are given in the main text. In this illustration, $\xi_1^0(t) = \xi_2^0(t) = 1$ and $\xi_1(a) = \xi_2(a) = 1, \xi_3(a) = \cdots = \xi_6(a) = 0$.

(1972, 1974) gave a "functional" version of the limit theorem, that is, a version that immediately implies the corresponding convergence for functionals of the processes, such as the maximum. Our next goal is to discover what the analogous results are for general branching processes.

4.1.2 Deterministic approximations of general processes

In this section, we turn to general branching processes in which reproduction can occur at various ages (see Sections 3.1 and 3.3). Consider a population with Z_0 ancestors at time 0. For $i = 1, \ldots, Z_0$, let $\xi_i^0(t)$ be the number of births to individual i by time t and let T_i^0 denote the time that the individual dies. [Of course, $\xi_i^0(t) = \xi_i^0(T_i^0)$ for $t > T_i^0$.] It is assumed that $\{[\xi_i^0(t), T_i^0], i = 1, \ldots, Z_0\}$ are independent and identically distributed. Similarly, for $j = 1, 2, \ldots$, let $\xi_j(a)$ denote the number of births by age a for the jth individual born *after* time zero and let T_j be the age at death for that individual. [Again, $\xi_j(a) = \xi_j(T_j)$ for $a > T_j$.] These definitions are illustrated in Figure 4.1. As in Section 3.3, we assume that $\{[\xi_j(a), T_j], j = 1, 2, \ldots\}$ are independent and identically distributed. In that section we considered a special case of this model, in which ancestors had age zero at time zero. Then the distributions of $\xi_i^0(a)$ and $\xi_j(a)$ coincide, as do those of T_i^0 and T_j. In the more general setup discussed here, distributions for offspring usually differ from those of the initial population.

Let $B(t)$ denote the total number of births in the population up to time t. Then, if η_j denotes the time of the jth birth,

$$B(t) = \sum_{i=1}^{Z_0} \xi_i^0(t) + \sum_{j=1}^{B(t)} \xi_j(t - \eta_j) , \tag{4.16}$$

which can be written as

$$B(t) = \sum_{i=1}^{Z_0} \xi_i^0(t) + \int_0^t \xi_{B(u)}(t - u) \, dB(u) . \tag{4.17}$$

(For simplicity, we disregard the possibility of simultaneous births here.) The population size at time t equals

$$
\begin{aligned}
Z_t &= \sum_{i=1}^{Z_0} 1_{[0,T_i^0)}(t) + \sum_{j=1}^{B(t)} 1_{[0,T_j)}(t - \eta_j) \\
&= \sum_{i=1}^{Z_0} 1_{[0,T_i^0)}(t) + \int_0^t 1_{[0,T_{B(u)})}(t - u) \, dB(u) .
\end{aligned}
\tag{4.18}
$$

We define

$$\mu^0(t) = \mathbb{E}[\xi_i^0(t)] \qquad \mu(t) = \mathbb{E}[\xi_i(t)] \tag{4.19}$$

and

$$S^0(t) = \mathbb{P}\{T_i^0 > t\} \qquad S(t) = \mathbb{P}\{T_i > t\} , \tag{4.20}$$

in which we assume that $\mu^0(t)$ and $\mu(t)$ are finite for all t. The reproduction function $\mu(t)$ is introduced in Section 3.3, and the survival function $S(t)$ is 1 minus the life-span distribution function, $L(t)$. Setting $b(t) = \mathbb{E}[B(t)]$ and $m(t) = \mathbb{E}[Z_t]$, we have

$$b(t) = Z_0 \mu^0(t) + \int_0^t \mu(t - u) \, db(u) , \tag{4.21}$$

and

$$m(t) = Z_0 S^0(t) + \int_0^t S(t - u) \, db(u) . \tag{4.22}$$

The solution of Equation (4.21) can be obtained by setting $b_0(t) = Z_0 \mu^0(t)$ and iterating

$$b_{k+1}(t) = Z_0 \mu^0(t) + \int_0^t \mu(t - u) \, db_k(u) . \tag{4.23}$$

The sequence $\{b_k(t)\}$ increases monotonically, and $b(t) = \lim_{k \to \infty} b_k(t)$, where $b(t)$ may be infinite. If, however, there exists an age $a > 0$ such that the expected number of offspring up to that age, $\mu(a) < 1$, then $b(t) < \infty$ for all t. This is certainly a reasonable assumption in any biological application.

As in Section 4.1.1, the Law of Large Numbers suggests the approximations

$$\frac{B(t)}{Z_0} \approx \bar{b}(t) = \frac{b(t)}{Z_0} \qquad \frac{Z_t}{Z_0} \approx \bar{m}(t) = \frac{m(t)}{Z_0} , \qquad (4.24)$$

where $\bar{b}$ satisfies Equation (4.21), with Z_0 replaced by 1, and $\bar{m}$ is given by Equation (4.22), with Z_0 replaced by 1 and b replaced by $\bar{b}$. This approximation is, indeed, valid in the sense that $B(t)/Z_0$ converges to $\bar{b}(t)$ and Z_t/Z_0 converges to $\bar{m}(t)$ as Z_0 goes to infinity.

Limit theorems of this type can be found in Kurtz (1983) and Solomon (1987). The limiting model is essentially the standard continuous-time demographic model (e.g., Keyfitz 1977).

4.2   Discrete-Time Dynamical Systems as Population Models

F.C. Klebaner

Deterministic discrete-time processes are characterized by a recurrence equation $x_{n+1} = f(x_n)$ that specifies the relation between the value of a state at time n, x_n, and its value one time step later, x_{n+1}. Such processes are used as models for changes in population size or density. An example is the Ricker model introduced in Section 1.4. The (asymptotic) dynamics of these models is described in many textbooks on biological systems (e.g., Case 2000). We thus include only a short survey here, to connect the field with branching processes and to acquaint those unfamiliar with the area with some basic concepts. Others may well skip Section 4.2.1 and proceed to Section 4.2.2, in which we relate these models to branching processes.

4.2.1   Dynamics of deterministic models

Recall the formulation in Section 4.1 of the Ricker model

$$z_{n+1} = m z_n e^{-b z_n} , \qquad (4.25)$$

where z_n denotes population size. It can be reformulated in terms of population density by the transformation $x_n = z_n/K$, where K is the area occupied by the population, and rescaling the parameter b. This yields a model of the form

$$x_{n+1} = x_n R(x_n) , \qquad (4.26)$$

which is the general form of density-dependent models that we consider here. Note that in this section we use "density" in a very loose way. If K is an area, x_n is, indeed, a density in the strict sense, but in the following we also consider cases in which K represents a different constant, such as carrying capacity in the logistic model (see below). In either case, we refer to x_n as a density.

The function $R(x)$ is the individual reproduction function. This function typically depends on the availability of resources and, therefore, it is reasonable to assume that, as population density increases, $R(x)$ tends to zero (the reader may verify that in the Ricker model this is indeed so). This is called negative density dependence.

In the Ricker model, and the other examples given in this section, $R(x)$ tends to a positive constant as x approaches zero, which signifies that at low population densities there is no resource limitation and the population grows at a constant rate independent of its density. The branching process equivalent of this situation is the Galton–Watson process. There are also models, however, in which a negative density dependence occurs at low densities too. This so-called "Allee effect" (Allee 1931) can, for instance, result from the difficulty of finding suitable mates when population size is small.

In ecological models a limitation on population size is often assumed, usually called a *carrying capacity* (of the environment for the population in question). When the population is far from its carrying capacity it reproduces at a constant rate, and the reproduction rate declines when the population approaches its carrying capacity and resources become exhausted. In this case, we can let K represent the carrying capacity. The simplest such dependence is linear, with x denoting the density $0 \le x \le 1$,

$$R(x) = r(1 - x) \,. \tag{4.27}$$

This is how the logistic model for population dynamics is obtained,

$$x_{n+1} = r x_n (1 - x_n) \,. \tag{4.28}$$

It is the most studied scheme from a mathematical perspective (see Thompson and Stuart 1986), and represents a prototype of simple models that exhibit complex behavior (May 1976).

Another famous density-dependent population approach uses the relation

$$x_{n+1} = r x_n / (1 + a x_n)^b \,, \tag{4.29}$$

which was applied by Hassell *et al.* (1976) to compare population dynamics of 28 species of insects (see also Smith 1974).

Iterating the recurrence equation

$$x_{n+1} = f(x_n) = f(f(x_{n-1})) \tag{4.30}$$

and so on, we obtain that the density in the nth generation is given by the nth iterate of the function f evaluated at the initial population density x_0,

$$x_n = f^{(n)}(x_0) \,, \tag{4.31}$$

where $f^{(n)}$ means f taken n times.

Behavior of the iterates of functions such as the logistic or Ricker is studied within non-linear discrete-time dynamics and chaos theory. Typically, the functions have a shape parameter (r in the logistic model) and, depending on the value of this parameter, the iterates of f converge to a fixed point (a stable fixed point), or they oscillate between a finite number of points (convergence to a stable cycle), or they exhibit chaotic behavior (which means that their positions for large n are described by a distribution function, rather than a limited set of predetermined points).

A point x is called a *fixed point* for f if $f(x) = x$. Clearly, for functions of the form $f(x) = xR(x)$ the point $x = 0$ is a fixed point, but there may be others. What happens if the population density is very low, near to zero, but positive? Development in this case depends on the stability of the point zero. A fixed point x^* is called *stable* or *attracting* if for all the initial points x_0 near x^*, the iterates $f^{(n)}(x_0)$ converge to x^* as $n = 1, 2, \ldots$ increases.

Alternatively, if there is an interval that includes x^* such that for some n the iterate x_n is outside this neighborhood, the fixed point x^* is called *unstable* or *repelling*. A sufficient condition for the stability of a fixed point x^* is that $|f'(x^*)| < 1$, and for it to be unstable the condition is $|f'(x^*)| > 1$. In the case $|f'(x^*)| = 1$, x^* may be attracting or repelling. For a necessary and sufficient condition for a fixed point to be attracting, see, for example, Theorem 2.2.1 in Sharkovskii *et al.* (1993).

If 0 is a repelling fixed point, f has another fixed point x^* that may also be either attracting or repelling. If it happens to be attracting, the long-term iterates converge to it, and if it is repelling, a cycling behavior occurs. For example, a cycle of period 2 means that there are two points, x_1^* and x_2^*, such that $f(x_1^*) = x_2^*$ and $f(x_2^*) = x_1^*$. This cycle is attracting if for large even values of n the iterates x_n are in the vicinity of a point x_1^* and if for large odd values of n the iterates x_n are in the vicinity of a point x_2^*. We can also describe a cycle of period 2 by means of fixed points of the twice-iterated function, $f^{(2)}$. If f has a two-cycle $\{x_1^*, x_2^*\}$, then $f^{(2)}$ has two fixed points x_1^* and x_2^*. The cycle is stable or attracting if these fixed points are stable. For a cycle to be attracting it is enough that

$$|f^{(2)\prime}(x_1^*)| < 1 \,. \tag{4.32}$$

Using the chain rule of differentiation we find that $f^{(2)\prime}(x_1^*) = f'(f(x_1^*))f'(x_1^*) = f'(x_2^*)f'(x_1^*)$. Thus, a sufficient condition for a cycle to be attracting is given by

$$|f'(x_1^*)f'(x_2^*)| < 1 \,. \tag{4.33}$$

Of course, a cycle of period d and its stability are defined similarly.

A large class of dynamical systems has asymptotically periodic trajectories: a dynamical system is called *simple* if each of its trajectories is periodic or asymptotically periodic. Moreover, there is a class of simple dynamical systems in which the stable cycle is unique and trajectories (x_n) are attracted to it for almost all initial points x_0 *(ibid)*.

Example 4.1 We examine the behavior of the iterates in the logistic model $x_{n+1} = rx_n(1 - x_n)$. The function $f(x) = rx(1 - x)$ has a single fixed point 0 if $r \le 1$, which is attracting, and $x_n \to 0$ as $n \to \infty$ for any x_0. For $1 < r \le 3$, the fixed point zero becomes repelling and another fixed point appears, $x^* = 1 - 1/r$. This point is attracting if $r < 3$, as $|f'(x^*)| < 1$, $(f'(x^*) = 2 - r)$. If $r = 3$, x^* is still attracting, although $|f'(2/3)| = 1$. If $r \le 3$, then $x_n \to x^*$ as $n \to \infty$ for any $x_0 \neq 0, 1$. When $r > 3$, then $x^* = 1 - 1/r$ is repelling as $|f'(x^*)| > 1$. For values of r in the range $3 < r < 1 + \sqrt{6} \approx 3.449$, f has a stable cycle of period 2. These points are determined as roots of $f^{(2)}(x) = x$. For all x_0 outside an (actually finite) exceptional set, x_n converges to this cycle.

When r increases further, $1 + \sqrt{6} \leq r$, the stable two-cycle becomes unstable and a stable cycle of period 4 is created. For all points x_0 outside an exceptional set (which contains the fixed points, cycles, and their pre-images), $f^{(n)}(x_0)$ converges to this four-cycle. This phenomenon (appearance of stable cycles of higher powers of two instead of unstable ones) is known as period doubling bifurcation, and continues until r reaches some value $r_c \approx 3.569\ldots$, at which point no stable trajectories of longer periodicity exist, and the system displays no simple dynamics.

For any $r < r_c$ there is a stable cycle of period 2^k (k depends on r), and $f^{(n)}(x_0)$ converges to this cycle for all x_0 except those that go to repelling cycles of periods 2^i, $i = 0, 1, \ldots, k - 1$. The value r_c is known as the value for the onset of chaos. For $r > r_c$ there are infinitely many cycles, all of which may be repelling. For certain values (*periodic windows*) of the parameter the system admits attracting cycles of periods not restricted to the powers of 2.

When $r = 4$, the long-term behavior of $f^{(n)}(x_0)$ is described by the probability distribution

$$\frac{1}{\pi \sqrt{x(1 - x)}} \, , \tag{4.34}$$

that is, for large n the probability of finding $f^{(n)}(x_0)$ in an interval $[a, b]$ is given by

$$\int_a^b \frac{dx}{\pi \sqrt{x(1 - x)}} \, . \tag{4.35}$$

$$\diamond \diamond \diamond$$

Remark. For the Ricker model, $f(x) = xe^{r-x}$, a bifurcation to a cycle of period 2 occurs at $r = 2$, then further from a 2-cycle to a 4-cycle at 2, 2.526, etc., and the value of onset of chaos is $r_c \approx 2.692$.

$$\diamond \diamond \diamond$$

4.2.2 Density dependent branching processes and dynamical systems

Deterministic models of the form $x_{n+1} = f(x_n)$ are macroscopic, they give a rule according to which the whole population evolves. Branching processes, however, are microscopic, built upon the individual behavior of population members and determined by their offspring distribution.

As we have seen, though, the macroscopic models allow feedback in the form of the effects of population density on growth, and may exhibit periodic behavior, which does not appear in branching processes. The bridge between the two approaches is provided by branching processes with a similar feedback (i.e., population-size- and density-dependent branching processes).

In branching processes dependent on population size, the distribution of offspring numbers depends on the size of the population z, and in density-dependent branching processes the distribution of offspring depends on the population density, or concentration z/K, where K is a parameter such as (but not necessarily) the carrying capacity. If this parameter is fixed, there is no difference between the two types of models, but when it becomes large, density-dependent models may simplify and allow approximations. Indeed, for large values of K they reduce to deterministic dynamical systems plus a small noise. The dynamical system part represents the deterministic approximation and the noise admits a Gaussian approximation.

This Gaussian approximation is different from the diffusion approximation of branching processes, treated in the preceding sections. In diffusion approximations the time is scaled to arrive at a continuous-time process, but in this case time remains discrete and the limiting process is not continuous, but jumps as the corresponding dynamical system.

As indicated in Section 2.6, density-dependent branching processes are defined in the same way as the classic Galton–Watson process, except that offspring distributions are allowed to depend on population density. This can be written as

$$Z_{n+1} = \sum_{j=1}^{Z_n} \xi_{j,n}(Z_n/K) \,, \tag{4.36}$$

to indicate that the distribution of $\xi_{j,n}(Z_n/K)$ is dependent on Z_n/K. The offspring numbers themselves are independent, with the common distribution being that of $\xi(x)$ if $Z_n/K = x$.

Example 4.2 Consider branching processes that occur in polymerase chain reactions (PCRs). PCR is a stepwise procedure in molecular biology whereby in each step DNA molecules either remain or are replaced by two copies. It is further described in Section 7.5.

The reaction can be modeled as a (single-type) Galton–Watson branching process, in which each individual has one or two offspring in the next generation. The probability of the latter event is usually termed the *efficiency* in the present connection. It is natural from the experimental setup that the efficiency of the reaction should decrease.

Under classic, so-called Michaelis–Menten kinetics, largely valid for enzymatic reactions, it follows that the probability of successful copying (two offspring) is given by

$$p(z) = \frac{K}{K+z} \,, \tag{4.37}$$

where z is the number of molecules and K is the Michaelis–Menten constant of the reaction. Initially, the efficiency is close to 1, since K is large compared to z.

The result is a density-dependent binary splitting Galton–Watson process, whereby the alternative to splitting is remaining into the next generation (experiment cycle), or equivalently giving birth to one offspring. The offspring number $\xi(z)$ takes values 1 and 2 with probabilities $1 - p(z)$ and $p(z)$, respectively.

$$\diamond \diamond \diamond$$

As shown in Section 4.2.1, it is more convenient to consider the density process $x_n^K = Z_n/K$, which evolves according to

$$x_{n+1}^K = \frac{1}{K} \sum_{j=1}^{Kx_n^K} \xi_{j,n}(x_n^K) \,. \tag{4.38}$$

We show that this model is a stochastic analog of the deterministic model $x_{n+1} = f(x_n)$ with a suitable function f. Indeed, denote by $R(x) = \mathbb{E}[\xi(x)]$ the mean offspring number when the population density is x. By subtracting and adding it within the sum, we have

$$x_{n+1}^K = \frac{1}{K} \sum_{j=1}^{Kx_n^K} R(x_n^K) + \frac{1}{K} \sum_{j=1}^{Kx_n^K} [\xi_{j,n}(x_n^K) - R(x_n^K)] , \qquad (4.39)$$

or

$$x_{n+1}^K = x_n^K R(x_n^K) + \eta_n^K = f(x_n^K) + \eta_n^K , \qquad (4.40)$$

where $f(x) = xR(x)$ and

$$\eta_n^K = \frac{1}{K} \sum_{j=1}^{Kx_n^K} [\xi_{j,n}(x_n^K) - R(x_n^K)] \qquad (4.41)$$

is random. This random term is small for large values of K, essentially by the Law of Large Numbers. Let x_n be the nth iterate of $f(x) = xR(x)$, starting from $x_0 = Z_0/K$. The results below state that, for large values of the carrying capacity, the process x_n^K is approximated by the deterministic sequence x_n, with the difference $x_n^K - x_n$ being approximately normal with mean zero and variance of order $1/K$.

In the next two theorems we assume that the function $f(x) = xR(x)$ has a continuous derivative and that the variances $\sigma^2(x)$ of offsprings $\xi(x)$ are bounded by some constant, say $\sigma^2(x) \leq C$.

Theorem 4.1 (Consistency theorem) *For any fixed n, as $K \to \infty$, $x_n^K \to x_n$ in probability.*

The proof uses induction on n and Chebyshev's inequality,

$$\mathbb{E}[(\eta_n^K)^2] = \mathbb{E}[\mathbb{E}[(\eta_n^K)^2|x_n^K]] = \frac{1}{K}\mathbb{E}[\sigma^2(x_n^K)] \leq C/K \to 0 . \qquad (4.42)$$

Theorem 4.2 (Fluctuation theorem) *Assume, in addition, that the third absolute moments $\mathbb{E}[(\xi(x) - R(x))^3]$ are bounded. Then, for any fixed n, as $K \to \infty$, $(x_n^K - x_n)\sqrt{K}$ converges in distribution to a normal random variable $N(0, D_n^2)$, where $D_0 = 0$ and D_n is defined by the recurrence relation*

$$D_{n+1}^2 = x_n\sigma^2(x_n) + [f'(x_n)D_n]^2 . \qquad (4.43)$$

The proof of this result follows by induction on n and an analysis of characteristic functions, and can be found in Klebaner (1993) under less stringent assumptions. In Klebaner and Nerman (1994) these results are established not only for a single fixed n, but also for any collection of times $n_1, n_2, \ldots, n_k$ (which corresponds to the convergence of processes, or the functional version of the limit theorem). Watkins (2000) generalized these results to multi-type processes (structured populations) and referred to them as consistency and fluctuation theorems in biology.

Example 4.2 (*continued*) We can express the probability of successful division in PCR as a function of the so-called dimensionless reduced concentration $x = z/K$ (Schnell and Mendoza 1997a),

$$p(x) = \frac{1}{1+x} \, . \tag{4.44}$$

Since the expected number of offspring per individual is

$$m(x) = 1 - p(x) + 2p(x) = 1 + 1/(1+x) \, , \tag{4.45}$$

we obtain that the population density process has the representation

$$x_{n+1}^K = f(x_n^K) + \eta_n^K \, , \tag{4.46}$$

with $f(x) = xm(x) = x/(1+x)$. The reproduction variance can be checked easily to satisfy

$$\sigma^2(x) = p(x)(1 - p(x)) = x/(1+x)^2 \, , \tag{4.47}$$

obviously bounded. Thus, for large values of K, the sequence $x_n^K = Z_n/K$ can be approximated by the deterministic sequence x_n obtained as the nth iterate of f starting at x_0. This is further developed in Jagers and Klebaner (2003).

4.3   Branching Processes and Structured Population Dynamics

M. Gyllenberg and P. Jagers

4.3.1   Introduction

Most of the classic deterministic population models developed by Lotka, Volterra, and others during the "golden age" of theoretical ecology in the 1930s are concerned with a single homogeneous population or with the interaction between several homogeneous populations. In particular, these models are based on the assumption that, at least on average, all individuals in a population behave identically with respect to reproduction, survival, exploitation of resources, competition, and other processes of importance for the dynamics of the population. They are deterministic versions of single-type Markov branching processes, or modifications of such processes.

In reality, matters are more complex. The reproductive behavior is age specific and depends on nutrition, and the same applies to survival. A predator is hardly likely to catch a prey that is larger, stronger, and quicker than itself. The list could be continued almost forever. In many cases, differences between individuals make a difference for the resultant dynamics. Moreover, it may be important to predict the *composition* and not only the size of a population (think, for instance, of human demography in which estimates of the future age distribution of the population influence socio-economic decision making). In branching processes, such considerations lead to multi-type processes with type spaces of varying complexity. In deterministic approaches, they take us to the realm of *structured populations* (Metz and Diekmann 1986). In this section a short introduction to the modeling of structured populations as developed by Diekmann *et al.* (1993, 1998, 2001, 2003) and Gyllenberg *et al.* (1997) is given, and a comparison made between deterministic modeling and the approach based on branching processes.

As a historical footnote, as we have pointed out, branching processes evolved from the simple Galton–Watson processes into processes that mirror more aspects of the real world. Structured population dynamics developed from straightforward differential equation models of growth into the general theory presented here. Thus, the backgrounds are different both in topic (extinction versus growth) and methods (probability versus differential equations). However, from the different starting points the two, in essence, have converged gradually, even though the notation and methods may differ. So as to acquaint the reader with the formalism of structured populations, this section is mainly formulated in the vein of the latter.

The level of mathematical abstraction is higher than in the rest of this book – the corresponding general theory for branching processes in continuous time with types of a completely general form has only been sketched in Chapter 3 – but some readers might find it interesting to see a modern and general formulation of structured population dynamics and compare it to that of branching processes.

4.3.2 Modeling structured populations

Individual state and processes at the individual level.   When modeling structured populations the starting point is to single out those characteristics that are essential for the development, survival, and reproductive behavior of an individual. Typically, these may include continuous quantities such as age, size, and spatial location, but often also discrete quantities such as sex and life-history stage (e.g., egg, larva, pupa, and adult in insects) are important. All these quantities are collected into a vector, which is called the *state of the individual*. The set of all conceivable individual states is called the *individual state space* and is denoted by $\mathcal{X}$. As an example, if the population is structured by age and size of individuals, $\mathcal{X}$ is a subset of the positive quadrant of $\mathbf{R}^2$ (see Figure 4.2).

The next task is to model processes at the individual level. Processes to be modeled are development of the individual state (i.e., growth, aging, etc.), survival, and reproduction (how many offspring and with what individual state at birth). We assume that the future development of an individual is determined completely, at least in a stochastic sense, by its present individual state. Thus, modeling at the individual level amounts to the following: given the present state of an individual, describe the probability of that individual still being alive and having an individual state in a given subset of the individual state space at any future time, and also give a similar description of its offspring. It is this Markovian property that allows us to call the elements of $\mathcal{X}$ *states*.

It is important to realize that the dynamical processes mentioned above depend on the environmental conditions. Here, we do not think of the environment as solely determined by exogenously given factors like temperature, but of something that not only affects, but also is affected by, the population. To qualify as

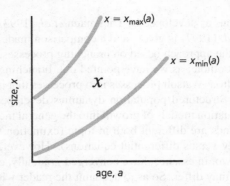

Figure 4.2 The individual state space $\mathcal{X}$ for a population structured by age and size of individuals. The boundary of $\mathcal{X}$ consists of the curves $x = x_{min}(a)$ and $x = x_{max}(a)$, which give for each $a \geq 0$ the minimum and maximum possible size of individuals of age a, and the line segment $[(0, x) : x_{min}(0) \leq x \leq x_{max}(0)]$, which represents the states at birth.

an environmental condition, the variable must be such that all the interactions between individuals (e.g., competition for food) are described by it. Describing this feedback mechanism is also part of the modeling.

If the feedback loop is cut and the environmental condition assumed to be given as a function of time, a *linear* model is obtained, in which there are no interactions between individuals. In stochastic terminology, individuals live and reproduce independently of each other, and a classic branching process obtains. The linearity at the level of expectations is an analytic expression of the underlying independence.

For branching processes, the concept of states and state variables is introduced in Section 2.5. Types can then be viewed either as "states at birth" or as describing the whole path through various states during a life. There is a substantial mathematical literature on branching random walks, branching diffusion, and so-called superprocesses, in which the objects of study are random measures and their development under "branching" [see Dynkin (1991) and Dawson (1993) for overviews]. However, these are Markov processes in real time; that is, they presume exponentially distributed life spans, and have not been developed with a focus on biological issues (which is not to say that they could not become useful in biology). In general branching processes, with arbitrary life spans and birth that occurs as a point process, little attention has been paid to modeling the individual's path through a state space.

Population state. Once the mechanisms at the individual level are specified, the next task is to lift the model to the population level: what we want to study are phenomena at that level. This is simply a matter of bookkeeping (i.e., keeping track of all the individuals in the population). The state of the population is, by definition, the *distribution* of individual states within the population. Mathematically speaking, the population state is a measure m of the individual state space $\mathcal{X}$, with the interpretation that $m(A)$ is the expected number of individuals with an individual state in the subset A of $\mathcal{X}$. Observe that the population state m contains

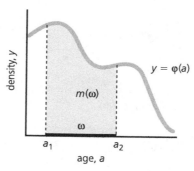

Figure 4.3 The age–density φ and the population state m for an age-structured population.

information about both the *size* of the population [indeed, $m(\mathcal{X})$ is the (expected) total number of individuals in the population] and its *composition* (relative frequencies of different individual states).

From here on, we do not spell out "expected" when talking about population numbers, even though in the present theory these need not be integer valued, and are, indeed, often expressed through continuous densities. However, this is a deterministic theory, and the distinction between expectation of entities and the entities themselves is not maintained.

Even though a measure is conceptually simpler than a density, many people with a background in physics or physics-inspired applied mathematics find densities more familiar. We therefore briefly give the connection between measures and densities. For simplicity, we restrict this to the one-dimensional case $\mathcal{X} = [0, \infty)$ (think of age-structured problems). The relation between the age density φ and the population state is now

$$m(A) = \int_A \varphi(a) \, da, \quad A \subset \mathcal{X}. \tag{4.48}$$

In particular, if $A = [a_1, a_2]$, we obtain that $\int_{a_1}^{a_2} \varphi(a) \, da$ is the number of individuals with age between a_1 and a_2 (see Figure 4.3).

In a branching process formulation, the "bookkeeping" amounts to adding all the individuals present with states in A to obtain, say, Z^A, which is random if the number of individuals or their states are. Then,

$$m(A) = \mathbb{E}[Z^A]. \tag{4.49}$$

The question is then "only" to determine the probabilities that tell us how to compute the expectation. This, of course, depends on the random mechanisms at the individual level, and one point of structured population dynamics is to use general formulations that do not require independence between individuals and also allow the environment to vary from the very beginning. We proceed to this now.

Environmental interaction variable as input. It is natural to start the model formulation by focusing upon the linear problem in which, as described above, the

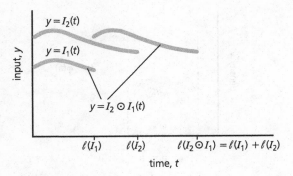

Figure 4.4 The concatenation $I_2 \odot I_1$ of I_1 and I_2.

environment condition or interaction variable I is a given function of time. We also call the interaction variable the *input* of the system. As is well-known from the theory of dynamical systems, one cannot be sure beforehand that, say, a differential equation has a solution for all time – solutions may "blow up" in finite time. One of the first tasks in the study of differential equations is to prove well-posedness, that is, the existence and uniqueness of *local* solutions, and to find conditions that guarantee a solution that is *global* (i.e., existing at all times). We therefore consider inputs I defined on half-open intervals $[0, \ell(I))$, where $\ell(I)$ – in this section – is the *length* or duration of the input I. However, if a population has evolved under the input I_1 and then continues to evolve under another input I_2, we must be able to describe the combined effect. We are therefore led to the following definition of *concatenation* of inputs: the concatenation $I_2 \odot I_1$ of I_1 and I_2 is defined on the interval $[0, \ell(I_1) + \ell(I_2))$ by

$$(I_2 \odot I_1)(t) = \begin{cases} I_1(t) & \text{for } 0 \le t < \ell(I_1) \\ I_2(t - \ell(I_1)) & \text{for } \ell(I_1) \le t < \ell(I_1) + \ell(I_2) . \end{cases} \tag{4.50}$$

Figure 4.4 illustrates the definition of concatenation.

Notice the order of the inputs I_1 and I_2 in the concatenation $I_2 \odot I_1$, which to some readers may seem illogical, but which conforms to common mathematical practice (compare the composition $f \circ g$ of two functions, where you also read from right to left: the function g acts first and only then does f).

4.3.3   Linear structured population models with input

The basic modeling ingredients.   We model individual development and survival as a Markov process with death as an absorbing state. We therefore have to prescribe the transition probabilities of this process. In fact, as our first basic model ingredient we take u_I, which is defined as follows:

$u_I(x, A)$ is the probability that, given the input I, an individual in state $x \in \mathcal{X}$ at a certain time is still alive $\ell(I)$ time units later, and then has its state in A.

As our second ingredient we take the object Λ_I which is similar to u_I but describes reproduction. Λ_I is called the *reproduction kernel* and is defined in the following way:

$\Lambda_I(x, A)$ is the expected number of offspring, with state-at-birth in A, produced by an individual, in state $x \in \mathcal{X}$ at a certain time, within the time interval of length $\ell(I)$ following that time, given the input I.

Whereas individual states for branching processes and transitions between them are discussed only perfunctorily in Section 2.5, the Λ_I for constant I is, of course, nothing but the μ introduced in Section 3.3.2, $\mu(s, A \times B)$ being the expected number of children with types in the set A, born to an s-type mother while she is of age in B.

The interpretation of the ingredients u_I and Λ_I requires that certain consistency relations must hold. The first one is the familiar Chapman–Kolmogorov equation, which in the present notation takes the form

$$u_{I_2 \odot I_1} = u_{I_2} \times u_{I_1} \tag{4.51}$$

for all inputs I_1 and I_2. Here, we have used the $\times$ product, which is defined by

$$\left(k^1 \times k^2\right)(x, A) = \int_{\mathcal{X}} k^1(y, A) k^2(x, dy) . \tag{4.52}$$

The Chapman–Kolmogorov equation, Equation (4.51), is a convenient and compact way to express mathematically what only amounts to common sense, but takes many words to explain. Consider an individual with individual state x at time zero and assume that it develops under the influence of the input I_1. The probability that it is still alive and has individual state in the infinitesimal set dy at time $\ell(I_1)$ is $u_{I_1}(x, dy)$. Now reset the clock to zero and assume that the individual continues to develop under the influence of the input I_2. Assuming that the intermediate state was y, the probability of the individual state being in A at the end of the input interval is $u_{I_2}(y, A)$. Summing over all possible intermediate states y we obtain

$$\int_{\mathcal{X}} u_{I_2}(y, A) u_{I_1}(x, dy) = \left(u_{I_2} \times u_{I_1}\right)(x, A) . \tag{4.53}$$

However, this should, of course, be equal to the probability of having an individual state in A if the individual develops from the starting point to the end point under the influence of the concatenation of I_1 and I_2. The Chapman–Kolmogorov equation expresses precisely this fact.

The transition probability u_I and the reproduction kernel Λ_I are related by the following consistency relation:

$$\Lambda_{I_2 \odot I_1} = \Lambda_{I_1} + \Lambda_{I_2} \times u_{I_1} . \tag{4.54}$$

The interpretation is similar to that of the Chapman–Kolmogorov equation. The first term on the right-hand side of Equation (4.54) represents the offspring born to an individual during the interval $[0, \ell(I_1)]$ when it was under the influence of the input I_1, and the second term represents the offspring born during the interval $[\ell(I_1), \ell(I_1) + \ell(I_2))$. The $\times$ product of Λ_{I_2} with u_{I_1} reflects that at time $\ell(I_1)$ the individual no longer has its initial state, but has developed according to the transition probability u_{I_1}. The sum of these two terms must, of course, be equal to the total (cumulative) offspring $\Lambda_{I_2 \odot I_1}$ produced during the whole interval $[0, \ell(I_1) + \ell(I_2))$.

As indicated, the transition probabilities u_I and the reproduction kernel Λ_I lend themselves directly to an interpretation in terms of the life and reproduction of a branching individual. Its state evolves in a Markov process with the given transition probabilities and it gives birth according to a reproduction point process, with expectation given by Λ_I. However, the latter does not suffice to describe reproduction: we not only need expectation, but also the whole distribution. Once this has been made precise, we can proceed to defining the process Z_t^A, which gives the number of individuals at any time $t \geq 0$ with states in a set A from an initial condition and the behavior of individuals.

This is completely parallel to the definition of general branching processes. Only questions of existence, that is, whether the assumptions made are consistent, become more involved. These matters are too complex for this book in the case of independent and (given the type) identically distributed reproduction processes, as pointed out in the introduction to Chapter 3. In the present case the problems of existence of the underlying processes become horrendous, and structured population dynamics contents itself by showing that the arising expectations are well defined.

The generation expansion.   As explained in the preceding subsection, the reproduction kernel Λ_I represents all the offspring born to an individual given the input I. Using solely this ingredient we can actually compute the corresponding object, not for direct offspring, but for grandchildren. In the general case, the formula is notationally, but not conceptually, rather complicated, so we refrain from displaying it. Instead, we represent it symbolically as

$$\Lambda_I^{2*} = (\Lambda * \Lambda)_I \ . \tag{4.55}$$

Again, we refer to the article of Diekmann *et al.* (2001) for a precise definition of the $*$ operation in a structured population dynamics context. The same operation, of course, appears in branching processes (e.g., Jagers 1989). The choice of the symbol "$*$," which is the usual symbol for convolution (see below) is no accident. We illustrate this with the simplest possible example: age-structured population dynamics with constant input. By the definition of age a, all individuals are born with the same state, $a = 0$. We therefore have

$$\Lambda_I(0, A) = L(\ell(I))\delta_0(A) \ , \tag{4.56}$$

where δ_0 is the point mass concentrated at the origin and $L(a)$ is the expected number of offspring born to an individual until it reaches age a. The reproduction kernel, thus, is characterized essentially by the scalar function L. The cumulative number of grandchildren of an individual until it reaches age a is given by the convolution

$$(L * L)(a) = \int_0^a L(a - u)L(du) \tag{4.57}$$

and

$$\Lambda_I^{*2}(0, A) = (L * L)(\ell(I))\delta_0(A) . \tag{4.58}$$

The general case is simply a straightforward extension of this simple case.

One can, of course, continue in the same manner to obtain higher generations. The threefold convolution of the reproduction kernel with itself gives the great-grandchildren, etc. The *clan* (i.e., the totality of all descendants of a given ancestor) is obtained by summing over all generations,

$$\Lambda_I^c = \sum_{n=1}^{\infty} \Lambda_I^{n*} . \tag{4.59}$$

In Equation (4.59), $\Lambda_I^{1*} = \Lambda_I$ and $\Lambda_I^{n*} = \left(\Lambda^{(n-1)*} * \Lambda\right)_I$ for $n \geq 2$.

The positivity of the family Λ_I guarantees that Equation (4.59) has a meaning in any case, but additional conditions on Λ_I (e.g., a reproduction delay that does not allow newborns to give birth) guarantee that the sum converges to something finite (Diekmann *et al.* 1998).

Λ_I^c has the same interpretation as Λ_I, but now refers to the whole clan. Since every member of the clan is either a child of the ancestor or a child of a member of the clan, or, alternatively, either a child of the ancestor or a member of the clan of a child of the ancestor, we obtain the consistency relation

$$\Lambda_I^c = \Lambda_I + \left(\Lambda * \Lambda^c\right)_I = \Lambda_I + \left(\Lambda^c * \Lambda\right)_I . \tag{4.60}$$

Alternatively, one can view Equation (4.60) as an equation in which the Λ_I are the data and Λ_I^c the unknown. The discussion above tells us that the generation expansion (4.59) defines a solution of this equation. One can show that under mild conditions this solution is unique (Diekmann *et al.* 1998).

Now let $u_I^c(x, A)$ be $u_I(x, A)$ plus the expected number of descendants (i.e., children, grandchildren, great-grandchildren, etc.) of an individual, initially of individual state x, that are still alive and have individual state in A, $\ell(I)$ time units later. This verbal description of u_I^c can be formalized as

$$u_I^c = u_I + \left(u * \Lambda^c\right)_I . \tag{4.61}$$

Note that to obtain the clan kernels we only have to solve one equation, namely Equation (4.60). Once Λ_I^c has been solved from this equation, u_I^c is obtained from the explicit Equation (4.61).

The branching process that underlies the generation expansion is the embedded Galton–Watson process of Section 3.1, in its multi-type version, so that $\zeta_n^A(I)$ is the number of individuals of the nth generation with types in A, born up to time

$\ell(I)$. Thus,

$$\Lambda_I^{n*} = \mathbb{E}[\zeta_n^A(I)] , \tag{4.62}$$

and the (expected) clan is the expectation of

$$\sum_{n=0}^{\infty} \zeta_n^A(I) . \tag{4.63}$$

The dynamical system at the level of the population.   As defined, the population state is the distribution of individual states and can therefore be represented by a measure m on the individual state space $\mathcal{X}$. The dynamical system T_I that describes the dynamics at the population level should, therefore, be such that, given the initial population state m_0 and the input I, $T_I\, m_0$ is the population state at time $\ell(I)$. The population at time $\ell(I)$ consists of those individuals present in the initial population that are still alive plus all living descendants of the initial population. For each individual, this clan is given by u_I^c. Summing over all individuals present, initially we obtain the composition of the population at time $\ell(I)$,

$$(T_I\, m_0)(A) = \int_{\mathcal{X}} u_I^c(x, A) m_0(dx) . \tag{4.64}$$

The mathematical task is now to prove that Equation (4.64) indeed defines a dynamical system at the level of the population. By this we mean that if T_{I_1} is first applied to the initial state m_0 and then T_{I_2} to the resultant state, the same state should be obtained as that obtained by the direct application of $T_{I_2 \odot I_1}$ to the initial state. In symbols:

$$T_{I_2}\, T_{I_1} = T_{I_2 \odot I_1} . \tag{4.65}$$

Equation (4.65) is the analog of the Chapman–Kolmogorov equation at the level of the population. Indeed, to prove that Equation (4.65) holds true, it has to be shown that u_I^c satisfies the Chapman–Kolmogorov equation

$$u_{I_2 \odot I_1}^c = u_{I_2}^c \times u_{I_1}^c . \tag{4.66}$$

Diekmann *et al.* (1998) proved Equation (4.66). It turned out that the shortest way to prove Equation (4.66) is first to prove the clan analog of the consistency Equation (4.54):

$$\Lambda_{I_2 \odot I_1}^c = \Lambda_{I_1}^c + \Lambda_{I_2}^c \times u_{I_1}^c . \tag{4.67}$$

The need to prove that the dynamical system is well defined arises precisely because we have not proved that our assumptions yield a well-defined probabilistic structure for the underlying branching process. However, as pointed out in Section 3.3, this should be intuitively clear (anyway, it is a mathematical problem outside the realm of this book). In the same spirit, we have not discussed the existence and uniqueness matters mentioned above.

Conclusions. Before we continue, let us recapitulate the situation for the linear-structured population problem with input. The data or ingredients of the problem are two kernels u_I, which describes individual development and survival, and Λ_I, which describes individual reproduction. Using solely these ingredients we *construct*, through the generation expansion (4.59), the corresponding objects at the level of the clan. Simple bookkeeping allows us to lift the system to the level of the population. The system (4.64) is, indeed, a dynamical system and it is the unique solution of the population problem in the sense that the clan kernel Λ_I^c is the only kernel that satisfies the consistency relation (4.60).

4.3.4   State at birth

In Section 4.3.3 we made no assumptions concerning the sets A in which $\Lambda_I(x, A)$ takes on non-zero values, that is, we allowed newborns to have any state in $\mathcal{X}$. In most models the set $\mathcal{X}_b$ of possible states at birth is considerably smaller than $\mathcal{X}$. For instance, if age a is a component of the individual state, every individual state in $\mathcal{X}_b$ has $a = 0$ (recall Figure 4.2). Often $\mathcal{X}_b$ is actually a finite set, an example being an age- and sex-structured model in which $\mathcal{X}_b = \{(0, \text{female}), (0, \text{male})\}$.

Since every individual present in the population is either part of the initial population or is born after the initial time, we can base our bookkeeping solely on the newborns. This usually facilitates the analysis considerably. In the case of a finite number of states at birth, the problem essentially reduces to a finite dimensional problem. What is meant precisely by this is made clear in Section 4.3.5. This is also true when the number of possible states at birth is finite in a stochastic sense. For instance, if the states at birth are distributed according to a fixed probability distribution independent of the parent, we arrive at a one-dimensional problem.

In certain models, most notably size-structured models of populations of cells that reproduce by fission (see Section 3.3; Diekmann *et al.* 1984; Gyllenberg 1986), there are infinitely many sizes at birth, which depend on the size of the mother cell, but before a cell can reproduce it has to pass a threshold size. This threshold size serves as a *renewal state* very similar to the state at birth, and the bookkeeping can be based on the flux of individuals through this renewal state (Diekmann *et al.* 1998; Jagers 2001).

Basing the bookkeeping on newborns makes the connection with multi-type branching processes obvious. An individual's state at birth is determined (at least in a stochastic sense) by its mother and the state at birth determines the future behavior of the individual, given the input. The state at birth therefore corresponds precisely to the type of an individual, as defined in Chapter 2.3.

4.3.5   Constant environment

Now we consider the case of constant inputs. This situation can arise in two different ways. Firstly, the model may be linear and time independent. This is usually a good approximation in the initial phase of an invasion when any density-dependent effect has not begun to operate and the individuals are still independent. Secondly, the original model may be non-linear, but the population has reached a steady state. At a steady state the environmental condition is, of course, constant in time.

We denote constant inputs by $\bar{I}$. With a slight abuse of the notation, we use the same symbol $\bar{I}$ to denote the constant value the input takes on. Being constant, $\bar{I}$ is defined for all times: $\ell(\bar{I}) = \infty$.

When the input is constant, the time of birth does not matter, so to determine whether the population grows, declines, or remains constant we can work either in real time or consider the population from a generation perspective. It follows that we need not bother about individual state development as described by $u_{\bar{I}}$; we only need to consider the reproduction kernel $\Lambda_{\bar{I}}$. Hence, in this constant environment case, the structured population theory is really another formulation of the expected values of general branching processes.

Consider, now, a population state m concentrated on $\mathcal{X}_b$. The state m thus represents a collection of newborns. They need not have been born at the same time, but we think of them as belonging to the same generation. We now define the *next generation operator* $W_{\bar{I}}$ by

$$W_{\bar{I}} m = \Lambda_{\bar{I}} \times m , \tag{4.68}$$

or, in more detail, by

$$\left(W_{\bar{I}} m \right)(A) = \int_{\mathcal{X}_b} \Lambda_{\bar{I}}(x, A) m(dx), \quad A \subset \mathcal{X}_b . \tag{4.69}$$

The measure $W_{\bar{I}} m$ gives the distribution of states at birth of all the children to those individuals characterized by the state-at-birth distribution m. Notice that, since $\ell(\bar{I}) = \infty$, we are, indeed, considering life-long production of offspring. Hence $W_{\bar{I}} m$ describes the next generation, and

$$\left(W_{\bar{I}} m \right)(A) = \int_S \mu(s, A \times \mathbb{R}_+) m(ds) = \mu(m, A \times \mathbb{R}_+) , \tag{4.70}$$

in the notation of Section 3.3.2.

The parallel between the two approaches means that the Perron–Frobenius theory discussed earlier is also relevant here. Though notation and arguments differ, the essentials remain the same: the Malthusian parameter, the stable state, the type distribution, etc. A notable difference is that structured population dynamics, through its deterministic philosophy, regards the critical case as evolving toward a stationary limit, whereas the underlying stochastic process dies out. In the former interpretation, limits of expectations of critical populations are much more interesting, and should often mirror pseudo-stationarities in the stochastic world (see the discussion in Sections 6.8 and 6.9). This is the background for the search for fixed points in Section 4.3.6.

4.3.6 Non-linear structured population models

In Section 4.3.3 we point out that, under natural assumptions, the ingredients u_I and Λ_I uniquely determine a linear dynamical system T_I at the level of the population. This result was obtained under the condition that the input I is given.

In this section we consider the situation in which the input I is not given beforehand, but fed back into the system from an output. In this way we account for how the population affects its own environment and model interactions between individuals.

We need one more ingredient, namely a function denoted by γ that describes how individuals contribute to the value of the environmental condition. This contribution is assumed to be state specific (i.e., γ is assumed to be a function of the individual state). A population with state m therefore produces an output

$$H(m) = \gamma \times m = \int_X \gamma(x)m(dx) . \tag{4.71}$$

More complicated output maps are, of course, conceivable. H could, for instance, be non-linear or depend on the input I. From a stochastic viewpoint a problem arises here, as all relations are between expected entities. Thus, as an example, the reproduction intensity at a certain time could be a function of the expected population size, but not of the actual one. Also, if $b(n)$ is an individual birth intensity in a population of size n, as we have seen repeatedly, $\mathbb{E}[b(Z_t)]$ could be quite different from $b(\mathbb{E}[Z_t])$.

Consider now a population initially with state m_0 and assume that the input is given by I. We let $\rho(t)I$ denote the restriction of I to the interval $[0, t)$. At time t the population state is $T_{\rho(t)I}m$ and the value of the output is $H(T_{\rho(t)I}m)$. The $\times$ product allows a neat representation of the input–output map P_{m_0},

$$P_{m_0}(I) = \gamma \times u^c_{\rho(\cdot)I} \times m_0 . \tag{4.72}$$

Note that both sides of Equation (4.72) are functions of time. The dot on the right-hand side denotes the position of the suppressed time variable.

We now obtain a non-linear model by closing the feedback loop with the requirement that input equals output,

$$I = P_{m_0}(I) . \tag{4.73}$$

We have thus arrived at a *fixed point problem*:

Given the ingredients u_I, Λ_I, and γ and the initial population state m_0, show that the input–output map P_{m_0} defined by Equation (4.72) has a unique fixed point I_{m_0}. The dynamical system that describes the time evolution of the population state is then given by

$$S(t, m_0) = T_{\rho(t)I_{m_0}}m_0 \tag{4.74}$$

and we say that the problem has been solved.

Diekmann *et al.* (2001) gave several sets of sufficient conditions for the existence of a unique solution to the non-linear population problem, but these are too technical to be reproduced here. The proof of existence and uniqueness of solutions is *constructive* based on successive approximations. One starts with an initial guess of what the input I could be. Call this input I_0. Then, the corresponding output $P_{m_0}(I_0)$ is computed and fed back in as input I_1. This procedure is repeated indefinitely to obtain a sequence of successive approximations (inputs) $I_0, I_1, I_2, \ldots$. The sufficient conditions alluded to above are such that they guarantee the convergence of this sequence to some I that then necessarily satisfies Equation (4.73) (i.e., is a fixed point of the input–output map).

5

Extinction

5.1 The Role of Extinction in Evolution

Extinction plays an important role in evolution. From the fossil record we know that many species have become extinct. Raup (1991) estimates that only one in a thousand species that have ever lived on earth still exist today. He concludes that all species have a relatively low risk of extinction most of the time, but that rare intervals of a vastly higher risk of extinction occur throughout evolutionary history. These so-called mass extinctions, in which a large percentage of the then-existing species disappears, seem to be separated by periods with lengths of the order of 10^8 years. Raup argues that extinctions of widespread species probably involve extreme global environmental conditions not normally experienced by species (i.e., to which they are not adapted). Mass extinctions involve huge environmental stresses that cut across ecological lines.

During the past 100 years many species have become extinct. Although extinction apparently is an inescapable factor of life on earth, many recent extinctions probably involve human-related causes. Some people have even voiced the opinion that we may be heading for a new mass extinction because of human interference (see Lawton and May 1995). To avoid extinction, it is important to be able to estimate the vulnerability of species and to have some impression as to which factors contribute to extinction risk and what can be done about them.

On a less dramatic but nevertheless important scale, there are local extinctions of species or loss of genetic diversity from local populations. During the past decades a whole new body of theory has been developed, meta-population theory, which focuses on the relation between the dynamics of local populations and of larger groups of subpopulations. Elements in this are vulnerability of small populations, extinction times, and rescue effects through migrations between subpopulations.

Extinction is not necessarily negative. Darwinian evolution involves a continuous adaptation of species (i.e., the replacement of resident populations by more successful mutants). Sometimes humanity may even strive for extinction, as in the eradication of smallpox.

Furthermore, 1 minus the extinction risk equals the chance of establishment success, for instance invasion success into new habitats. So the study of extinction risk also gives us information about, for example, the best strategies to introduce biological control agents, or to re-introduce a species into an area where it had become extinct.

5.2   Extinction or Explosion: The Merciless Dichotomy

Many populations that surround us seem in balance with their environments. Their sizes are fairly stable, or if they vary they do so in a predictable cyclic manner (famous examples of this are certain predator–prey systems), so there is stability over longer time periods. If a population has settled down to such stability, it is said to be in its *stationary* phase; the population itself is also called stationary. Mathematically, such cases appear as limits: as an illustration, consider a population that started from a number of newborn individuals. Initially, its size remains essentially constant, after which there is a period of rapid growth as these individuals mature to reproductive age. A new period of relative constancy follows, and so on, until the age distribution levels off – actually approaches the stable age distribution, which we return to later.

It is thus not a futile hope that in the critical cases, in which, on average, each individual is replaced by one child, the population tends to stationarity. In deterministic models this actually occurs. Much deterministic theory is concerned with the stationary situation, which often has a simpler structure than the transitory phases that precede it. Also, it is thought to depict the long-run behavior of actual populations. A closer look at the model, however, discloses that only completely deterministic population models possess stationary states. (As do infinite populations – to a biologist such concepts may seem strange, but one way to approximate very large populations spread out in space might be through infinite populations, finite in any habitat, with individuals moving between the infinitely many habitats.) As soon as we accept the most minute variation in offspring numbers between individuals in the same circumstances, it turns out that there can be no finite, eternal populations at all, and hence no truly stationary states. What can occur is a sort of quasi-stationarity: even if we are all bound to extinction on a cosmological time scale, there may be plateaus of what is, for all human purposes, a stationary phase. (If you feel that all this is self-evident, you are welcome to skip the following, possibly somewhat over-fussy, argument.)

Thus, consider a population with, for a more precise setting and for simplicity, just one single type of individual that evolves in discrete time, say by seasons. It need not be a branching process in any strict sense of the word; the following argument applies to very general structures, even though phrased in population dynamics terminology.

Clearly, in the absence of any resource or space limitations, a population can die out or it can grow beyond all bounds. The latter is, indeed, the precise meaning of saying that size "tends to infinity." Conversely, if the population size does not surpass all given bounds (and also does not become extinct), there must exist a level to which it keeps returning, over and over again. This is illustrated in Figure 5.1.

This little exercise in the meaning of words reveals three logically possible cases:

1. The population dies out;
2. The population grows beyond all limits;

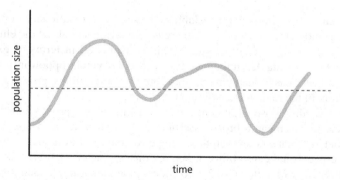

Figure 5.1 Example of a population that repeatedly reaches the same size.

3. There are numbers $0 < A < B$ such that the population size lies between A and B infinitely often.

Now focus upon the third case. A little thought shows that this apparent option is, indeed, infeasible or, at least, incompatible with just a tiny leeway for individual variation in reproduction. Indeed, if there is such individual-to-individual variation the population size cannot be constant. Hence, in case (3) population size can increase as well as decrease. However, if population size can decrease, individuals can remain childless. And if that is the case, there should be a non-zero (albeit tiny) risk of none of the at most B possible parents in the population producing any children, at least if there is some independence between individuals.

This is made more precise through the very weak independence property given in Assumption 5.1.

Assumption 5.1 *Consider any individual alive. Whatever be the history and present situation of the population, and whatever be the future destinies of the progeny of all other individuals, there is a number $\pi > 0$ such that the probability that the daughter population of the individual ultimately dies out is greater than π.*

Given Assumption 5.1, there is a number $p(B) \geq \pi^B > 0$ such that whenever the population has a size between A and B, the probability of future extinction exceeds $p(B)$. Since we return to this situation infinitely often, sooner or later luck fails, and the population becomes extinct. Hence, at a closer scrutiny case (3) is impossible. Clearly, it is the existence of $p(B)$ rather than π that is the crucial assumption. Just a tiny possibility for individual variation in reproduction leads to $p(B)$, and including it in the definition of the concept of a "proper population" (since any decent physical population certainly satisfies it), we conclude:

Metatheorem 5.1 *Any proper population under any environmental regime, and any sensible form of collaboration and/or competition or independence between individuals, resource dependence, etc., ultimately either becomes extinct or increases to infinity.*

Metatheorem 5.1 contradicts one established idea, that of stationary populations. However, as indicated in Section 5.1, there is also awareness about the elusiveness of life in population biology. Usually, this is expressed in terms of extinction as ultimately unavoidable: unbounded growth is no viable option in a bounded world. Thus, whereas in mathematics there are two possibilities, growth to infinity or extinction, in the real world only the latter occurs.

Of course, the loose argument given constitutes no proof – indeed, by a "metatheorem" we mean a broad assertion, neither strictly stated nor rigorously proved, that summarizes an insight we hope to convey. For a stricter argument, and other aspects on the instability of populations, see Jagers (1992). (The theorem there that replaces Metatheorem 5.1 considers potentially unbounded populations that satisfy a very weak Markovian property. Assume that for any population size x there is a number $\delta(x) > 0$ such that the conditional extinction probability in the future is at least $\delta(x)$, whatever the prehistory of a population, if only its present size does not exceed the level x. Then, the population must either die out or its size tends to infinity, as generations pass.)

To quote Halley and Iwasa (1998), "In the end all populations die out. However, to understand how this comes about and how quickly is a serious problem." The next two sections are devoted to these matters. They need not, of course, be studied in the extreme generality of Metatheorem 5.1. However, for branching style processes much is known, in particular for the single-type Galton–Watson case and related structures.

5.3 Extinction and Generating Functions

The idealized unbounded populations found in pure thought thus either grow beyond all limits or die out. When does one or the other occur? In the complete generality of the preceding section, where interaction and variation in time are involved, there is certainly no all-embracing answer. However, for simple Galton–Watson processes there is. And as we remarked in the introduction to Chapter 2, the successive generation sizes in a quite general continuous- or discrete-time branching process constitute a Galton–Watson process, the embedded Galton–Watson or generation process. Furthermore, any population – even one in which mothers can give birth repeatedly and at any time or (fertile) age, as in human populations – dies out if and only if there is an empty generation. Indeed, if a population dies out, then only a finite number of individuals are ever born into the population. Hence, generations must eventually also be empty (i.e., the generation counting process finally becomes 0). From the reverse aspect, if the generation process dies out, only finitely many individuals can come into existence. Since each has a finite life span, the population itself must die out, sooner or later. This basic fact deserves to be stated as a theorem (which, indeed, holds not only for branching processes in the strict sense of the word, but also very generally):

Theorem 5.1 *A general branching process with finite individual life spans dies out if and only if the embedded Galton–Watson process of generations does so.*

Thus, to establish the extinction probability for Galton–Watson processes is the same as doing it for any branching process, even though times to extinction are, of course, not the same, since generations tend to spread out in chronological time, unless reproduction and survival are seasonal.

A handy tool in determining the extinction probability of single-type Galton–Watson processes is the generating function of the reproduction distribution, already used by Galton and Watson (1875) for that purpose. It is usually denoted by f and viewed as a function of a real variable $s \in [0, 1]$:

$$f(s) = \mathbb{E}[s^\xi] = \sum_{k=0}^{\infty} \mathbb{P}(\xi = k)s^k = \sum_{k=0}^{\infty} p_k s^k, \ 0 \le s \le 1 , \qquad (5.1)$$

in terms of a random variable ξ, which gives the offspring of an individual, or in terms of its distribution $p_0, p_1, p_2, \ldots$. We refer to it as the *reproduction generating function*.

For the benchmark processes

$$f(s) = \mathbb{E}[s^\xi] = \sum_{k=0}^{\infty} \mathbb{P}(\xi = k)s^k = q + ps^2 , \qquad (5.2)$$

in the binary splitting case, and

$$f(s) = \sum_{k=0}^{\infty} qp^k s^k = \frac{q}{1 - ps} , \qquad (5.3)$$

for geometric offspring size distribution.

The generating function of the number of individuals in the nth generation of a Galton–Watson process is

$$f_n(s) = \mathbb{E}[s^{Z_n}] = \sum_{k=0}^{\infty} \mathbb{P}(Z_n = k)s^k . \qquad (5.4)$$

If there is only one ancestor, then clearly $f_0(s) = s, 0 \le s \le 1$, and the generating function of the first generation is simply the reproduction generating function, $f_1 = f$. Furthermore, the conditional expectation argument used in Section 2.2 yields

$$\begin{aligned} f_n(s) &= \mathbb{E}[s^{Z_n}] = \mathbb{E}[\mathbb{E}[s^{Z_n}|Z_{n-1}]] \\ &= \mathbb{E}[\mathbb{E}[s^{\xi_1+\xi_2+\cdots+\xi_{Z_{n-1}}}|Z_{n-1}]] = \mathbb{E}[\mathbb{E}[s^{\xi_1}s^{\xi_2}\cdots s^{\xi_{Z_{n-1}}}|Z_{n-1}]] , \end{aligned} \qquad (5.5)$$

where ξ_i means the number of children of the ith individual in the $n - 1$th generation. All these numbers have the same distribution and generating function f, and are independent of themselves and of Z_{n-1}. Hence, we can proceed and replace the conditional expectation of the product by the product of conditional expectations, and obtain

$$f_n(s) = \mathbb{E}[\mathbb{E}[s^{\xi_1}|Z_{n-1}]\mathbb{E}[s^{\xi_2}|Z_{n-1}]\cdots\mathbb{E}[s^{\xi_{Z_{n-1}}}|Z_{n-1}]] = \mathbb{E}[f(s)^{Z_{n-1}}]$$
$$= f_{n-1}(f(s)) = \cdots = f(f(\cdots(f(s))\cdots)) = f(f_{n-1}(s)) \,. \tag{5.6}$$

(However, here independence is crucial, as opposed to the same argument being used to derive $\mathbb{E}[Z_n] = m^n$ in Chapter 2!)

The classic relations between generating function derivatives and expectations (see the Appendix) can be used to recover the formulas derived for population means and variances, derived in Section 2.2 under less stringent independence conditions.

In any probability textbook you can further learn that not only does the probability distribution determine its generating function, but the converse is also true: if you know the generating function of a random variable, you know its distribution. In principle, thus Equation (5.6) above describes all distributional properties of the nth generation size.

The step from knowledge in principle to explicit formulas is often prohibitively long. As mentioned, one useful consequence of Equation (5.6) is the form of population means and variances. Another is that

$$f_n(0) = \mathbb{P}(Z_n = 0) \,, \tag{5.7}$$

as you can see by inserting $s = 0$ into

$$f_n(s) = \sum_{k=0}^{\infty} \mathbb{P}(Z_n = k)s^k \,, \tag{5.8}$$

which removes all terms but the first. Now, if one generation is empty, the same must be true for the next. Therefore

$$f_n(0) = \mathbb{P}(Z_n = 0) \le \mathbb{P}(Z_{n+1} = 0) = f_{n+1}(0) \,. \tag{5.9}$$

It follows that the sequence

$$Q_n = \mathbb{P}(\text{extinction by generation } n) = \mathbb{P}(Z_n = 0) = f_n(0), \ n = 1, 2, \ldots, \tag{5.10}$$

must increase to the extinction probability, which we denote by Q,

$$\lim_{n\to\infty} Q_n = Q \,. \tag{5.11}$$

Since

$$Q_n = f_n(0) = f(f_{n-1}(0)) = f(Q_{n-1}) \tag{5.12}$$

and the function f is continuous, it follows that $Q = f(Q)$. In these definitions and arguments we have assumed throughout that the population starts from one single ancestor. If $Z_0 = N$, then the extinction probability of the whole population is Q^N, since all N independent families must die out.

The fundamental equation for Q was deduced long ago by Galton and Watson (1875). Quite intriguingly, they erroneously concluded that, since the equation is

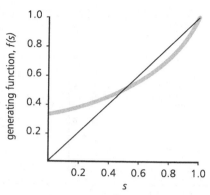

Figure 5.2 Example of a generating function of an offspring distribution with $m > 1$.

solved by 1, $f(1) = 1$, Q must be 1, and extinction thus certain. The truth is that if the process is supercritical (i.e., $m > 1$), there is also another smaller positive root of the equation, which is the correct extinction probability (Figure 5.2).

Thus, the process has a positive chance of escaping extinction, and then growing to infinity. In Chapter 6 we show that this occurs at the famous exponential rate, claimed already by Malthus (1798), and before him by Euler – for expected population sizes this is also obtained in Section 2.2. We now summarize the above analysis of extinction:

Theorem 5.2 *The probability of extinction in a single-type Galton–Watson process with one ancestor is the smallest non-negative root Q of the equation $f(x) = x$. It is less than 1 if and only if $m > 1$ [but for the degenerate case of any individual obtaining exactly one child, $\mathbb{P}(\xi = 1) = 1$, where $Q = 0$, of course].*

Example 5.1 For binary splitting, we must solve $q + px^2 = x$ with the result that $Q = q/p$ provided $p > 1/2$ and 1 otherwise. For geometrically distributed offspring, the equation is solved equally easily,

$$q/(1 - xp) = x \tag{5.13}$$

yields the same result, $Q = q/p$ for $p > 1/2$. In this case $m = p/q$ so that $Q = 1/m$ also. A process with a mean reproduction of, say, 1.2 therefore has a population doubling time of four generations [if $Z_0 = N$, $\mathbb{E}[Z_4] = N(1.2)^4 = N \times 2.07$], and an extinction probability higher than 80% ($1/1.2 = 0.83$).

This means that in a large population, which by the Law of Large Numbers exhibits a growth similar to the expected growth, nevertheless, a vast majority of the separate family lines becomes extinct (see Figure 5.3). In scientific literature this was first noted by Malthus, for human populations. Toward the end of the first volume of his famous *Essay on the Principle of Population*, he notes that 379 out of 487 bourgeois families in the city of Berne died out between 1583 and 1783. In our terminology this means that we should estimate the extinction probability to be at least $379/487 \approx 0.75$, but, if anything, higher, since Malthus' data give extinction only after some six generations, and $Q_6 < Q$.

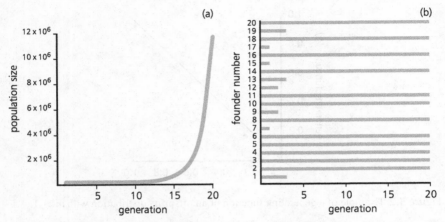

Figure 5.3 Simulation results of a branching process with extinction probability of 1/2. The offspring distribution is geometric, with expectation 2. (a) Total size of a population with 20 founders. (b) Survival times of each of the processes started by one founder during 20 generations.

When Galton and, before him, Bienaymé (1845) observed that noble family names tended to disappear in spite of the rapid population increase of the times, they could well have observed only this paradoxic combination of rapid average growth and frequent extinction of families. Indeed, it is easy to construct branching processes with, say, $m = 1.2$, Q around 0.8, and realistic numbers for the probabilities of various numbers of male children (carrying the family name on in 19th century France or England).

In evolution, even higher extinction frequencies are claimed. In Section 5.1 we mentioned that Raup (1991) asserts that 99.9% of all species that have ever existed are now extinct; indeed, he claims an extinction risk of 0.999. (Of course, in Raup's analysis other circumstances are relevant, such as invasion by new species, competition, habitat finiteness, and environmental change and catastrophes, continental drift, and so on.)

$$\diamond \diamond \diamond$$

The relation between generating functions of subsequent generations, based on the explicit rule in Equation (5.6), thus turns out to be very efficient in determining extinction probability. Unfortunately, it is less so when it comes to other properties of the process. Even though the process of composing generating functions, $f(f(\ldots f(s) \ldots))$, can be computerized, it is difficult and implicit.

Just check one of our benchmark processes, binary splitting:

$$f_1(s) = f(s) = q + ps^2, \ f_2(s) = q + p(q + ps^2)^2, \ f_3(s) = \cdots . \qquad (5.14)$$

Proceeding further is a temptation easily resisted.

At this juncture our other benchmark process presents its pleasant properties. Its reproduction process and mean are

$$f(s) = \frac{q}{1 - ps} \text{ and } f'(1) = m = p/q . \qquad (5.15)$$

Box 5.1 Iteration of generating functions for geometric distributions

Since

$$1 - f(s) = 1 - \frac{q}{1 - ps} = \frac{p(1 - s)}{1 - ps} , \tag{a}$$

simple manipulations yield

$$\frac{1}{1 - f(s)} - \frac{1}{m(1 - s)} = \frac{1 - ps}{p(1 - s)} - \frac{q}{p(1 - s)} =$$

$$\frac{1 - q - ps}{p(1 - s)} = \frac{p(1 - s)}{p(1 - s)} = 1 . \tag{b}$$

Substitution of $f_{n-1}(s)$ for s in Equation (b) and using Equation (5.6) yields

$$\frac{1}{1 - f_n(s)} - \frac{1}{m(1 - f_{n-1}(s))} = 1 , \tag{c}$$

or

$$\frac{1}{1 - f_n(s)} = 1 + \frac{1}{m(1 - f_{n-1}(s))} = 1 + \frac{1}{m} + \frac{1}{m^2(1 - f_{n-2}(s))} = \cdots , \tag{d}$$

which finally leads to

$$\frac{1}{1 - f_n(s)} = 1 + \frac{1}{m} + \left(\frac{1}{m}\right)^2 + \cdots + \left(\frac{1}{m}\right)^{n-1} + \left(\frac{1}{m}\right)^n \frac{1}{1 - s} =$$

$$\frac{m^n - 1}{m^{n-1}(m - 1)} + \frac{1}{m^n(1 - s)} . \tag{e}$$

The relation in Equation (5.16) follows immediately from this.

The former can be iterated – see Box 5.1 – to yield

$$1 - f_n(s) = \frac{m^n(m - 1)(1 - s)}{m(m^n - 1)(1 - s) + m - 1} , \tag{5.16}$$

and, in particular,

$$1 - Q_n = 1 - f_n(0) = \frac{m^n(m - 1)}{m^{n+1} - 1} , \tag{5.17}$$

which facilitates the calculation of many entities, such as the time to extinction.

5.4 Time to Extinction in Simple Processes

In Section 5.3 we claim that the extinction probability Q is 1 if $m \leq 1$. In the subcritical case this is proved easily, and we can even obtain a sharp bound. Indeed, by Markov's inequality

$$\mathbb{P}(Z_n > 0) = \mathbb{P}(Z_n \geq 1) \leq \mathbb{E}[Z_n] = Nm^n , \tag{5.18}$$

where N is the number of founders of the population. And certainly

$$\lim_{n \to \infty} m^n = 0 , \tag{5.19}$$

if $m < 1$.

This estimate allows interesting conclusions about the development of subcritical populations. Indeed, suppose that we wish to find the time (generation) when the survival probability becomes less than some small number r. Clearly,

$$Nm^n \leq r \tag{5.20}$$

holds if

$$n \geq \frac{\ln N - \ln r}{|\ln m|} , \tag{5.21}$$

given that $\ln m < 0$.

For example, if $m = 0.95$ and $r = 0.01$, starting populations of size $N = 1, 2, \ldots, 100$ yield $n \geq 90, 104, \ldots, 180$. In other words, the probability that such a population starting from a single mutant should survive more than, say, 90 generations is less than 0.01. If, instead, the population is initiated by 20 invaders, it takes 149 generations until the survival probability, or rather its bound, passes the level 0.01.

To say that the bound $1 - Q_n \leq Nm^n$ is "sharp" means that

$$1 - Q_n \to 0 \tag{5.22}$$

at the same rate as m^n. In general, the proof of this is rather delicate, but in the geometric case, Equation (5.17) tells us that for $m < 1$ and $Z_0 = 1$

$$\mathbb{P}(Z_n > 0) = 1 - Q_n \sim (1 - m)m^n , \tag{5.23}$$

as $n \to \infty$.

A similar assertion holds for all branching processes that satisfy a mild condition (Theorem 5.3) on reproduction – so mild that it is impossible to imagine biological populations not fulfilling it.

Theorem 5.3 *Consider a subcritical Galton–Watson process that starts from $Z_0 = N$ individuals. It is always true that $\mathbb{P}(Z_n > 0) \leq \mathbb{E}[Z_n] = Nm^n$. If the reproduction distribution satisfies*

$$\mathbb{E}[\xi \ln(\xi + 1)] < \infty , \tag{5.24}$$

there is a constant $0 < c_N \leq N$ such that

$$\mathbb{P}(Z_n > 0) \sim c_N m^n , \tag{5.25}$$

as $n \to \infty$. If, also, the reproduction variance $\sigma^2 < \infty$,

$$Nm^n \geq \mathbb{P}(Z_n > 0) \geq \frac{N(1 - m)m}{\sigma^2(1 - m^n) + m^{n+1}(1 - m)}\left(1 - \frac{(N - 1)m^n}{2}\right)m^n$$
$$\approx N(1 - m)m^{n+1}/\sigma^2 . \tag{5.26}$$

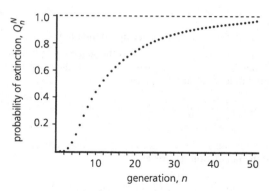

Figure 5.4 Probability of extinction at or before generation n for a process with a geometric offspring distribution with $m = 0.95$ and $N = 10$ founders.

The theorem contains two approximations. The strict meaning of the first is that

$$\lim_{n \to \infty} \mathbb{P}_N(Z_n > 0)/m^n = c_N , \qquad (5.27)$$

where we indicate the dependence on the starting size of the population by suffices throughout. In the second approximation, the terms of size m^{2n} or smaller are disregarded.

For a complete proof of Theorem 5.3, see standard mathematical textbooks. Here, we contend ourselves with deducing the inequalities in Equation (5.26) and showing that $\mathbb{P}_1(Z_n > 0)/m^n$ decreases as n grows, so that c_1 must exist.

For the interpretation of the limiting constant in the case of one ancestor, note that

$$\frac{\mathbb{P}_1(Z_n > 0)}{m^n} = \frac{\mathbb{P}_1(Z_n > 0)}{\mathbb{E}_1[Z_n]} = \frac{1}{\mathbb{E}_1[Z_n|Z_n > 0]} . \qquad (5.28)$$

Hence, under the conditions of Theorem 5.3,

$$1/c_1 = \lim_{n \to \infty} \mathbb{E}_1[Z_n|Z_n > 0] . \qquad (5.29)$$

Thus, c_1 is one over the long-run mean size of subcritical populations, from one ancestor, provided they have not died out.

The distribution function of the time τ to extinction is found easily (we identify generations and time), since extinction occurs before time n if and only if the nth generation is empty (see Figure 5.4),

$$\mathbb{P}_N(\tau \leq n) = \mathbb{P}_N(Z_n = 0) = Q_n^N . \qquad (5.30)$$

Thus, the conclusion, Equation (5.26), of Theorem 5.3 can be reformulated as

$$\frac{N(1-m)m}{\sigma^2(1-m^n) + m^{n+1}(1-m)} \left(1 - \frac{(N-1)m^n}{2}\right)m^n$$

$$\leq \mathbb{P}_N(\tau > n) \leq Nm^n , \qquad (5.31)$$

Box 5.2 Deduction of the survival chance approximation

Turning now to the proofs, we first show that the sequence in Equation (5.28) does not increase. Indeed, the process survives to time n if the offspring of at least one child of the initial individual survives to time n, that is

$$\mathbb{P}_1(Z_n > 0) = \mathbb{P}_1 \left(\bigcup_{l=1}^{Z_1} \{ Z_{n-1}^{(i)} > 0 \} \right), \tag{a}$$

where $Z_{n-1}^{(i)}$ is the number of individuals at time n that stem from the ith child of the ancestor of the population. However,

$$\mathbb{P}_1 \left(\bigcup_{l=1}^{Z_1} \{ Z_{n-1}^{(i)} > 0 \} \right) = \mathbb{E}_1 \left[\mathbb{P} \left(\bigcup_{l=1}^{Z_1} \{ Z_{n-1}^{(i)} > 0 \} \mid Z_1 \right) \right]$$

$$\leq \mathbb{E}_1 \left[\sum_{l=1}^{Z_1} \mathbb{P}_1 \left(Z_{n-1}^{(i)} > 0 \mid Z_1 \right) \right], \tag{b}$$

and by the Wald identity (see the Appendix) we have

$$\mathbb{E}_1 \left[\sum_{l=1}^{Z_1} \mathbb{P}_1 \left(Z_{n-1}^{(i)} > 0 \mid Z_1 \right) \right] = \mathbb{E}_1 [Z_1] \, \mathbb{E}_1 \left[\mathbb{P}_1(Z_{n-1} > 0) \right] = m \mathbb{P}_1(Z_{n-1} > 0) \quad \text{(c)}$$

since all the children behave independently and the survival probability up to time n of a daughter population equals the survival probability up to time $n - 1$ of the original population.

Combining Equation (b) and Equation (c), we obtain

$$c_1 \leq \frac{\mathbb{P}_1(Z_n > 0)}{m^n} \leq \frac{\mathbb{P}_1(Z_{n-1} > 0)}{m^{n-1}}. \tag{d}$$

In other words, the expected population size, given non-extinction,

$$\mathbb{E}_1[Z_n \mid Z_n > 0] = m^n / \mathbb{P}_1(Z_n > 0) \tag{e}$$

increases toward $1/c_1$ as generations pass.

If $Z_0 = N > 1$, the arguments above fail. However, if $Q_n = \mathbb{P}_1(Z_n = 0)$ is the probability of extinction in the first n generations, with one single ancestor, then

$$\mathbb{P}_N(Z_n > 0) = 1 - Q_n^N \leq N(1 - Q_n) = N \mathbb{P}_1(Z_n > 0), \tag{f}$$

which shows that $c_N \leq c_1 N$. In the geometric case equality holds here, $c_N = (1 - m)N$, as can be calculated from Equation (5.17).

More generally, in the biological case of finite reproduction variance $\mathrm{Var}_N[Z_n] < \infty$ also. A useful, though not so well known inequality, ascribed to Lyapunov (see the Appendix), shows that

$$\mathbb{P}(X > 0) \geq \frac{(\mathbb{E}[X])^2}{\mathbb{E}[X^2]} \tag{g}$$

for any random variable $X \geq 0$. The choice of $X = Z_n$ and $Z_0 = 1$ gives

$$\mathbb{P}_1(Z_n > 0) \geq \frac{(\mathbb{E}_1[Z_n])^2}{\mathbb{E}_1[Z_n^2]} = \frac{(\mathbb{E}_1[Z_n])^2}{\mathrm{Var}_1[Z_n] + (\mathbb{E}_1[Z_n])^2} = \frac{m^{2n}}{\mathrm{Var}_1[Z_n] + m^{2n}}. \tag{h}$$

continued

Box 5.2 *continued*

From Chapter 2, Equation (2.16),

$$\text{Var}_1[Z_n] = \frac{\sigma^2 m^n (m^n - 1)}{m(m - 1)} . \tag{i}$$

Thus

$$\begin{aligned}
\mathbb{P}_1(Z_n > 0) &\geq \frac{m^{2n+1}(m - 1)}{\sigma^2 m^n (m^n - 1) + m^{2n+1}(m - 1)} \\
&= \frac{(1 - m)m^{n+1}}{\sigma^2(1 - m^n) + m^{n+1}(1 - m)} .
\end{aligned} \tag{j}$$

Now, suppose that the population is initiated by $N \geq 1$ individuals. In the Appendix we derive the inequality

$$1 - (1 - x)^k \geq kx - \frac{k(k - 1)}{2}x^2 \tag{k}$$

for non-negative integers k and $0 \leq x \leq 1$. This yields

$$\begin{aligned}
\mathbb{P}_N(Z_n > 0) = \; & 1 - (1 - \mathbb{P}_1(Z_n > 0))^N \geq N\mathbb{P}_1(Z_n > 0)\left(1 - \frac{N-1}{2}\mathbb{P}_1(Z_n > 0)\right) \\
\geq \; & \frac{N(1 - m)m^{n+1}}{\sigma^2(1 - m^n) + m^{n+1}(1 - m)}\left(1 - \frac{(N - 1)m^n}{2}\right) \\
= \; & \frac{N(1 - m)m}{\sigma^2(1 - m^n) + m^{n+1}(1 - m)}\left(1 - \frac{(N - 1)m^n}{2}\right)m^n ,
\end{aligned} \tag{l}$$

where we used the fact that $\mathbb{P}_1(Z_n > 0) \leq m^n$. Thus, letting $n \to \infty$,

$$c_N \geq N(1 - m)m/\sigma^2 , \tag{m}$$

in the general situation.

or approximately, disregarding terms of order m^{2n},

$$N(1 - m)m^{n+1}/\sigma^2 \leq \mathbb{P}_N(\tau > n) \leq Nm^n . \tag{5.32}$$

Later we show how this can be applied in practical examples.

Another extremely useful approximate relation, valid under the conditions of Theorem 5.3, is

$$\mathbb{E}_N[\tau] \sim \frac{\ln N}{|\ln m|}, \quad N \to \infty . \tag{5.33}$$

Its importance lies in that it requires only the initial, or present, size of the population and the offspring mean to make assertions about the expected survival time. The formula was first derived for continuous-time Markov branching processes (Pakes 1989). We establish a stronger result, which also shows the accuracy of the asymptotic relation Equation (5.33). Its proof may be skipped by the not so mathematical reader.

Theorem 5.4 *Consider a Galton–Watson process with mean $m < 1$ and starting from $Z_0 = N$ individuals, and assume that the logarithmic moment condition of Theorem 5.3 holds. Then, for $N \geq 3$, the time τ to extinction satisfies*

$$\left(\frac{\ln N - \ln \ln N}{|\ln m|} - 1\right)\left(1 - \frac{1}{Nc_1}\right) \leq \mathbb{E}_N[\tau] \leq \frac{\ln N}{|\ln m|} + \frac{2 - m}{1 - m}, \tag{5.34}$$

where

$$c_1 \geqslant (1 - m)\, m \big/ \sigma^2 \tag{5.35}$$

if $\sigma^2 < \infty$ [see Equation (m) in Box 5.2].

Proof. From Equation (d) in Box 5.2,

$$c_1 m^n \leq \mathbb{P}_1(Z_n > 0) = \mathbb{P}_1(\tau > n) \leq m^n, \tag{5.36}$$

and

$$\mathbb{P}_N(\tau > n) \leq N m^n. \tag{5.37}$$

Set

$$\phi(N) = \frac{\ln N}{|\ln m|}, \qquad \psi(N) = \frac{\ln \ln N}{|\ln m|}. \tag{5.38}$$

Observe that $N m^{\phi(N)} = N m^{-(\ln N)/\ln m} = 1$,

$$\exp\left\{-c_1 N m^{\phi(N) - \psi(N)}\right\} = \exp\left\{-c_1 m^{-\psi(N)}\right\} = \exp\left\{-c_1 \ln N\right\} = 1/N c_1. \tag{5.39}$$

Further (see the Appendix),

$$\mathbb{E}_N[\tau] = \sum_{n=0}^{\infty} \mathbb{P}_N(\tau > n), \tag{5.40}$$

and therefore, by Equation (5.37),

$$\mathbb{E}_N[\tau] \leq \sum_{0 \leq n < \phi(N)} \mathbb{P}_N(\tau > n) + N \sum_{n \geq \phi(N)} m^n$$

$$\leq \phi(N) + 1 + \frac{N m^{\phi(N)}}{1 - m} = \frac{\ln N}{|\ln m|} + \frac{2 - m}{1 - m}. \tag{5.41}$$

However,

$$\mathbb{E}_N[\tau] \geq \sum_{0 \leq n < \phi(N) - \psi(N)} \mathbb{P}_N(\tau > n) = \sum_{0 \leq n < \phi(N) - \psi(N)} (1 - \mathbb{P}_N(\tau \leq n)). \tag{5.42}$$

Clearly,

$$\mathbb{P}_N(\tau \leq n) = \mathbb{P}_1^N(\tau \leq n) = (1 - \mathbb{P}_1(\tau > n))^N$$

$$\leq e^{-N\mathbb{P}_1(\tau > n)} \leq e^{-c_1 N m^n}, \tag{5.43}$$

where we used first the inequality $1 - x \leq e^{-x}$, $x > 0$ and then Equation (5.36). Hence, for $0 \leq n \leq \phi(N) - \psi(N)$,

$$
\begin{aligned}
\mathbb{P}_N(\tau > n) &\geq \mathbb{P}_N(\tau > \phi(N) - \psi(N)) \\
&= 1 - \mathbb{P}_N(\tau \leq \phi(N) - \psi(N)) \\
&\geq 1 - e^{-c_1 m^{\phi(N) - \psi(N)}} \geq 1 - \frac{1}{N c_1} .
\end{aligned}
\tag{5.44}
$$

Therefore,

$$
\begin{aligned}
\min_{0 \leq n \leq \phi(N) - \psi(N)} \mathbb{P}_N(\tau > n) &\geq (\phi(N) - \psi(N) - 1)\left(1 - \frac{1}{N c_1}\right) \\
&\geq \left(\frac{\ln N - \ln \ln N}{|\ln m|} - 1\right)\left(1 - \frac{1}{N c_1}\right)
\end{aligned}
\tag{5.45}
$$

Combining Equation (5.41) and Equation (5.45) gives Equation (5.34).

◇ ◇ ◇

Example 5.2 We consider an example borrowed from Caswell *et al.* (1999). They studied the threats posed by a decline of survival probabilities for the North Atlantic right whale, within the framework of the following model.

A female right whale may produce 0, 1, or 2 females the following year. It is assumed that the death of a parent results in the death of a calf in the first year. Thus, a female at time n produces no offspring if she dies before $n + 1$, one offspring (herself) if she survives without reproducing female offspring, and two offspring (herself and her calf) if she survives and gives birth to a female calf. Generation length is then 1 year. Let p be the survival probability and μ be the probability of begetting a female calf. The reproduction generating function of the process becomes

$$
f(s) = 1 - p + p(1 - \mu)s + p\mu s^2 ,
\tag{5.46}
$$

with mean $m = p(1 - \mu) + 2p\mu = p(1 + \mu)$. Caswell *et al.* (1999), using different sources, give the estimates in Table 5.1 for p, μ, and, as a result, for m.

Table 5.1 Estimates of Caswell *et al.* (1999) for North Atlantic right whales.

	$\mu = 0.051$	$\mu = 0.038$
$p = 0.94$	$m = 0.988$	$m = 0.976$

Applying Equation (5.21) and Equation (1) in Box 5.2 to the data, we obtain the lower estimates in Table 5.2 for the number n of generations (years) that the population of whales (now at around 150 female members) can survive with probability higher than 0.99 and the upper estimates for the number of generations within which the population will die out with a probability greater than 0.99.

In particular, this shows that, provided reproduction conditions remain the same in the future, under the worst-case scenario the whale population will die out within 400 years with a probability of more than 99 percent.

Table 5.2 Estimates of survival and extinction for North Atlantic right whales using Equation (5.21) and Equation (e) in Box 5.2.

	m	0.988	0.976
Survival with probability ≥ 0.99 for at least n years	$n \geq$	357	177
Extinction with probability ≥ 0.99 within at most n years	$n \leq$	796	395

For the North Atlantic right whales we obtain the estimates in Table 5.3, by means of Equation (5.33), for the expected time to extinction in the subcritical situation (bottom line of Table 5.3). These figures agree with the results given in Caswell *et al.* (1999), showing through direct calculations that $\mathbb{E}[\tau|Z_0 = 150] \approx 191$ if $p = 0.94$ and $\mu = 0.038$ and, therefore, $m = 0.976$. Our method is, however, robust in the sense that it is built upon Equation (5.33), so the result does not depend on the particular form of the reproduction distribution.

Table 5.3 Estimates of expected time to extinction for North Atlantic right whales from Equation (5.33).

m	0.988	0.976
$\mathbb{E}[\tau|Z_0 = 150] \approx$	415	206

◇ ◇ ◇

5.5 Multi-type Processes

In Section 2.3.1 we show that the average numbers of individuals of different types in any indecomposable non-periodic branching process behave "nicely." However, what can be said about the extinction probability of multi-type processes?

First, recall that periodic processes can be transformed easily into non-periodic ones. As in Section 2.3.1, we thus only consider non-periodic processes. From Section 2.3 we know that their properties crucially depend on the Perron root (the largest positive eigenvalue), ρ, of the mean matrix M,

$$\mathbb{E}\left[Z_n^T\right] = \mathbb{E}\left[Z_0^T\right] M^n \sim \mathbb{E}\left[Z_0^T\right] \rho^n A , \qquad (5.47)$$

where Z_n is a d-dimensional vector that specifies the numbers of individuals of different types in the nth generation, and A is a $d \times d$ matrix with elements $a_{hj} = v_h u_j$, which consists of products of the elements of the normed eigenvectors u and v. These are defined by

$$u^T M = \rho u^T , \; Mv = \rho v , \; \sum_{j=1}^{d} u_j v_j = 1 , \; \sum_{j=1}^{d} u_j = 1 . \qquad (5.48)$$

From this, it is seen that in a subcritical process, where $\rho < 1$, the expected population size decreases over generations. In fact, any indecomposable subcritical branching process dies out. Indeed, if the population was initiated by one individual of type h, the chance that it has not died out by time n is

$$\mathbb{P}_h(Z_{n1} + Z_{n2} + \cdots + Z_{nd} > 0) = \mathbb{P}_h(Z_{n1} + Z_{n2} + \cdots + Z_{nd} \geq 1) . \qquad (5.49)$$

(We use suffixes to indicate starting conditions in this way, wherever needed.)
Now, recall from Section 2.3.1 that all coordinates of v are strictly positive, and
hence so is $v^* = \min_{1 \leq i \leq d} v_i$. Obviously

$$\mathbb{P}_h(Z_{n1} + Z_{n2} + \cdots + Z_{nd} \geq 1)$$
$$\leq \mathbb{P}_h(Z_{n1}v_1 + Z_{n2}v_2 + \cdots + Z_{nd}v_d \geq v^*) \tag{5.50}$$

and, by the Chebyshev inequality (see the Appendix), the right-hand side is smaller
than or equal to

$$\mathbb{E}_h[Z_{n1}v_1 + Z_{n2}v_2 + \cdots + Z_{nd}v_d]/v^* =$$
$$= \sum_{j=1}^{d} \mathbb{E}_h[Z_{nj}]v_j/v^* = \sum_{j=1}^{d} m_{hj}^{(n)} v_j/v^* , \tag{5.51}$$

where $m_{hj}^{(n)}$ are elements of the matrix M^n. However,

$$M^n v = M^{n-1}(Mv) = \rho M^{n-1} v = \cdots = \rho^n v , \tag{5.52}$$

or, in coordinate form,

$$\sum_{j=1}^{d} m_{hj}^{(n)} v_j = \rho^n v_h . \tag{5.53}$$

Hence

$$\mathbb{P}_h(Z_{n1} + Z_{n2} + \cdots + Z_{nd} > 0) \leq \rho^n v_h/v^* , \tag{5.54}$$

and, if $\rho < 1$, the right-hand side of this inequality tends to 0 as $n \to \infty$. It can
be shown that critical processes ($\rho = 1$) also die out, provided, of course, that at
least one type runs the risk of having no offspring.

Supercritical processes, $\rho > 1$, have a positive chance of persistence. For
multi-type processes the extinction probability can also be calculated by means of
the probability generating function. This is now a vector-valued function $f(s) =
f(s_1, \ldots, s_d)$ of dimension d, with its hth element equaling

$$f_h(s_1, \ldots, s_d) = f_h(s) = \mathbb{E}_h[s_1^{\xi_1} s_2^{\xi_2} \ldots s_d^{\xi_d}] , \tag{5.55}$$

where ξ_j is the number of offspring of type j and the suffix h serves to indicate the
mother's type.

Let Q be a vector with elements Q_h equal to the probability that a population
started from a type h ancestor becomes extinct, $\mathbb{P}_h(Z_n \to 0)$. Then

$$Q = f(Q), \quad \text{with} \quad 0 \leq Q_h < 1, h = 1, \ldots, d . \tag{5.56}$$

For supercritical branching processes, exactly one vector Q satisfies these conditions. The derivation of this result is very similar to that for single-type processes
given in Section 5.4 (Athreya and Ney 1972).

Example 5.3 Many parasites can enter resting phases during which they do not reproduce.
This is done by forming so-called spores. Spores are relatively robust against environmental

influences, such as attacks by a host's immune system. In this form, parasites can be transmitted easily to new hosts. We can study the effect of spore formation on the growth rate of a parasite population within one host by a branching process with two types: an active state (1) and a resting state (2). If a parasite is in state (1) it runs a risk c of dying, it can enter state (2) with chance p, and it can reproduce by binary splitting with chance $r = 1 - p - c$. A spore [state (2)] is transmitted to another host with probability e. It stays as a spore in the same host with probability q and becomes active within the host with chance $a = 1 - q - e$.

The reproduction mean matrix of this process becomes

$$M = \begin{pmatrix} 2r & p \\ a & q \end{pmatrix},$$ (5.57)

which has as its largest eigenvalue

$$\rho = \frac{1}{2} \left(q + 2r + \sqrt{(q + 2r)^2 - 4(2rq - pa)} \right).$$ (5.58)

Furthermore, the reproduction generating functions are given by

$$f_1(s_1, s_2) = c + rs_1^2 + ps_2, \quad f_2(s_1, s_2) = as_1 + qs_2 + e.$$ (5.59)

When $p = 0.2$, $q = 0.8$, $c = 0.3$, and $e = 0.1$, the process is supercritical, since $\rho = 1.07$. Solving the equations for $f_1(Q_1, Q_2) = Q_1$ and $f_2(Q_1, Q_2) = Q_2$ yields $Q_1 = 0.8$ and $Q_2 = 0.9$, so with the chosen parameter values the parasite population has a survival chance of 0.2 if the host is infected by a parasite in the active state, and a survival chance of 0.1 if the initiator was a spore.

5.6 Slightly Supercritical Populations

We have already mentioned the appearance of an advantageous gene in a resident population, seemingly, or for the time being, stationary, as a natural topic for analysis in terms of branching processes. Of particular interest is the not so clear-cut situation in which the mutant has only a slight advantage over the existing population. What are the chances that it will establish itself? Can they be described broadly in terms valid for a wide class of situations, independently of the small details of reproduction, usually unknown? Such questions lead to the study of survival probabilities of branching processes with reproduction means a little larger than 1, the resident population being thought of as critical.

A sequence of authors has studied these, from Haldane (1927) to Ewens (1968), Eshel (1981), Hoppe (1992), and Athreya, to whom the final results belong. We describe first the single- and then the multi-type case, and finally give an example by Law and Dieckmann (1998) that deals with the merger of lineages in evolution. For proofs we refer to Athreya (1992, 1993).

5.6.1 Single-type processes

As a first step, we derive a lower bound for the survival probability, which is valid for any supercritical Galton–Watson process (with finite variance) – for critical and subcritical processes this is also true, but trivial. The derivation is an enchanting short walk in the world of elementary calculus, so please come along.

Write $P = 1 - Q$ for the survival probability of a Galton–Watson process so that, in the usual notation,

$$P = 1 - f(1 - P) = \mathbb{E}[1 - (1 - P)^\xi] . \qquad (5.60)$$

Taking $x = P$ and $k = \xi$ in Equation (k) in Box 5.2 leads to

$$\mathbb{E}[1 - (1 - P)^\xi] \geq \mathbb{E}[\xi P - \xi(\xi - 1)P^2/2] , \qquad (5.61)$$

or

$$P \geq mP - \mathbb{E}[\xi(\xi - 1)]P^2/2 . \qquad (5.62)$$

Dividing by P (known to be strictly positive in the supercritical case) and rearranging gives

$$P \geq \frac{2(m - 1)}{\mathbb{E}[\xi(\xi - 1)]} . \qquad (5.63)$$

Furthermore,

$$\mathbb{E}[\xi(\xi - 1)] = \sigma^2 + m(m - 1) , \qquad (5.64)$$

and thus, for *any* Galton–Watson process with reproduction mean $m = 1 + \varepsilon > 1$ and variance σ^2, we can conclude that

$$P \geq \frac{2\varepsilon}{\sigma^2 + m\varepsilon} . \qquad (5.65)$$

Of course, this is also true for $\varepsilon = 0$.

To gain a feeling for the accuracy of the bound, consider the case of binary splitting, $f(s) = q + ps^2$, $q + p = 1$, with $\mathbb{E}[\xi] = m = 2p > 1$ and $\mathbb{E}[\xi(\xi-1)] = f''(1) = 2p$. As noted in Example 5.1, the probability of extinction Q equals q/p. Therefore, the survival probability P is

$$P = 1 - Q = \frac{p - q}{p} = \frac{2p - 1}{p} = \frac{2(2p - 1)}{2p} = \frac{2(m - 1)}{\mathbb{E}[\xi(\xi - 1)]} . \qquad (5.66)$$

Thus, despite its simplicity, the Estimate (5.65) is *sharp* in the sense that there is no smaller bound valid for all Galton–Watson processes, and it is natural to suspect that for small ε, indeed,

$$P \approx \frac{2\varepsilon}{\sigma^2 + m\varepsilon} \approx \frac{2(m - 1)}{\sigma^2} . \qquad (5.67)$$

This is a (generalized) form of Haldane's approximation from 1927 – Haldane's version was $P \approx 2(m - 1)$, valid for Poisson reproduction, where $\sigma^2 = m \approx 1$.

To obtain it in a strict form, consider Galton–Watson processes with reproduction generating functions

$$f^{(\varepsilon)}(s) = \mathbb{E}[s^{\xi^{(\varepsilon)}}] = \sum_{k=0}^{\infty} \mathbb{P}(\xi^{(\varepsilon)} = k)s^k , \quad \mathbb{E}[\xi^{(\varepsilon)}] = 1 + \varepsilon \geq 1 . \qquad (5.68)$$

Usually, we use the notation ξ for the offspring number, whatever its distribution. In this final part of the subsection, we index it by ε, since the very variation of the latter, approaching 0, is in focus. Thus, we assume that $m = 1 + \varepsilon$ is only "slightly" larger than 1.

However insignificant the supercriticality, there is a positive survival chance $P = 1 - Q$, and P is the unique positive solution of

$$f^{(\varepsilon)}(1 - P) = 1 - P , \quad \text{or} \quad 1 - f^{(\varepsilon)}(1 - P) = P . \tag{5.69}$$

In real life the exact distribution of $\xi^{(\varepsilon)}$ is usually unknown and, thus, so is the function $f^{(\varepsilon)}(s)$. As indicated, it would be useful to express P, at least approximately, in terms of a few natural and more easily determined characteristics of $\xi^{(\varepsilon)}$, rather than as a solution of the complicated Equation (5.69). This is what Equation (5.67) aspires to and Theorem 5.5 accomplishes:

Theorem 5.5 *Assume that the reproduction generating functions*

$$f^{(\varepsilon)}(s) = \mathbb{E}[s^{\xi^{(\varepsilon)}}] , \varepsilon \geq 0 , \tag{5.70}$$

are such that $\mathbb{E}[\xi^{(\varepsilon)}] = 1 + \varepsilon$ *and for some* $\varepsilon_0 > 0$ *and* $\delta > 0$

$$\sup_{0 \leq \varepsilon \leq \varepsilon_0} \mathbb{E}[(\xi^{(\varepsilon)})^{2+\delta}] < \infty . \tag{5.71}$$

Then

$$P = \frac{2\varepsilon}{\mathbb{E}[\xi^{(\varepsilon)}(\xi^{(\varepsilon)} - 1)]} + o(\varepsilon) , \quad \text{as } \varepsilon \to 0 . \tag{5.72}$$

If, further,

$$\sigma_\varepsilon^2 = \text{Var}[\xi^{(\varepsilon)}] \to \sigma_0^2 > 0 , \tag{5.73}$$

as $\varepsilon \to 0$, *then*

$$P = 2\varepsilon/\sigma_0^2 + o(\varepsilon) . \tag{5.74}$$

5.6.2 Multi-type populations

Survival probabilities for slightly supercritical multi-type populations are much more delicate; indeed this whole subsection is mathematically more heavy than those preceding it. Its essence is the last line of the theorem, and shows how the survival probability approximation theorem generalizes to a multi-type setting.

Recall (see Section 5.5) that if Q_h is the probability of extinction of a d-type population started by an individual of type $h \in \{1, 2, \ldots, d\}$ and $P_h = 1 - Q_h$, then

$$P = 1 - f(1 - P) , \tag{5.75}$$

where P denotes the vector $(P_1, \ldots, P_d)$. Now, assume that the reproduction generating functions depend on a small parameter $\varepsilon > 0$, but no longer spell this out with a superscript, neither on the vector ξ nor on the generating function,

$$f_h(s_1, \ldots, s_d) = f_h(s) = \mathbb{E}[s_1^{\xi_{h1}} \cdots s_d^{\xi_{hd}}] \,. \tag{5.76}$$

We denote the reproduction mean matrix by

$$M_\varepsilon = \left(m_{hj}(\varepsilon) \right) \,. \tag{5.77}$$

(Strictly speaking, the notation here and in the following does not conform to earlier usage, as in Section 5.5: the ancestral type h determines the probability measure and hence the distribution of the number of children of the various types, which is indicated there by indexing expectations and probabilities. Here, a suffix is placed on the random variable itself, even though this may seem less logical, as the random variable is defined in the same way, only its distribution changes.)

Suppose that for any $\varepsilon > 0$, M_ε is positively regular, which means an integer $n(\varepsilon) \geq 1$ exists such that $M_\varepsilon^{n(\varepsilon)}$ has all entries strictly positive, the maximal eigenvalue $\rho(\varepsilon)$ of M_ε is positive, and

$$\rho(\varepsilon) \to 1 \quad \text{as} \quad \varepsilon \to 0 \,. \tag{5.78}$$

Let $u(\varepsilon) = (u_1(\varepsilon), \ldots, u_d(\varepsilon))^T$ and $v(\varepsilon) = (v_1(\varepsilon), \ldots, v_d(\varepsilon))^T$ be the left and right eigenvectors, respectively, of M_ε with eigenvalues $\rho(\varepsilon)$ that satisfy

$$u_h(\varepsilon) > 0 \,, \quad v_h(\varepsilon) > 0 \,, \quad \sum_h u_h(\varepsilon) v_h(\varepsilon) = \sum_h u_h(\varepsilon) = 1 \,, \tag{5.79}$$

and, finally, let

$$B(\varepsilon) = \sum_{h=1}^{d} \sum_{j=1}^{d} \sum_{k=1}^{d} u_h(\varepsilon) v_j(\varepsilon) \mathbb{E}[\xi_{hj}(\xi_{hk} - \delta_{jk})] v_k(\varepsilon) \,, \tag{5.80}$$

where $\delta_{jk} = 1$ if $j = k$, and $\delta_{jk} = 0$ otherwise. After some algebra, this can be written as

$$B(\varepsilon) = \sum_h u_h \text{Var}[\sum_j v_j \xi_{hj}] + \rho(\rho - 1) \sum_j u_j v_j^2 \,, \tag{5.81}$$

where the dependence upon ε is omitted. When the latter tends to 0, the parameter ρ approaches 1, so that $B(\varepsilon)$ can be approximated by the first term of the expression.

The multi-type analogue of Theorem 5.5 is given in Theorem 5.6.

Theorem 5.6 *Assume that any limit matrix of the matrices $\{M_\varepsilon\}$ is positively regular,*

$$\liminf_{\varepsilon \to 0} B(\varepsilon) > 0 \,, \tag{5.82}$$

and for some $\varepsilon_0 > 0$ and $\delta > 0$

$$\sup_{h, 0 \leq \varepsilon \leq \varepsilon_0} \mathbb{E}[(\xi_{h1}^2 + \cdots + \xi_{hd}^2)^{1+\delta}] < \infty \,. \tag{5.83}$$

Then, as $\varepsilon \to 0$, $Q_h = Q_h(\varepsilon) \to 1$, *and*

$$P_h(\varepsilon) = 1 - Q_h(\varepsilon) = \frac{2(\rho(\varepsilon) - 1)}{B(\varepsilon)} v_h(\varepsilon) + o(\varepsilon). \tag{5.84}$$

Again, provided $B(\varepsilon) \to B(0)$ and also the eigenvector $v(\varepsilon) \to v(0)$, we can conclude that

$$1 - Q_h(\varepsilon) = \frac{2(\rho(\varepsilon) - 1)}{B(0)} v_h(0) + o(\varepsilon). \tag{5.85}$$

Example 5.4: Evolution of symbiosis. It is commonly believed that in the course of evolutionary history there have been several occasions in which different species merged to form symbioses that eventually evolved into single new species. One example, of many others, is the evolution of chloroplasts of eukaryotic plants, which is thought to have started by mergers between photosynthetic Gram-negative bacteria and early eukaryotes. Law and Dieckmann (1998) examined the conditions under which such evolution from facultative to obligate symbiosis may occur.

They considered two species, one exploiter and one victim. Each individual can occur in free-living form or in symbiosis with an individual of the other species. In the latter case they form what has been called a *holobiont*. In free-living form, the two species use separate resources, whereas in the holobionts a transfer of resources occurs between the partners. The exploiter species restricts the flow of its resources to its partner more than the victim does. Law and Dieckmann examined, in each species, the co-evolution of traits that determine the transfer of resources to the partner and the probability of coupled birth events of the partners in symbiosis.

It was assumed that mutations of small effect could occur in either species, but are rare and far apart, so that the population dynamics seems, and in the relevant time scale essentially is, in equilibrium when a new mutant is introduced. This is the situation sketched at the beginning of this section, in which the initial growth of a mutant population can be considered to take place in a constant background environment determined by the resident population. Thus, the probability of invasion success can be calculated from the results of this section.

In the Law–Dieckmann model two types of mutants are possible:

- In free-living form; and
- Within a holobiont.

Those in free-living form can die (with probability d), reproduce by splitting (probability b), or enter into symbiosis with a free-living individual of the other species (probability e). The holobiont can die (with probability μ) or it can release its partners (with probability α). A mutant within a holobiont can reproduce separately (probability b_f), in which case it produces one free-living individual, or both partners can produce a new holobiont together (probability b_s). The reproduction vectors and corresponding probabilities are given, and the dependence on the mutant trait, x, is made explicit in Tables 5.4 and 5.5.

Besides the birth rates in symbiosis, the mortality risk of free-living mutants is assumed to depend on the mutant trait value, since there might be a cost to adaptation to the symbiont form. Parameters b, μ, and α are supposedly unaffected by the trait value, whereas as long as the mutant is rare e can be assumed to depend only on the resident trait value. [Relations

Table 5.4 Offspring of free-living mutants.

(ξ_{11}, ξ_{12})	Probability
$(0, 0)$	$d(x)$
$(1, 0)$	$1 - b - d(x) - e$
$(2, 0)$	b
$(0, 1)$	e

Table 5.5 Offspring of mutants in symbiosis.

(ξ_{21}, ξ_{22})	Probability
$(0, 0)$	μ
$(1, 0)$	α
$(0, 1)$	$1 - \mu - \alpha - b_f(x) - b_s(x)$
$(1, 1)$	$b_f(x)$
$(0, 2)$	$b_s(x)$

between parameter values and trait values were obtained from a physiological resource-based approach, for which we refer to Law and Dieckmann (1998).]

The reproduction mean matrix is

$$M_x = \begin{pmatrix} 1 + b - d(x) - e & e \\ \alpha + b_f(x) & 1 + b_s(x) - \mu - \alpha \end{pmatrix} . \tag{5.86}$$

Calculation of its dominant eigenvalue $\rho(x)$ and of the corresponding normed eigenvectors $v(x), u(x)$ is tedious, but straightforward. Furthermore, from the tables given above it can be found that

$$\mathbb{E}[\xi_{11}(\xi_{11} - 1)] = 2b , \; \mathbb{E}[\xi_{12}(\xi_{12} - 1)] = \mathbb{E}[\xi_{21}(\xi_{21} - 1)] = 0 ,$$

$$\mathbb{E}[\xi_{11}\xi_{12}] = 0 , \; \mathbb{E}[\xi_{22}\xi_{21}] = b_f(x) ,$$

$$\mathbb{E}[\xi_{22}(\xi_{22} - 1)] = 2b_s(x) . \tag{5.87}$$

Substitution of these expressions in Equation (5.84) gives

$$B(x) = u_1(x)v_1^2(x)2b + 2u_2(x)v_1(x)v_2(x)b_f(x) + u_2(x)v_2^2(x)2b_s(x) . \tag{5.88}$$

By Theorem 5.6 the chance that a free-living mutant can invade the resident population successfully is thus approximately

$$P_f(x) = \frac{2v_1(x)(\rho(x) - 1)}{B(x)} , \tag{5.89}$$

whereas, if the mutation originates in a holobiont, the invasion chance is approximately

$$P_s(x) = \frac{2v_2(x)(\rho(x) - 1)}{B(x)} . \tag{5.90}$$

From these expressions, the so-called selection differentials can be calculated by taking derivatives at the point where x equals the resident value x_r. In terms of our formulation this means that we consider entities as functions of $\varepsilon = x - x_r$, and differentiate with respect to ε at the origin.

The differentials indicate the direction in which selection pressure operates and are used to calculate an approximation to the mean path of evolution [for further details, see Dieckmann and Law (1996) and Law and Dieckmann (1998)]. The outcome is that the merger of two initially separate lineages can, indeed, evolve, even if resource transfer is entirely unidirectional from victim to exploiter. The critical feature is that for each species there are more deaths than births in the free-living state, which is reversed in symbiosis. Such situations can be achieved either by increased benefits of symbiosis to each partner or by costs to the free-living state, incurred by adaptive defense in the symbiotic state of either partner.

5.7 Accounting for Time Being Continuous

5.7.1 Continuous-time Markov branching

Continuous-time Markov branching processes were introduced in Section 3.2.1. As seen there, many properties of such processes, like the behavior of the expected population size, are similar to those of Galton–Watson branching processes. However, the analogy extends much further.

For instance, by Chebyshev's inequality

$$\mathbb{P}\left(Z_t > 0\right) = \mathbb{P}\left(Z_t \geq 1\right) \leq \mathbb{E}\left[Z_t\right] = e^{\lambda(m-1)t} \tag{5.91}$$

if the population started from one ancestor. Thus, if $m < 1$,

$$\lim_{t \to \infty} \mathbb{P}\left(Z_t > 0\right) = 0 . \tag{5.92}$$

As in Section 5.3, generating functions can provide more detailed information. They are defined in a standard way:

$$f_t(s) = \mathbb{E}\left[s^{Z_t}\right] = \sum_{k=0}^{\infty} \mathbb{P}\left(Z_t = k\right) s^k, \ 0 \leq s \leq 1 , \tag{5.93}$$

where it is assumed that $f_0(s) = s$, that is, the population is initiated by a single individual at time $t = 0$. Since individuals do not age (Section 3.2.1), we can always think of them as newly born, and for any $u > 0$

$$Z_{t+u} = Z_{t,1} + Z_{t,2} + \cdots + Z_{t,Z_u} , \tag{5.94}$$

where the random variables $Z_{t,j}, \ j = 1, \ldots, Z_u$ are independent and each of them has the same distribution as Z_t. Hence,

$$f_{t+u}(s) = \mathbb{E}\left[s^{Z_{t+u}}\right] = \mathbb{E}\left[\mathbb{E}\left[s^{Z_{t+u}} | Z_u\right]\right] = \mathbb{E}\left[\mathbb{E}\left[s^{Z_{t,1}+Z_{t,2}+\cdots+Z_{t,Z_u}} | Z_u\right]\right]$$

$$= \mathbb{E}\left[\prod_{j=1}^{Z_u} \mathbb{E}\left[s^{Z_{t,j}} | Z_u\right]\right] = \mathbb{E}\left[(f_t(s))^{Z_u}\right] = f_u(f_t(s)) . \tag{5.95}$$

In view of the symmetry between t and u,

$$f_{t+u}(s) = f_u(f_t(s)) = f_t(f_u(s)) \ . \tag{5.96}$$

This corresponds to Equation (5.6) for Galton–Watson processes.

Now, write $f(s) = \mathbb{E}[s^\xi]$ for the generating function of the offspring number ξ. If T denotes life span, and u is small, then Z_u is either ξ (if the ancestor died by u – but none of her children had the time to do so, as should be true for small u) or 1 (meaning that the ancestor survived up to time u, and thus had not had any children yet, since this is a splitting process). More formally,

$$f_u(s) \approx s\mathbb{P}(T > u) + \mathbb{E}[s^\xi]\mathbb{P}(T \le u) = se^{-\lambda u} + f(s)(1 - e^{-\lambda u}) \ . \tag{5.97}$$

Replacing s by $f_t(s)$ in this, we obtain

$$\begin{aligned}
f_{t+u}(s) = f_u(f_t(s)) &\approx f_t(s)e^{-\lambda u} + f(f_t(s))(1 - e^{-\lambda u}) \\
&= f_t(s) + (f(f_t(s)) - f_t(s))(1 - e^{-\lambda u}) \ .
\end{aligned} \tag{5.98}$$

Since $1 - e^{-\lambda u} \approx \lambda u$ for small u,

$$f_{t+u}(s) - f_t(s) \approx \lambda u \, (f(f_t(s)) - f_t(s)) \tag{5.99}$$

for small u. After obvious manipulations

$$\frac{f_{t+u}(s) - f_t(s)}{u} \approx \lambda \, (f(f_t(s)) - f_t(s)) \ . \tag{5.100}$$

Now let $u \to 0$, to conclude that

$$\frac{\partial f_t(s)}{\partial t} = \lambda(f(f_t(s)) - f_t(s)), \quad f_0(s) = s \ . \tag{5.101}$$

In particular, $s = 0$ yields $Q(t) = \mathbb{P}(Z_t = 0) = f_t(0)$ and

$$\frac{dQ(t)}{dt} = \lambda \, (f(Q(t)) - Q(t)) \ . \tag{5.102}$$

Generally, the differential Equation (5.101) is not solved easily. However, for one of our benchmark processes, binary splitting,

$$f(s) = q + ps^2 = 1 - \frac{m}{2} + \frac{m}{2}s^2 = s - (m-1)(1-s) + \frac{m}{2}(1-s)^2 \ , \tag{5.103}$$

with $m = 2p$, and Equation (5.101) turns into

$$\frac{\partial f_t(s)}{\partial t} = \lambda \left(-(m-1)(1 - f_t(s)) + \frac{m}{2}(1 - f_t(s))^2 \right) \ , \tag{5.104}$$

with the initial condition $f_0(s) = s$. The solution is

$$f_t(s) = 1 - \frac{e^{\lambda(m-1)t}(1-s)}{1 + m\left(e^{\lambda(m-1)t} - 1\right)(1-s)/2(m-1)} \ , \tag{5.105}$$

if $m \neq 1$, and

$$f_t(s) = 1 - \frac{1-s}{1 + \lambda t (1-s)/2},$$ (5.106)

if $m = 1$. Hence, if $m < 1$, then as $t \to \infty$,

$$\mathbb{P}(Z_t > 0) = 1 - f_t(0) \sim \frac{2(1-m)}{2-m} e^{\lambda(m-1)t}.$$ (5.107)

Similar asymptotic relations are valid generally. More precisely, assume that the reproduction variance is finite. (Actually, it suffices to suppose that

$$\mathbb{E}[\xi \ln(\xi+1)] < \infty,$$ (5.108)

but this will not impress biological readers!) Let a subcritical continuous-time Markov process start with $Z_0 = N$ individuals. Then there exists a constant $0 < c \leq N$ such that

$$\mathbb{P}(Z_t > 0) \sim c e^{-\lambda(1-m)t}, \quad t \to \infty.$$ (5.109)

We do not formulate the remaining counterparts of Galton–Watson properties, mentioned in Section 5.4. The reader can do this by replacing m by $e^{\lambda(m-1)}$ and n by t throughout.

5.7.2   General processes

For single-type general branching processes, survival probabilities exhibit the same long-run behavior as in Equation (5.109), but the methods of deriving it are very different. The Chebyshev inequality remains the same,

$$\mathbb{P}(Z_t > 0) = \mathbb{P}(Z_t \geq 1) \leq \mathbb{E}[Z_t],$$ (5.110)

and also, from Section 3.3,

$$\mathbb{E}[Z_t] \sim e^{\alpha t}(1-\ell)/\alpha\beta$$ (5.111)

in all well-behaved subcritical cases with one ancestor. Here, as opposed to the supercritical situation stressed in Section 3.3, $\alpha < 0$, so that we obtain, indeed, an exponential decrease, and since $\ell > 1$ for subcritical populations, the right-hand side is also positive, as it ought to be. Thus, we have found a proper upper bound on the survival probability. The cumbersome task is to establish the lower bound and exact asymptotics. Anyhow, under quite natural conditions, including a general process version of Condition (5.108), it holds true that

$$\mathbb{P}(Z_t > 0) \sim N c e^{-|\alpha|t}$$ (5.112)

for $Z_0 = N$ and some constant c, $0 < c < (\ell - 1)/|\alpha|\beta$, as $t \to \infty$.

As for other such assertions, we refer to the mathematical literature for proofs (Jagers 1975).

5.8 Population Size Dependent Processes

F.C. Klebaner

Recall that branching processes dependent on the population size are defined as Galton–Watson processes, but allowing that the offspring distribution may depend on the size of the mother's generation,

$$Z_{n+1} = \sum_{i=1}^{Z_n} \xi_{i,n}(Z_n) . \tag{5.113}$$

Again, we use the somewhat inadvertent notation style commented upon in Section 5.6: here, the parentheses remind us of the dependence in distribution of reproduction on Z_n. In the same vein, denote by $\xi(z)$ a random variable with the reproduction distribution

$$\mathbb{P}(\xi(z) = k) = p_k(z), \ k = 0, 1, 2, \dots , \tag{5.114}$$

which is in force when the population size is z.

We address the extinction or explosion dichotomy first. If, for some population size z, $p_1(z) = 1$, then once the population reaches this size, each member has one single offspring and the population size remains z forever. Barring this (quite artificial) possibility, extinction or explosion occurs, as in Theorem 5.7.

Theorem 5.7 *Assume that for all z, $p_1(z) < 1$. Then*

$$\mathbb{P}(Z_n \to \infty \ or \ Z_n \to 0) = 1 . \tag{5.115}$$

Provided $\mathbb{P}(\xi(z) = 0) = p_0(z) > 0$ for all z, this is a consequence of the Metatheorem 5.1 [in its stricter formulation, take $\delta(x) = min_{1 \le z \le x} p_0(z)^z$]. Actually, the situation is simpler here because the process is a Markov chain. By general theory, it is enough to establish that any state $z \ne 0$ is transient, meaning that having started with a population of size z, the probability of returning to the same size in the future is less than 1. This is clear, since there is always a positive probability of dying out; indeed, it is at least $p_0(z)^z$ at level z and the population cannot recover from 0. The result remains true without the simplifying (but natural) assumption that $p_0(z) > 0$ for all z, but the proof needs some extra concepts [see Fujimagari (1976); Klebaner (1984)].

It is important to have some information on the extinction probability. This is not simple, but in some cases there are straightforward results. For example, if

$$m(z) = \sum_k k p_k(z) \le 1 \tag{5.116}$$

for all z, so that the population is subcritical or critical throughout, it dies out (barring the degenerate case mentioned). However, if it is supercritical with bounded variances, the extinction probability is less than 1.

The first statement is easy to see from the result on the extinction or explosion. First,

$$\mathbb{E}[Z_n] = \mathbb{E}[\mathbb{E}[Z_n|Z_{n-1}]] = \mathbb{E}[m(Z_{n-1})Z_{n-1}] \leq \mathbb{E}[Z_{n-1}] . \tag{5.117}$$

Second, we know that Z_n tends to some limit Z_∞, which is either 0 or ∞. So we can use a result on mathematical expectations of positive variables, which states that the expectation of the limit does not exceed the limit of expectations (Fatou's lemma, see the Appendix) to conclude that in the case of (sub)critical reproduction $\mathbb{E}[Z_\infty] \leq \mathbb{E}[Z_0]$. Since the latter is finite, we must have that $\mathbb{P}(Z_\infty = \infty) = 0$ and thus $\mathbb{P}(Z_\infty = 0) = 1$. In other words, extinction is certain.

Before we give general results on extinction, here is a simple result on generating functions that sometimes yields a useful bound on the extinction risk.

Theorem 5.8 *Let $f_z(s)$ be the generating function of offspring distribution when the population size is z. Assume that for all z and some $u > 0$*

$$\int_0^1 (f_z(s))^z s^{u-1} \, ds \leq \frac{1}{z+u} . \tag{5.118}$$

Then, the extinction probability does not exceed $u/(z_0 + u)$, where z_0 is the non-random starting population.

The proof is that for a non-negative random variable X with generating function $f(s)$

$$\mathbb{E}\left[\frac{1}{X+u}\right] = \mathbb{E}\left[\int_0^1 s^{X+u-1} \, ds\right] = \int_0^1 f(s)s^{u-1} \, ds , \tag{5.119}$$

obtained by interchanging the order of expectation and integral.

Together with the condition of the theorem, this yields

$$\mathbb{E}\left[\frac{1}{Z_n+u} \Big| Z_{n-1} = z\right] = \int_0^1 (f_z(s))^z s^{u-1} \, ds \leq \frac{1}{z+u} . \tag{5.120}$$

Taking expectations of this results in

$$\mathbb{E}\left[\frac{1}{Z_n+u}\right] \leq \mathbb{E}\left[\frac{1}{Z_{n-1}+u}\right] , \tag{5.121}$$

which can be repeated to give

$$\mathbb{E}\left[\frac{1}{Z_n+u}\right] \leq \mathbb{E}\left[\frac{1}{Z_0+u}\right] = \frac{1}{z_0+u} . \tag{5.122}$$

As $n \to \infty$, either $Z_n \to \infty$, so that $1/(Z_n + u) \to 1/(Z_\infty + u) = 0$ or else $Z_n \to 0$ and $1/(Z_n + u) \to 1/u$. Thus, by Fatou's lemma (see the Appendix),

$$\frac{1}{z_0+u} \geq \liminf_{n\to\infty} \mathbb{E}\left[\frac{1}{Z_n+u}\right] \geq \mathbb{E}\left[\liminf_{n\to\infty} \frac{1}{Z_n+u}\right] = \mathbb{P}(Z_n \to 0)/u . \tag{5.123}$$

Example 5.5: Near-critical binary splitting. The theorem applies to binary splitting where the probability of division is $p(z) = 1/2 + 1/(2z)$, and $q(z) = 1 - p(z)$ is the probability of no children, if population size is z. Thus, $m(z) = 1 + 1/z$ and the process is supercritical, but approaches criticality as z increases.

The probability generating function of the offspring number is $f_z(s) = q(z) + p(z)s^2$. We take $u = 2$ in Equation (5.118) to obtain

$$
\begin{aligned}
\mathbb{E}\big[\frac{1}{Z_{n+1}+2}|Z_n = z\big] &= \int_0^1 (q(z) + p(z)s^2)^z s\, ds \\
&= 1/2 \int_0^1 (q(z) + p(z)y)^z\, dy \quad (s^2 = y) \\
&= 1/(2p(z)) \int_{q(z)}^1 t^z\, dt \quad (t = q(z) + p(z)y) \\
&\le 1/(2p(z)) \int_0^1 t^z\, dt = 1/(2p(z)(z+1)) \\
&= 1/(z + 2 + 1/z) < 1/(z+2)\,,
\end{aligned}
\tag{5.124}
$$

which proves that Condition (5.118) is valid with $u = 2$. This means that the survival probability is at least 1/3 if $Z_0 = 1$, and close to 1 if the initial population is large, in sharp contrast to the strictly critical case. Indeed, $Z_0 = 1000$ yields an extinction probability less than 2 promille.

◇ ◇ ◇

The following result is a classification theorem of Höpfner (1985). Recall that $m(z)$ is the mean of offspring distribution when population size is z and write $v(z) = \mathbb{E}[\xi(z)(\xi(z) - 1)]$. The letters c, C, M, N denote positive constants.

Theorem 5.9 *First assume that $m(z) \le 1 + c/z$ and $\sigma^2 - M/z \le v(z) < \infty$, for all $z > N$. Then $\mathbb{E}[\xi^2(z)] \le C$ and $\sigma^2 > 2c$ imply $Q = 1$, and $\mathbb{E}[\xi^3(z)] \le C$ and $\sigma^2 = 2c$ also imply $Q = 1$.*

Now assume that $1 + c/z \le m(z) < \infty$ and $\sigma^2 + M/z \ge v(z)$, for all $z > N$, and that $\sigma^2 < 2c$. Then $Q < 1$.

In the framework of a more general growth model, Kersting (1986) gives the best possible results in this vein. Since the concepts used are too advanced for this book, the interested reader is referred to the article. These were applied by Klebaner (1990) to obtain conditions for extinction or survival in multi-type population size dependent Galton–Watson processes.

5.9 Effects of Sexual Reproduction

G. Alsmeyer

To examine the effect of sexual reproduction on extinction probabilities, we turn to the Galton–Watson process with mating, which is introduced in Section 2.8. Recall that in this model the nth generation consists of F_n females and M_n males, who form $Z_n = \zeta(F_n, M_n)$ couples where F_n and M_n are random variables and ζ is a deterministic function, called a *mating function*. Each couple produces offspring independently of all other couples and according to the same distribution. Thus, let for each couple $p_{j,k}$ denote the probability of producing j female and k male children. With X_k and Y_k denoting, respectively, the number of female and male offspring of the kth couple of the nth generation (labeled in arbitrary fashion), we arrive at

$$F_{n+1} = \sum_{k=1}^{Z_n} X_k \quad \text{and} \quad M_{n+1} = \sum_{k=1}^{Z_n} Y_k \tag{5.125}$$

for $n \geq 0$, where the (X_k, Y_k) are independent and identically distributed. This is the familiar structure for the Galton–Watson branching process with the one, but important, difference that here the summation ranges over the number of couples of the preceding generation. Choosing the "asexual" mating function $\zeta(x, y) = x$, we see that Z_n just equals the number of females in the nth generation ($Z_n = F_n$ for all $n \geq 0$) and is, indeed, a classic branching process.

From a mathematical viewpoint it is desirable to restrict the class of offspring distributions to facilitate explicit computations. Daley (1968) gave two alternative possibilities:

Assumption 5.2 *Conditionally on the total number of offspring, their sex is determined at random (analogously to flipping a, possibly biased, coin). More formally, if $\hat{p}_{j+k}$ denotes the probability that a couple produces $j + k$ children, and if the probability that a child is female equals θ, then*

$$p_{j,k} = \binom{j+k}{j} \theta^j (1-\theta)^k \hat{p}_{j+k} \tag{5.126}$$

for all $j, k \geq 0$.

Assumption 5.3 *Another possibility is that the numbers of male and female offspring are independent with a possibly different distribution. In this case*

$$p_{j,k} = p_j^F p_k^M \tag{5.127}$$

for all $j, k \geq 0$, where p_j^F and p_k^M denote the probabilities that a couple has, respectively, j female and k male children.

The mechanism that corresponds to Assumption 5.2 is most common in mammals, and occurs in humans with $\theta = 0.5$. Assumption 5.3 may be reasonable in situations with environmental sex determination, such as temperature dependence in many reptiles [for examples, see Bull (1983)]. The two mechanisms are equivalent for Poisson-distributed numbers of male and female offspring, so Assumption 5.3 with $(p_j^F)_{j \geq 0}$ and $(p_j^M)_{j \geq 0}$ Poisson distributions with means m^F and m^M, respectively, is equivalent to Assumption 5.2 with the Poisson distribution $(\hat{p}_j)_{j \geq 0}$ with mean $m^F + m^M$, and $\theta = m^F / (m^F + m^M)$.

Let us stipulate without further discussion that hereafter $(p_{j,k})_{j,k \geq 0}$ always satisfies Assumption 5.2 or 5.3 and, further, that there is a positive probability of producing offspring that are of one sex only, that is,

$$\max(p_{0,\bullet}, p_{\bullet,0}) > 0, \tag{5.128}$$

where $p_{0,\bullet} = \sum_{k \geq 0} p_{0,k}$ and $p_{\bullet,0} = \sum_{j \geq 0} p_{j,0}$ denote the respective probabilities that a couple has only male or only female offspring. This condition holds automatically under Assumption 5.2 because $0 < \theta < 1$, and is equivalent to $\max(p_0^F, p_0^M) > 0$ under Assumption 5.3. As before, we assume that the mating function ζ is common sense and superadditive (see Section 2.8).

5.9.1 Criticality

We now turn to the fundamental question of finding conditions that guarantee certain ultimate extinction of a Galton–Watson process with mating $(Z_n)_{n \geq 0}$. To be more precise, let

$$Q_j = \mathbb{P}(Z_n = 0 \text{ eventually} | Z_0 = j) \tag{5.129}$$

denote the extinction probability given $j \geq 1$ ancestor couples. Then the question in its most ambitious form may be restated as: Is there an intuitive condition for $Q_1 = Q_2 = \cdots = 1$, as for the simple Galton–Watson process, where we know that certain extinction occurs if, and only if, each individual produces at most one child on average and has a positive chance of having no children?

The following example, from Hull (1982), shows that one cannot expect an equally simple answer for processes with sexual reproduction. Consider the mating function $\zeta(x, y) = 0$ if $x = 0$ or $y = 0$, and $\zeta(x, y) = x + y - 1$ otherwise. Let $p_{j,k}$ be of the form of Equation (5.126) for some $0 < \theta < 1$ and with $(\hat{p}_j)_{j \geq 0}$ defined through $\hat{p}_3 = 1$, and hence $\hat{p}_j = 0$ otherwise. Then, every couple has exactly three children. Nonetheless, extinction occurs if, for some $n \geq 0$, all couples of the nth generation produce only female or only male offspring. By comparison with a process of an inbreeding population in which couples are formed only by children of the same parents, Hull showed that $Q_j < 1$ for all $j \geq 1$ and any choice of θ (see Theorem 5.10 and its proof in Box 5.3). This may come as a surprise because

$$m = \mathbb{E}[Z_1 | Z_0 = 1] = 2(1 - \theta^3 - (1 - \theta)^3) \tag{5.130}$$

is strictly less than 1 if $\theta = 0.8$. However, as pointed out later by Bruss (1984), more relevant here are the *average unit reproduction means*

$$m_j = \frac{1}{j} \mathbb{E}[Z_{n+1} | Z_n = j], \quad j \geq 1, \tag{5.131}$$

which give the mean population growth rates per generation for the various levels j. For the simple Galton–Watson process this is disguised by the lucky coincidence that m_j does not depend on j. In the given example,

$$m_j = \frac{3j - 1}{j}(1 - \theta^{3j} - (1 - \theta)^{3j}), \tag{5.132}$$

which, for any choice of θ, increases to 3 as j tends to infinity. In the case $\theta = 0.8$, we thus see that the population, when originating from one ancestor couple, can actually survive because, with positive probability, it eventually reaches a level at which the growth becomes supercritical ($m_j > 1$, for all sufficiently large j).

It is quite intuitive, and actually confirmed by the following result of Daley *et al.* (1986), that this latter observation holds true more generally.

Theorem 5.10 *For a Galton–Watson process $(Z_n)_{n \geq 0}$ with a common sense, superadditive mating function ζ, the average reproduction means m_j are convergent to the limit $m_\infty = \sup_{k \geq 1} m_k$. Furthermore, $m_\infty \leq 1$ implies certain extinction for any initial population size (i.e., $Q_1 = Q_2 = \cdots = 1$), while in the case $m_\infty > 1$ (ultimate supercriticality) the population survives with positive probability for a sufficiently large initial population size, in fact $1 > Q_{i_0} \geq Q_{i_0 + 1} \geq \cdots$ for some positive integer i_0.*

For those readers who wonder whether there are examples of ultimately supercritical processes with common sense superadditive mating functions that die out if the initial population size is too small, we note that this happens, for instance, if the mating function ζ is chosen such that $\zeta(x, y) = 0$ whenever x or y is less than some (arbitrarily chosen) threshold. Other, less trivial examples can also be given, but further discussion is omitted because the biological relevance of any such example seems doubtful. We add in support of the latter statement that, whenever the considered population has a positive chance of increase at any given level i, formally stated as $\mathbb{P}(Z_{n+1} > i | Z_n = i) > 0$ for all $i \in I\!N$, then ultimate supercriticality implies a positive chance of survival for *all* initial population sizes, so $i_0 = 1$ and $1 > Q_1 \geq Q_2 \geq \cdots$. Since, by the Strong Law of Large Numbers (see the Appendix), $\bar{X}_j = j^{-1} \sum_{i=1}^{j} X_i$ and $\bar{Y}_j = j^{-1} \sum_{i=1}^{j} Y_i$ tend to the average numbers of female and male children per couple, m^F and m^M, writing

$$m_\infty = \lim_{j \to \infty} \mathbb{E}\big[\zeta\big(\sum_{i=1}^{j} X_i, \sum_{i=1}^{j} Y_i\big)\big]/j = \lim_{j \to \infty} \mathbb{E}\big[\zeta\big(j\bar{X}_j, j\bar{Y}_j\big)\big]/j , \qquad (5.133)$$

it should not be surprising that one can show that

$$m_\infty = \lim_{j \to \infty} \mathbb{E}\big[\zeta\big(jm^F, jm^M\big)\big]/j = r(m^F, m^M) \qquad (5.134)$$

for a suitable function r (see Daley *et al.* 1986, Lemma 2.3). We note in passing the technical point that m^F and m^M need not be integers, but that, by linear interpolation, $\zeta(x, y)$ can always be defined for all pairs (x, y) of non-negative numbers without losing superadditivity. For the examples given in Section 2.8 this is clear anyway. Although it often may be hard to determine r explicitly, there are many examples of ζ, including ours, for which this is easy. In fact, $\zeta(x, y) = \min(x, dy)$ implies $r = \zeta$ and $m_\infty = \min(m^F, dm^M)$ and $\zeta(x, y) = x \min(1, y)$ implies $r(x, y) = x$ and $m_\infty = m^F$.

Box 5.3 Proof of Theorem 5.10

In the following we present the main arguments of the proof, without technicalities. The first observation to make is that

$$\mathbb{P}(Z_{n+1} = j | Z_n = i) = \mathbb{P}\Big(\zeta\big(\sum_{k=1}^{i} X_{n+1,k}, \sum_{k=1}^{i} Y_{n+1,k}\big) = j\Big) \tag{a}$$

and, since the mating function ζ is monotonic in each argument, this implies

$$\mathbb{P}(Z_1 > k | Z_0 = i) = \mathbb{P}\Big(\zeta\big(\sum_{j=1}^{i} X_j, \sum_{j=1}^{i} Y_j\big) > k\Big) \leq \mathbb{P}\Big(\zeta\big(\sum_{j=1}^{i+1} X_j, \sum_{j=1}^{i+1} Y_j\big) > k\Big)$$

$$= \mathbb{P}(Z_1 > k | Z_0 = i + 1) \tag{b}$$

for all $i, k \in \mathbb{N}_0$. So the probability of exceeding a size k in the next generation forms an increasing function of the current population size. A Markov chain with this property is called *stochastically monotone*. By an easy inductive argument, one can prove that Equation (b) generalizes to

$$\mathbb{P}(Z_n > k | Z_0 = i) \leq \mathbb{P}(Z_n > k | Z_0 = i + 1) \tag{c}$$

for all $i, k = 0, 1, 2, \ldots$ and $n = 1, 2, \ldots$, which in turn yields the important fact that the extinction probability Q_i is a decreasing function of the initial population size i. Namely, by letting n tend to infinity in Equation (c),

$$1 - Q_i = \lim_{n\to\infty} \mathbb{P}(Z_n > 0 | Z_0 = i) \leq \lim_{n\to\infty} \mathbb{P}(Z_n > 0 | Z_0 = i + 1) = 1 - Q_{i+1} \tag{d}$$

for all $i \in \mathbb{N}_0$. For a more intuitive comparison argument, suppose the population starts with $i + 1$ ancestor couples ($Z_0 = i + 1$). Choose an arbitrary subset of i couples and denote by $(Z'_n)_{n\geq 0}$ the process based on this subset, hence $Z'_0 = i$. Then the Z_1 couples that form the first generation of the original population are those formed by the offspring of the i ancestor couples of the subpopulation plus, generally, some more because of the one additional ancestor couple in the original population and the monotonicity of the mating function. This shows $Z'_1 \leq Z_1$ and finally leads to the conclusion that $Z'_n \leq Z_n$ for all $n \geq 0$ when repeating the argument for the subsequent generations. Since the extinction probabilities of $(Z_n)_{n\geq 0}$ and $(Z'_n)_{n\geq 0}$ are Q_{i+1} and Q_i, respectively, the inequality $Q_i \geq Q_{i+1}$ follows as a consequence.

We now show that m_j converges to $m_\infty = \sup_{k\geq 1} m_k$. Indeed, from the definition in Equation (5.131)

$$(j + k)m_{j+k} = \mathbb{E}[Z_1 | Z_0 = j + k] = \mathbb{E}\Big[\zeta\big(\sum_{l=1}^{j+k} X_l, \sum_{l=1}^{j+k} Y_l\big)\Big] \tag{e}$$

and from the superadditivity of ζ, the result is larger than or equal to

$$\mathbb{E}\Big[\zeta\big(\sum_{l=1}^{j} X_l, \sum_{l=1}^{j} Y_l\big)\Big] + \mathbb{E}\Big[\zeta\big(\sum_{l=j+1}^{j+k} X_l, \sum_{l=j+1}^{j+k} Y_l\big)\Big]. \tag{f}$$

continued

Box 5.3 *continued*

From the independence and identical distribution of the offspring variables (X_l, Y_l), it follows that this equals

$$\mathbb{E}\Big[\zeta\Big(\sum_{l=1}^{j} X_l, \sum_{l=1}^{j} Y_l\Big)\Big] + \mathbb{E}\Big[\zeta\Big(\sum_{l=1}^{k} X_l, \sum_{l=1}^{k} Y_l\Big)\Big]$$

$$= \mathbb{E}[Z_1|Z_0 = j] + \mathbb{E}[Z_1|Z_0 = k] = jm_j + km_k \tag{g}$$

for all $j, k \geq 1$. Combining Equation (e) with Equation (g), we find that $(j + k)$ $m_{j+k} \geq jm_j + km_k$ for all $j, k \geq 1$, which implies that jm_j is superadditive. Applying standard results on superadditive functions to $(jm_j)_{j\geq 1}$ (e.g., Hille and Phillips 1957) then yields the asserted convergence of the m_j to $m_\infty = \sup_{k\geq 1} m_k$.

Suppose now that $m_\infty \leq 1$ and thus $m_j \leq 1$ for all $j \geq 1$. Then

$$\mathbb{E}[Z_{n+1}|Z_n = i] = im_i \leq i \tag{h}$$

holds for all $i, n \geq 0$. A stochastic sequence with this property is called a super-martingale (see the Appendix). A fundamental result from the theory of stochastic processes says that every non-negative supermartingale converges to a finite random variable, hence $Z_n \to Z_\infty$ (for any given initial population size). However, Z_∞ must then be identical to 0 by the extinction–explosion dichotomy (see Section 5.2), and so $Q_1 = Q_2 = \cdots = 1$, as asserted.

To see that $Q_i < 1$ for all sufficiently large i in the case $m_\infty > 1$ is more difficult and too technical to be presented here. However, a rather simple argument from Hull (1982) exists under the stronger condition $m_1 > 1$, and is again based on a comparison of $(Z_n)_{n\geq 0}$ with another process, a supercritical Galton–Watson process. Define $Z_0' = Z_0$ and then, recursively,

$$Z_n' = \sum_{j=1}^{Z_{n-1}'} \zeta(X_j, Y_j) \tag{i}$$

for $n \geq 2$. One may think of $(Z_n')_{n\geq 0}$ as describing an inbreeding population in which couples are formed according to the same mating function, but only by children of the same parents. The superadditivity of ζ implies

$$Z_1' = \sum_{j=1}^{Z_0} \zeta(X_j, Y_j) \leq \zeta\Big(\sum_{j=1}^{Z_0} X_j, \sum_{j=1}^{Z_0} Y_j\Big) = Z_1, \tag{j}$$

and then, inductively, $Z_n' \leq Z_n$ for all $n \geq 0$. Since all $\zeta(X_j, Y_j)$ are independent with the same distribution $(p_k)_{k\geq 0}$, say, $(Z_n')_{n\geq 0}$ is distributed as a simple Galton–Watson process with offspring distribution $(p_k)_{k\geq 0}$. It is further supercritical because $\mathbb{E}[\zeta(X_1, Y_1)] = \mathbb{E}[Z_1|Z_0 = 1] = m_1 > 1$. Consequently, $(Z_n')_{n\geq 0}$ survives with positive probability for any initial population size and so $(Z_n)_{n\geq 0}$ also does (i.e., $1 > Q_1 \geq Q_2 \geq \cdots$).

5.9.2 Sexual versus asexual reproduction: The extinction probability ratio

Given a large initial population size, how does mating affect the extinction probability as compared to the asexual case? This interesting and natural question appears to be a hard one from a mathematical point of view, which may be why only very few contributions to this subject are found in the literature, namely Daley *et al.* (1986) and Alsmeyer and Rösler (1996, 2002). For the classic basic branching process, the extinction probability Q_i given an initial size i satisfies $Q_i = Q_1^i$ and can be calculated exactly because Q_1 is found as the smallest solution in $[0, 1]$ to the equation $f(s) = s$, where $f(s) = \sum_{j \geq 0} p_j s^j$ denotes the generating function of the offspring distribution. Unfortunately, there is no such simple way to compute Q_i for Galton–Watson processes with mating, whatever the choice of the mating function.

Daley *et al.* (1986) suggest a finite Markov chain approximation, which is described roughly as follows. Let $(Z_n)_{n \geq 0}$ be a Galton–Watson process with super-additive mating function ζ and note that $(Z_n)_{n \geq 0}$ forms a temporally homogeneous Markov chain with transition matrix $P = (P_{ij})_{i,j \geq 0}$, that is,

$$P_{ij} = \mathbb{P}(Z_n = j | Z_{n-1} = i) \tag{5.135}$$

denotes the conditional probability that, at any time $n = 0, 1, 2, \ldots$, the population size changes from i to j. The state 0 is absorbing and thus $P_{00} = 1$. The extinction–explosion dichotomy further implies that, in the case of survival, the chain is asymptotically absorbed at ∞. Moreover, the latter is more and more likely to happen if the initial population size becomes large. Hence, the probability of extinction (absorption at 0) should only change very little if, for some integer N considerably larger than the initial state, $(Z_n)_{n \geq 0}$ is replaced with the finite Markov chain $(Z_n(N))_{n \geq 0}$, say, which evolves exactly as $(Z_n)_{n \geq 0}$ until a state $N + i$, $i \geq 1$, is hit, in which case the latter chain is absorbed at N. The extinction probabilities of both chains then only differ by the probability of the rare event that $(Z_n)_{n \geq 0}$ dies out after exceeding the high level N. However, extinction probabilities for the finite Markov chain $(Z_n(N))_{n \geq 0}$ can be obtained as the solutions to a *finite* system of linear equations.

To make this precise, fix a large integer N and let $(Z_n(N))_{n \geq 0}$ be defined as

$$Z_n(N) = Z_{\min(n, T(N))} = \begin{cases} Z_n, & \text{if } n < T(N) \\ Z_{T(N)}, & \text{if } n \geq T(N) \end{cases}, \tag{5.136}$$

where $T(N)$ is the first time k is such that $Z_k > N$. This chain has the transition matrix

$$P(N) = \begin{pmatrix} 1 & 0 & \ldots & 0 & 0 \\ P_{10} & P_{11} & \ldots & P_{1N} & 1 - \sum_{i=0}^{N} P_{1i} \\ \vdots & \vdots & \ddots & \vdots & \vdots \\ P_{N0} & P_{N1} & \ldots & P_{NN} & 1 - \sum_{i=0}^{N} P_{Ni} \\ 0 & 0 & \ldots & 0 & 1 \end{pmatrix}. \tag{5.137}$$

The extinction probabilities $Q_i(N) = \mathbb{P}(Z_n(N) = 0 \text{ eventually} | Z_0(N) = i)$, $i = 1, \ldots, n$, satisfy the system of linear equations

$$Q_i(N) = P_{i0} + \sum_{j=1}^{N} P_{ij} Q_j(N), \quad i = 1, \ldots, N, \tag{5.138}$$

which in matrix form reads

$$Q(N) = (I - R(N))^{-1} P_0(N), \tag{5.139}$$

where

$$Q(N) = \begin{pmatrix} Q_1(N) \\ \vdots \\ Q_N(N) \end{pmatrix}, \quad P_0(N) = \begin{pmatrix} P_{10} \\ \vdots \\ P_{N0} \end{pmatrix}, \quad R(N) = \begin{pmatrix} P_{11} & \cdots & P_{1N} \\ \vdots & \ddots & \vdots \\ P_{N1} & \cdots & P_{NN} \end{pmatrix}, \tag{5.140}$$

and I is the identity matrix. Note that

$$(I - R(N))^{-1} = I + R(N) + R(N)^2 + \cdots. \tag{5.141}$$

The following result from Daley *et al.* (1986) provides an estimate for $Q_i - Q_i(N)$ for $N \geq i$ and is stated without proof.

Theorem 5.11 *Given a Galton–Watson process* $(Z_n)_{n \geq 0}$ *with superadditive mating function,*

$$Q_i(i + j - 1) \leq Q_i \leq \min\left(1, \frac{Q_i(i + j - 1)}{1 - Q_j(i + j - 1)}\right) \tag{5.142}$$

for all $i, j \geq 1$.

Daley *et al.* (1986) used this finite chain approximation to compute the extinction probabilities Q_i of supercritical processes with monogamous or (unilateral) promiscuous mating functions for various initial generation sizes i. The numbers of female and male offspring per individual were assumed to be independent, with a Poisson distribution of mean 1.2, that is,

$$p_j^F = p_j^M = e^{-1.2} \frac{1.2^j}{j!} \quad j = 0, 1, 2, \ldots. \tag{5.143}$$

The simple (asexual) branching process with this offspring distribution has extinction probabilities Q^i for $i \geq 1$, where $Q = 0.6863$. These values can be compared to the respective extinction probabilities Q_i for the monogamous or (unilateral) promiscuous branching processes, which are clearly larger. Based on the numbers in Daley *et al.* (1986), Table 5.6 shows the values of the extinction probability ratio $R_i = Q_i / 0.6863^i$ for various initial generation sizes i.

In the monogamous case R_i apparently tends to infinity. Daley *et al.* (1986) note that there does not appear to be a simple way to find the precise asymptotic

Table 5.6 Extinction probability ratios for various initial generation sizes.

Initial generation size	Mating type Asexual	Monogamous	Promiscuous
i	0.6863^i	$R_i = Q_i/0.6863^i$	
1	0.6863	1.4530	1.2439
2	0.4710	2.0964	1.3161
3	0.3233	2.9938	1.3300
4	0.2219	4.2231	1.3308
5	0.1523	5.8779	1.3300
6	0.1045	8.0699	1.3292
10	0.2318×10^{-1}	25.0216	1.3296
20	0.5374×10^{-3}	204.1310	1.3295
40	0.2888×10^{-6}	2637.1191	1.3296
60	0.1552×10^{-9}	12847.9381	1.3293

behavior of R_i, but that, by a very rough heuristic argument based on the Central Limit Theorem, it seems plausible that $R_i \approx \exp(c\sqrt{i})$ for some $c > 0$ and sufficiently large i. They also point out that, in the promiscuous case, R_i seems to converge rapidly to about 1.33, but do not give a theoretical explanation for the particular value. However, convergence of R_i is quite plausible because the promiscuous process behaves exactly like the asexual process that pertains to the female subpopulation, as long as at least one male is born in each generation. An additional risk of extinction is caused only by the probability that a generation may have no male offspring at all, which becomes more and more unlikely for increasing initial population sizes. Based on these observations Alsmeyer and Rösler (1996, 2002) provide a deeper analysis of promiscuous processes with offspring distributions that satisfy Assumption 5.3. Although the mathematical details are far beyond the scope of this survey, as they involve potential theoretic aspects of branching processes, we summarize the major findings from these authors in Theorem 5.12.

So we consider a Galton–Watson process with (unilateral) promiscuous mating $(Z_n)_{n \geq 0}$ that has probabilities $p_{j,k} = p_j^F p_k^M$ of having j daughters and k sons, respectively, per couple. Since Z_n equals the number of females in the nth generation, as long as at least one male is alive it follows easily with Assumption 5.3 that the extinction probabilities Q_i depend on the male offspring distribution $(p_k^M)_{k \geq 0}$ only through p_0^M, the probability that a couple has no male offspring. Let $f(s) = \sum_{j \geq 0} p_j^F s^j$ be the generating function of the female offspring distribution $(p_j^F)_{j \geq 0}$, f_n its nth iterate (see Section 5.3), and Q the extinction probability of the associated simple Galton–Watson process, say $(F_n)_{n \geq 0}$, with this offspring distribution. Hence, Q is the smallest solution of $f(Q) = Q$ in [0, 1], and $f_n(s) \uparrow Q$ for each $s \leq Q$.

Theorem 5.12 *Suppose that $m^F = \sum_{j \geq 1} j p_j^F > 1$ and $\kappa = p_0^M < 1$.*
(a) The following assertions hold true for all $i \geq 1$:

1. *If $\kappa < p_0^F$, then*

$$1 \leq R_i \leq 1 + \frac{\kappa}{p_0^F} ; \tag{5.144}$$

2. *If $\kappa = p_0^F$, then*

$$1 + \frac{1 - Q}{1 + Q - p_0^F} \leq R_i \leq 2 ; \tag{5.145}$$

3. *If $p_0^F < \kappa < Q$, then*

$$1 + \frac{\kappa(1 - Q)}{\kappa Q + (1 - \kappa)p_0^F} \leq R_i \leq (n + 2)\left(\frac{1}{1 - \kappa} + \frac{p_0^F}{\kappa}\right), \tag{5.146}$$

where n is determined through $f_n(p_0^F) < \kappa \leq f_{n+1}(p_0^F)$;

4. *If $\kappa = Q$, then*

$$\frac{1 - Q}{Q(a_1 - Q) + (1 - Q)} \leq \frac{R_i}{a_i} \leq \frac{1}{1 - q} + \frac{p_0}{q}, \tag{5.147}$$

where $a_i = \mathbb{E}(\tau | \tau < \infty, F_0 = i)$ and $\tau = \inf\{n \geq 0 : F_n = 0\}$;

5. *If $\kappa > Q$, then*

$$1 \leq \frac{Q_i}{\kappa^i} \leq 1 + \frac{f(\kappa)}{\kappa - f(\kappa)} . \tag{5.148}$$

(b) If $0 < \kappa < Q$, then convergence of R_i does not hold in general, while

$$\lim_{i \to \infty} \ln\left(\frac{1}{f'(Q)}\right)\frac{R_i}{a_i} = \lim_{i \to \infty} \frac{R_i}{\ln i} = 1 \quad \text{if } \kappa = Q, \tag{5.149}$$

and

$$\lim_{i \to \infty} \frac{R_i}{\kappa^i} = 1 \quad \text{if } \kappa > Q. \tag{5.150}$$

The most intriguing result stated in Theorem 5.12 is that for $0 < \kappa < Q$ convergence of R_i fails to hold in general. This is even more surprising considering that all computational studies of R_i for this case indicate the contrary, namely a rapid convergence to some finite value, as in the above example studied by Daley *et al.* (1986); see Alsmeyer and Rösler (1996) for some examples. The disclosed phenomenon belongs to the class of so-called *near-constancy phenomena*, which also show up in other problems in the theory of branching processes (see also Biggins and Nadarajah 1993). It means that a considered sequence is seemingly convergent, but actually oscillates in a very small range (of the order 10^{-4} or smaller). The convergence results, Equation (5.149) and Equation (5.150), are much more appealing to intuitive thinking, their interpretation being that for $\kappa \geq Q$, the extinction of a population with large initial size is more likely to be caused by the disappearance of males than that of females.

Box 5.4 Lotka's data reconsidered by Hull

Lotka (1931a, 1931b) calculated the extinction probability of a male line of descent from one newborn male. Using data from a US census of 1920, he arrived at the conclusion that this risk equals 0.8715. His calculation, however, was based on an asexual branching process model. Recently, Hull (2001) corrected the estimate, using a branching process with monogamous mating. He used the offspring distribution given by Lotka and sex determination according to Assumption 5.2. From the same census data, he estimated that the probability of producing a female child, θ, equaled 0.485. With the numerical approach developed by Daley *et al.* (1986) and outlined above, he arrived at an estimated extinction probability of 0.9958, obviously much larger than Lotka's estimate. This would be a grim prospect for the survival of family names (the application that Lotka had in mind). As Hull notes, however, this analysis is based on the assumption of a single mating unit in the initial population. Moreover, a particular name that originates from a single family has a higher survival chance when other mating units produce females to act as mates for future generations, especially when males from that family are highly esteemed (Hull 1998).

5.10 Environmental Variation Revisited

In this section we reconsider the effect of environmental variation. Some examples of branching process models in varying environments were introduced in Section 2.9. We use the same notation here, that is, $m(k)$ denotes the expected number of offspring per individual in the kth reproduction cycle.

5.10.1 Deterministically varying environments

In Section 2.9 it is shown that

$$\mathbb{E}[Z_n] = \prod_{k=0}^{n-1} m(k)\mathbb{E}[Z_0] . \tag{5.151}$$

As mentioned there, this result is true whether or not reproduction of different individuals is independent. For the results in this subsection and the next, however, the independence assumption is necessary.

The generating function of the offspring distribution of individuals in the kth generation is defined as

$$f(k, s) = \mathbb{E}[s^{\xi(k)}] = \sum_{j=0}^{\infty} \mathbb{P}(\xi(k) = j)s^j = \sum_{j=0}^{\infty} p_j(k)s^j , \tag{5.152}$$

with $0 \le s \le 1$, where $\xi(k)$ denotes the number of offspring of an individual of the kth generation, and $p_j(k)$ the chance that this equals j.

Furthermore, we define the generating function of the number of individuals of generation k in a population that consists of one individual in generation

$r(0 \leq r \leq k)$ as

$$f_r(k, s) = \mathbb{E}[s^{Z_k}|Z_r = 1] = \sum_{j=0}^{\infty} \mathbb{P}(Z_k = j|Z_r = 1)s^j , \qquad (5.153)$$

with $0 \leq s \leq 1$. The conditional expectation argument yields

$$f_r(k, s) = \mathbb{E}[\mathbb{E}[s^{Z_k}|Z_{k-1}]|Z_r = 1]$$
$$= \mathbb{E}[\mathbb{E}[s^{\xi_1(k-1)+\xi_2(k-1)+\dots+\xi_{Z_{k-1}}(k-1)}|Z_{k-1}]|Z_r = 1] . \qquad (5.154)$$

Since the $\xi_i(k-1)$ are independent, this implies that

$$f_r(k, s) = \mathbb{E}\left[\mathbb{E}\left[s^{\xi_1(k-1)}|Z_{k-1}\right]\mathbb{E}\left[s^{\xi_2(k-1)}|Z_{k-1}\right] \times \cdots \right.$$
$$\left. \times \mathbb{E}\left[s^{\xi_{Z_{k-1}}(k-1)}|Z_{k-1}\right]|Z_r = 1\right] \qquad (5.155)$$

and, since they are distributed identically with generating function $f(k-1, s)$,

$$f_r(k, s) = \mathbb{E}\left[f(k-1, s)^{Z_{k-1}}|Z_r = 1\right]$$
$$= f_r(k-1, f(k-1, s)) . \qquad (5.156)$$

Repeated use of this argument results in

$$f_r(k, s) = f_r(r+1, f(r+1, \cdots f(k-1, s) \cdots)) . \qquad (5.157)$$

Furthermore,

$$f_r(r+1, s) = \mathbb{E}\left[s^{Z_{r+1}}|Z_r = 1\right] = \mathbb{E}\left[s^{\xi(r)}\right] = f(r, s) . \qquad (5.158)$$

Thus, finally,

$$f_r(k, s) = f(r, f(r+1, \cdots f(k-1, s) \cdots)) = f(r, f_{r+1}(k, s)) . \qquad (5.159)$$

With $r = 0$, $k = n$, and $s = 0$, Equation (5.153) reads

$$f_0(n, 0) = \mathbb{P}(Z_n = 0|Z_0 = 1) . \qquad (5.160)$$

Obviously, the extinction probability does not decrease over generations, and

$$f_0(n, 0) = \mathbb{P}(Z_n = 0|Z_0 = 1) \leq \mathbb{P}(Z_{n+1} = 0|Z_0 = 1) = f_0(n+1, 0) . \qquad (5.161)$$

It follows that the sequence

$$Q_n = \mathbb{P}(\text{extinction by generation } n|Z_0 = 1)$$
$$= \mathbb{P}(Z_n = 0|Z_0 = 1) = f_0(n, 0), n = 1, 2 \dots , \qquad (5.162)$$

must increase to the extinction probability $Q = \lim_{n\to\infty} Q_n$, and actually

$$Q = \lim_{n\to\infty} f_0(1, f(2, \cdots f(n-1, 0) \cdots)) . \qquad (5.163)$$

In contrast to the situation in constant environments, it is not easy to answer the question when $Q = 1$ and $Q < 1$ in general. However, there is a simple sufficient

Box 5.5 Survival chances with geometric reproduction

Assume that any individual of the kth generation produces j children, $j = 0, 1, 2, 3, 4, \ldots$, with probabilities

$$p_j(k) = q(k)p^j(k) . \tag{a}$$

The generating functions $f(k, s)$ are, in this case,

$$f(k, s) = \sum_{j=0}^{\infty} q(k)p^j(k)s^j = \frac{q(k)}{1 - p(k)s} . \tag{b}$$

Repeating the arguments of Section 5.3 in the slightly more cumbersome notation of varying environments, we see that

$$1 - f(k, s) = 1 - \frac{q(k)}{1 - p(k)s} = \frac{p(k)(1 - s)}{1 - p(k)s} , \tag{c}$$

so

$$\frac{1}{1 - f(k, s)} = \frac{1 - p(k)s}{p(k)(1 - s)} . \tag{d}$$

Since $m(k) = p(k)/q(k)$, this means that

$$\frac{1}{1 - f(k, s)} - \frac{1}{m(k)(1 - s)} = \frac{1 - p(k)s}{p(k)(1 - s)} - \frac{q(k)}{p(k)(1 - s)}$$
$$= \frac{1 - q(k) - p(k)s}{p(k)(1 - s)} = \frac{p(k)(1 - s)}{p(k)(1 - s)} = 1 . \tag{e}$$

With $k = 0$ and s replaced by $f_1(n, s)$,

$$\frac{1}{1 - f(0, f_1(n, s))} - \frac{1}{m(0)(1 - f_1(n, s))} = 1 . \tag{f}$$

However, Equation (5.159) with $r = 1$ and $k = n$ says that

$$f_1(n, s) = f(1, f(2, \ldots f(n-1, s) \ldots)) . \tag{g}$$

If we apply the same equation with $r = 0$ and $k = n$ on the left side, and Equation (g) for $f_1(n, s)$ substituted on the right side, we obtain

$$f_0(n, s) = f(0, f_1(n, s)) , \tag{h}$$

and so Equation (f) yields

$$\frac{1}{1 - f_0(n, s)} = 1 + \frac{1}{m(0)(1 - f_1(n, s))} . \tag{i}$$

Repeated use of this recurrence relation gives

$$\frac{1}{1 - f_0(n, s)} = 1 + \frac{1}{m(0)} + \frac{1}{m(0)m(1)(1 - f_2(n, s))} = \cdots$$
$$= 1 + (m(0))^{-1} + (m(0)m(1))^{-1} + \cdots$$
$$+ (m(0)m(1) \ldots m(n-2))^{-1}$$
$$+ (m(0)m(1) \ldots m(n-1)(1 - s))^{-1} , \tag{j}$$

and, in particular,

$$1 - Q_n = 1 - f_0(n, 0) = \left(1 + \sum_{k=0}^{n-1} \prod_{j=0}^{k} (m(j))^{-1} \right)^{-1} . \tag{k}$$

It follows that $Q_n \to 1$ if and only if the right-hand side of Equation (k) goes to 0 as $n \to \infty$.

condition, namely if

$$\lim_{n\to\infty} \prod_{k=0}^{n-1} m(k) = 0 , \tag{5.164}$$

then $Q = 1$. This can be derived with the help of Markov's inequality (see the Appendix),

$$\mathbb{P}(Z_n > 0|Z_0 = 1) = \mathbb{P}(Z_n \geq 1|Z_0 = 1)$$

$$\leq \mathbb{E}[Z_n|Z_0 = 1] = \prod_{k=0}^{n-1} m(k) . \tag{5.165}$$

In particular, if environments deteriorate monotonically in the mean, that is, if $m(k) \leq m(k+1)$, and if, furthermore, $\lim_{k\to\infty} m(k) = m < 1$, the process dies out in the long run.

Equation (5.163) can, in principle, be used to calculate the extinction probability Q. In many cases, however, this is not easy to do analytically, so we must resort to numerical methods. For one of our benchmark processes, namely processes with varying geometrical offspring distributions, an explicit expression of Q is derived in Box 5.5. If you do not wish to work through the derivation, read off the final expression for $1 - Q_n$, which is the n generation's survival probability.

What can we say generally about the asymptotic behavior of the population size in varying environments? A broad answer is, if $\prod_{j=0}^{n} m(j) \to \infty$, then under natural conditions and on the set of non-extinction the population grows (up to a random factor) as $\prod_{j=0}^{n} m(j)$. If, however, $\prod_{j=0}^{n} m(j) \to 0$, then, conditioned on non-extinction, the distribution of the number of individuals in the population stabilizes as time passes.

Example 5.6: Cell death and human embryo arrest. Human pre-implantation embryos that are produced in vitro exhibit a highly variable developmental potential. Only 25 percent of embryos that are transferred to patients 2 days after fertilization succeed in implanting, whereas the rest exhibit varying deficiencies. About 50 percent of embryos cultured in vitro arrest during the first week. Such losses probably result from a variety of factors during early development. To study the mechanism of embryo arrest, Hardy *et al.* (2001) constructed branching process models of cell division and death.

Since there are several indications that cell death does not occur in the first stages of embryonal development, they used a time-dependent branching process as their basic model. It is assumed that embryonic cells split into two daughter cells with probability $1 - \delta - \alpha_k$, die with probability α_k, where k denotes the generation number, or continue living, without splitting, with probability δ. Denote the stage at which cell death starts by n_s. It is assumed that before this generation $\alpha_k = 0$ and afterward $\alpha_k = \alpha$. Thus,

$$f(k, s) = (1 - \delta)s^2 + \delta s, \text{ for } 0 \leq k \leq n_s \tag{5.166}$$

$$f(k, s) = (1 - \delta - \alpha)s^2 + \delta s + \alpha, \text{ for } n_s < k . \tag{5.167}$$

The probability of embryo arrest (i.e., that all its cells have died by stage n_f, Q_{n_f}) equals 0 if $n_f < n_s$. Otherwise, we can use the relation

$$Q_{n_f} = f_0(0, f(1, \ldots f(n_f, 0) \ldots)) \tag{5.168}$$

to calculate this probability, by backward iteration,

$$f(n_f, 0) = \alpha, \text{ so } f(n_f - 1, f(n_f, 0)) = f(n_f - 1, \alpha), \tag{5.169}$$

and if $n_f - 1 > n_s$, this is $(1 - \delta - \alpha)\alpha^2 + \delta\alpha + \alpha$, whereas if $n_f - 1 \leq n_s$, it equals $(1 - \delta)\alpha^2 + \delta\alpha$. The resultant number is then substituted for s in the generating function $f(n_f - 2, s)$, etc. Parameter values were estimated from data on numbers of live and dead cells in embryos at different stages. Embryos were observed until stage 8. Hardy *et al.* (2001) give several estimates of the value of α, varying roughly between 0.14 and 0.5, whereas δ was in the range of 0.1 to 0.2. They found strong indications that the value of n_s was 2 generations. As an example, extinction probabilities for the case $\alpha = 0.3, \delta = 0.2, n_s = 2$ are

$$Q_0 = Q_1 = Q_2 = 0, Q_3 = 0.040, Q_4 = 0.078, Q_5 = 0.109,$$
$$Q_6 = 0.132, Q_7 = 0.15, Q_8 = 0.16, Q_9 = 0.17, Q_{10} = 0.18. \tag{5.170}$$

It is relatively easy to calculate ultimate extinction chances for this model, since from n_s onward the reproduction generating functions are the same. Thus, the ultimate extinction probability, if we start in generation 2 with one cell x, is calculated by solving

$$x = (1 - \delta - \alpha)x^2 + \delta x + \alpha, \tag{5.171}$$

which gives

$$x = \min\left(\frac{\alpha}{1 - \delta - \alpha}, 1\right). \tag{5.172}$$

If we start with one cell in generation 0, the extinction probability Q is

$$Q = f(1, f(2, x)) = (1 - \delta)((1 - \delta)x^2 + \delta x)^2 + \delta((1 - \delta)x^2 + \delta x). \tag{5.173}$$

Comparison of model predictions with data, however, revealed several inconsistencies. One of these is that, to explain the observed proportion of embryo arrest of about 50 percent at generation 8, α would have to be extremely high, about 0.5, whereas $\delta = 0.1$. Since $x = 1$ with this combination of parameter values, none of the embryos would survive. Their results indicated that different embryos had different cell death rates, which suggests that these rates are already "programmed" in the zygote. They examined this further with a model that allowed for inter-individual variation.

5.10.2 Random environments

If the environment is random, as in Section 2.9.2, then so is the reproduction distribution that the environment results in. In particular, both the mean offspring numbers $m_1, m_2, \ldots$ and the variances become random variables, each determined by the environment that happens to prevail during its particular season. Through the varying environments, the extinction probability turns random, but is then, of course, influenced by the environments in more than one year. The extinction probability is a function of the whole sequence of environments, or of the environmental *scenario*.

In symbols, this can be expressed as follows. Let $\epsilon_0, \epsilon_1, \epsilon_2, \ldots$ denote the environments during the initial, first, second, etc., reproduction seasons. Then m_0 is determined by ϵ_0 and m_2 by ϵ_2, but the extinction probability is of the form $Q(\mathcal{E})$, where $\mathcal{E} = (\epsilon_0, \epsilon_1, \ldots)$ is the whole sequence of environments. In other words, for a given scenario $\mathcal{E}$ and $Z_0 = 1$,

$$\mathbb{P}(Z_n \to 0 | \mathcal{E}) = Q(\mathcal{E}) . \tag{5.174}$$

If we take a further expectation, over all possible scenarios, we obtain the overall extinction probability,

$$Q = \mathbb{E}[Q(\mathcal{E})] . \tag{5.175}$$

The different sources of randomness are referred to as *demographic* (i.e., randomness given the environmental scenario) and *environmental*.

The first case of random environments to be treated was that of independent, identically distributed environments. Later literature turned to the more realistic setup of stationary environmental sequences (i.e., the situation in which the environments during any given period of seasons have the same distribution, whenever the period occurs). (Then a further condition of *ergodicity* is asked for. This implies that forms of the Law of Large Numbers hold, but for further information the reader is referred to literature on stochastic processes.) Special cases are sequences of identically distributed elements with a time-homogeneous Markovian dependence. This means that the distribution of ϵ_{k+1} depends only on ϵ_k among all other past environments, and the form of the dependence remains the same for all k.

Athreya and Karlin (1971a; see also the book by Athreya and Ney 1972) proved that branching processes in stationary and ergodic environments are bound to extinction if $\mathbb{E}[\ln m(\epsilon)] \leq 0$, if ϵ is the environment of a season. In other words, in this case $Q(\mathcal{E})$ always has the same value, 1, for all possible environmental sequences. They also showed that if $0 < \mathbb{E}[\ln m(\epsilon)] < \infty$, then $Q(\mathcal{E})$ is less than 1. In this case the branching process is supercritical and there is a positive chance of persistence in the long run.

How do we obtain information about the distribution of the extinction probability $Q(\mathcal{E})$ for supercritical processes? Denote the probability generating function of the offspring distribution of an individual in the environment ϵ by

$$f_\epsilon(s) = \mathbb{E}\left[s^\xi | \epsilon\right] = \sum_{j=0}^{\infty} s^j \mathbb{P}(\xi = j | \epsilon) . \tag{5.176}$$

For instance, if the environmental state can be described by a number so that $0 < \epsilon < \infty$ and offspring numbers have a Poisson distribution with expectation ϵ, given the environment,

$$\mathbb{P}(\xi = j | \epsilon) = e^{-\epsilon} \frac{\epsilon^j}{j!} \, , \tag{5.177}$$

then

$$f_\epsilon(s) = e^{-\epsilon(1-s)} \, . \tag{5.178}$$

Or, in the case of our benchmark processes, if $0 < \epsilon < 1$, and

$$\mathbb{P}(\xi = k | \epsilon) = (1 - \epsilon)\epsilon^k, k = 0, 1, 2, \dots \, , \tag{5.179}$$

or

$$\mathbb{P}(\xi = 2 | \epsilon) = \epsilon, \ \mathbb{P}(\xi = 0 | \epsilon) = 1 - \epsilon \, , \tag{5.180}$$

then

$$f_\epsilon(s) = \frac{1 - \epsilon}{1 - \epsilon s} \quad \text{or} \quad 1 - \epsilon + \epsilon s^2 \, . \tag{5.181}$$

Now let $\mathcal{E}_1 = (\epsilon_1, \epsilon_2, \dots)$ denote the environmental sequence in which environments come one year earlier than in $\mathcal{E}$, so that ϵ_1 is the environment during the starting rather than next season, etc. If the environment is stationary, it is only natural to expect that the distribution of $Q(\mathcal{E})$ should coincide with that of $f_{\epsilon_0}(Q(\mathcal{E}_1))$. In particular, in the case of independent and identically distributed environments, this implies that

$$\mathbb{E}[Q(\mathcal{E})] = \mathbb{E}[f_\epsilon(Q(\mathcal{E}))] \, , \tag{5.182}$$

where $\mathcal{E}$ and ϵ on the right-hand side are independent. Actually, then

$$
\begin{aligned}
Q = \mathbb{E}[f_\epsilon(Q(\mathcal{E}))] &= \mathbb{E}[\sum_{j=0}^{\infty} \mathbb{P}(\xi = j | \epsilon) Q(\mathcal{E})^j] \\
&= \sum_{j=0}^{\infty} \mathbb{E}[\mathbb{P}(\xi = j | \epsilon)] \mathbb{E}[Q(\mathcal{E})^j] \\
&= \sum_{j=0}^{\infty} \mathbb{P}(\xi = j) \mathbb{E}[Q(\mathcal{E})^j] = \mathbb{E}[f(Q(\mathcal{E}))] \, ,
\end{aligned}
\tag{5.183}
$$

where f is the overall generating function of reproduction (i.e., not conditionally upon the environment). Together with similar relationships for higher moments, this is the basis of several approximation methods.

Wilkinson (1969) was the first to derive a numerical method to calculate an approximation for $\mathbb{E}[Q(\mathcal{E})]$. Haccou and Iwasa (1996) give an approximation that is easier to calculate. Focus on cases like our benchmark or Poisson reproduction, where there is one environment $\tilde{\epsilon}$ determined by the requirement that its mean reproduction

$$\tilde{m} = e^{\mathbb{E}[\ln m(\epsilon)]} . \tag{5.184}$$

Let $\tilde{Q}$ be the extinction probability of a branching process in a constant environment with this mean $\tilde{m}$. Since this is a classic process, $\tilde{Q}$ can be found by (numerically) solving

$$\tilde{Q} = f_{\tilde{\epsilon}}(\tilde{Q}) . \tag{5.185}$$

The following approximation was derived:

$$\mathbb{E}[Q] \approx \tilde{Q} + \frac{\tilde{m}\left(1 - \tilde{Q}\right)}{2\left(1 - \tilde{m}\tilde{Q}\right)} \left(\frac{\tilde{m}\left(1 - \tilde{Q}\right)}{1 - \left(\tilde{m}\tilde{Q}\right)^2} - 1 \right) \mathrm{Var}[\ln m(\epsilon)] . \tag{5.186}$$

This works especially well for values of $\tilde{m}$ close to 1 and was shown to be more efficient than Wilkinson's approximation for the specific examined cases. An advantage of Wilkinson's method is, however, that it can also be used to approximate higher moments of $Q(\mathcal{E})$.

The stable distribution of $Q(\mathcal{E})$ can be generated through iterative methods. Haccou and Iwasa (1996) describe such an approach for independent identically distributed environmental states. Define

$$Q_{j+1} = f_{\epsilon_{j+1}}(Q_j) , \tag{5.187}$$

and then the method works as follows. Start with an array of arbitrary values of Q_0 larger than or equal to 0 and smaller than 1. Simulate an array of random values of ϵ and calculate the values of Q_1, etc. Continue to do this until the distribution of Q_j values is stable. In the examples given in the article this required about 500 iterations.

The method was generalized by Haccou and Vatutin (2003) to environmental sequences with Markovian dependence. Note that this iteration method works backward in time. Thus, if k denotes the generation number in the original branching process, then j in Equation (5.187) corresponds to $k + 1$ and $j + 1$ to k. When environmental states are sequentially dependent, as in the Markov case, the method should be adjusted. This can be done by generating random sequences of environmental states (with time running forward) beforehand and then reversing the sequences for use in the iterations.

That extinction probability $Q(\mathcal{E})$ is a random variable has some interesting consequences. It implies, for instance, that the probability that an initially small group of invaders can establish itself in a new environment depends on the way in which the invasion takes place. For instance, it can be shown that by invading sequentially rather than simultaneously, the establishment chance in randomly varying environments can be improved considerably. This result has many important implications (e.g., for dispersal, life history, and biological control strategies). Further details can be found in the articles cited.

6

Development of Populations

The extinction or explosion dichotomy, and the subsequent analysis of the role and character of extinction, left many questions unanswered. What is the form of the uninhibited growth of theoretical populations that do not die out? What determines its speed? Does the composition of the population stabilize, as numbers grow, by some Law-of-Large-Numbers effect? And what happens, in reality, as a result of the inevitable bounds imposed upon any population?

In the long run, we know there is no other way than extinction, but what happens before that event? What temporary stabilities, so-called *quasi-stationarities* would populations show in the contest between an intrinsic tendency toward expansion and exterior limitations? And what can the theoretical uninhibited growth tell us about the form of populations that develop for a long time in benign stable circumstances, which practically allow unlimited growth?

In more concrete terms, what can the mathematics of supercritical branching processes that do not die out, and of similar stochastic processes, tell us about doubling rates and distributions over ages, phases, cell mass, or DNA content in chemostats or other in vitro populations, or in young, quickly growing in vivo tumors?

Also, what can we conclude about population growth from observed age or body mass distributions in, say, caught fish or hunted moose or deer?

In this chapter we seek to answer such questions. First, we establish the Malthusian, exponential growth that is the mathematical alternative to extinction of freely reproducing populations in stable conditions. Then we proceed to the stabilization of composition (Section 6.2) and the meaning of reproductive value (Section 6.3). In Section 6.4, we briefly consider the development of populations bound for extinction (i.e., subcritical and critical processes), as well as supercritical processes that become extinct. Sections 6.5 to 6.7 deal with the consequences of dependences between individuals on population size effects. Still, we remain in principle in the realm of exponential growth or extinction, even though near-critical populations can show linear or polynomial growth.

Finally, in Sections 6.8 and 6.9, we say some words about the complicated phenomena that occur in the presence of quasi-stationarity, in processes that are supercritical whenever the population size is small, and in some sense subcritical when it is large.

6.1   Exponential Growth

6.1.1   Discrete time

In Section 2.2 we noted that single-type Galton–Watson processes satisfy

$$\mathbb{E}[Z_n] = \mathbb{E}[Z_0]\, m^n , \tag{6.1}$$

in terms of the mean reproduction m and expected initial population size $\mathbb{E}[Z_0]$. If the process is supercritical ($m > 1$), the expected population size thus grows exponentially, or in Malthus' words "goes on doubling itself at regular intervals."

However, what about the population size Z_n itself? We know already that a population has a positive chance $(1 - Q)$ of escaping extinction, and that if it does, its size tends to infinity. Population size certainly exhibits random fluctuations, but it is a beautiful mathematical result (Kesten and Stigum 1966a) that, under the logarithmic moment condition we first met in Theorem 5.3, long-run growth of Z_n is also exponential:

Theorem 6.1 *For supercritical Galton–Watson processes with $Z_0 = 1$*

$$\lim_{n \to \infty} Z_n / m^n = W \tag{6.2}$$

exists. Under the condition that

$$\mathbb{E}[\xi \ln(\xi + 1)] < \infty , \tag{6.3}$$

the expectation of W, $\mathbb{E}[W] = 1$. Furthermore, $W = 0$ precisely if the process dies out. Otherwise, it is a positive, continuous random variable.

This theorem sharpens the "merciless dichotomy" between unbounded growth or extinction to one between exponential growth or extinction for such processes. Similar logarithmic moment conditions are needed formally for the other theorems given in this subsection and the next. Since, however, they are always satisfied for models with biologically reasonable assumptions, we do not give their exact formulations. These can be found in any mathematically oriented branching processes book, such as Asmussen and Hering (1983), Athreya and Ney (1972), or Jagers (1975).

The size of a population that is not started by one, but by, say, k individuals, is the same as that of the sum of the sizes of k independent populations started by one individual. The asymptotic distribution of the size of such a population, normed by m^n, thus converges to that of the sum of k random variables that are independent with identical distributions equal to that of W.

That W, although it is positive in the case of non-extinction, is a random variable is quite natural. Populations that start with low values of Z_0 are small in the first stage of their development, and hence their growth is capricious. This is illustrated in Figure 6.1.

There is a multitude of different options of population development before the onset of the exponential growth of populations that do not become extinct. It

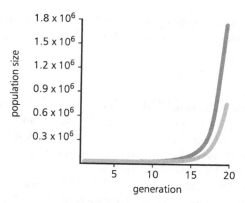

Figure 6.1 Growth curves of two exponentially increasing populations with the same number of founders.

is therefore not strange that W is a continuous random variable that can assume any positive value (i.e., has a strictly positive density function for all positive real values).

Unfortunately, it is difficult to say much more substantial about its distribution, in general. For one of our benchmark cases, geometric offspring, however, we can find the distribution of W explicitly. Indeed, the Laplace transform

$$\phi_n(u) = \mathbb{E}[e^{-uW_n}], \quad u \ge 0 \tag{6.4}$$

of $W_n = Z_n/m^n$ satisfies the recursive relation

$$\phi_n(u) = \mathbb{E}[\mathbb{E}[e^{-uW_n}|Z_1]] = \mathbb{E}[\mathbb{E}[e^{-uW_{n-1}/m}]^{Z_1}] = f(\phi_{n-1}(u/m)), \tag{6.5}$$

which is valid since, given Z_1, Z_n is the sum of Z_1 independent population sizes, each of which has the distribution Z_{n-1}, and Z_1 has the generating function

$$f(s) = E[s^{Z_1}], \tag{6.6}$$

as before. As $n \to \infty$, we obtain

$$\phi(u) = f(\phi(u/m)). \tag{6.7}$$

It can be shown that this equation is solved by a unique $\phi(u)$, which is the Laplace transform of a proper random variable. In the case of a geometric distribution, the equation has the form

$$\phi(u) = \frac{1}{1 + m(1 - \phi(u/m))}, \tag{6.8}$$

the relevant solution of which is

$$\phi(u) = \frac{1}{m} + \frac{m-1}{m}\frac{1}{1 + \dfrac{m}{m-1}u}. \tag{6.9}$$

The Laplace transform determines the distribution function uniquely, as can be checked in any standard book that deals with Laplace transforms, and in this particular case

$$\mathbb{P}(W \leq x) = \frac{1}{m} + \frac{m-1}{m}(1 - e^{-\frac{(m-1)x}{m}}) , \tag{6.10}$$

for any $x \geq 0$, and, of course, $\mathbb{P}(W \leq x) = 0$ if $x < 0$. Recalling that the extinction probability Q of the process with geometric distribution equals $1/m$ and that $1 - e^{-\frac{(m-1)x}{m}}$ is the distribution of a continuous random variable with mean $m/(m-1)$, we see that $\mathbb{E}[W] = 1$, which confirms the assertion of Theorem 6.1 in this regard, and for this particular case.

A further possibility might be to capture the distribution of W numerically. One option is to perform simulations. Indeed, for large n, $W_n = Z_n/m^n$ is close to W. However, an alternative, more efficient way is to obtain the Laplace transform, or moment-generating function, of W_n through Recursion (6.5). It starts from $\phi_0(u) = e^{-uZ_0}$, since $W_0 = Z_0$, assumed known, and stops at an n deemed large enough to yield

$$\phi_n(u) \approx \mathbb{E}[e^{-uW}] . \tag{6.11}$$

Remark. The mathematically knowledgeable reader should note that the first step to prove the theorem is a showcase application of the famous martingale theorem (cf. Williams 1991, or, for its statement only, the Appendix herein): since Galton–Watson processes are Markov and

$$\mathbb{E}[Z_{n+1}|Z_n] = mZ_n , \tag{6.12}$$

the random variables $W_n = Z_n/m^n$, $n = 0, 1, 2, \ldots$ constitute a non-negative martingale

$$\mathbb{E}[W_{n+1}|W_n] = W_n . \tag{6.13}$$

And non-negative martingales converge as $n \to \infty$.

◇ ◇ ◇

The exponential growth of the expected population size of *multi-type* Galton–Watson processes is established in Section 2.3. The martingale argument and proof of the logarithmic moment condition extends to supercritical indecomposable and non-periodic processes, so that the following can be proved in the notation of that section.

Theorem 6.2 *Consider a supercritical, indecomposable multi-type branching process. There then exists a random variable $W \geq 0$, such that, as $n \to \infty$,*

$$Z_n \sim W\rho^n u . \tag{6.14}$$

Under logarithmic moment conditions, $W = 0$ precisely in the cases where the population becomes extinct, and it has expectation v_h if there is just one ancestor of type h.

The theorem means that the number of individuals of type k, Z_{nk}, in a population with one ancestor of type h satisfies

$$Z_{nk} \sim W\rho^n u_k , \tag{6.15}$$

where, as in Section 2.3, u and v denote the normed eigenvectors of the mean matrix, and $\mathbb{E}[W] = v_h$. Another consequence is that the total population size $|Z_n|$ behaves as $W\rho^n$, since the u_k sum to one.

That W is not a vector implies that there are no random effects in the limit of the type distribution. Indeed, if we consider the ratio between the numbers of individuals of two types in generation n, say types i and k, then under non-extinction $W > 0$, and W and ρ^n cancel out so that

$$\lim_{n \to \infty} Z_{ni}/Z_{nk} = u_i/u_k , \tag{6.16}$$

exactly as for the ratio between their expectations. This shows that the composition over types must stabilize unless the population dies out, and provides a first hint of the very general stabilization properties of exponentially growing populations. The vector $(u_1, \ldots , u_d)$ defines the *stable type distribution*.

Since the expected relative contribution from a type h individual to the future population is proportional to v_h, this number is called the *reproductive value* of type h. However, be aware that in age-dependent models not only type, but also age influences the reproductive value of an individual (see Section 6.3). However, if age is modeled discretely, in the matrix population model tradition (see Section 2.4), so that the type is the age (in years or seasons), we have, of course, the present setting.

Now, the essential meaning of being indecomposable is that the population grows as one unit, since individuals of all types eventually have progeny of any type. However, growth at the same rate also occurs in decomposable processes if some subpopulation of "stem cell" type individuals grows and keeps generating other individuals who do not necessarily give birth to individuals of any other type, but also do not give birth to new populations that grow more quickly than the stem population (see Section 2.5).

In a more mathematical language such arguments amount to the possibility, maybe after some renumbering of types, that the expectation matrix M could be written as consisting of positive submatrices at the diagonal and under or above it and zeros for the rest, as in Section 2.3.1. If, further, the matrix has a simple largest eigenvalue greater than one, the convergence of $Z_n/\rho^n \to Wv$ as $n \to \infty$ continues to hold. The general result in this area is also given by Kesten and Stigum (1966b, 1967), and we refer to them for a proper formulation (see also Asmussen and Hering 1983).

As in the single-type case, a population that started from several ancestors, maybe of different types, behaves as the union of independent populations, each with one ancestor. For instance, in a population that begins with Z_{01} ancestors of type 1 and Z_{02} ancestors of type 2, Z_n/ρ^n converges in distribution to

$$v(W_{11} + W_{12} + \cdots + W_{1Z_{01}} + W_{21} + W_{22} + \cdots + W_{2Z_{02}}), \tag{6.17}$$

where W_{hj} are independent variables with the distribution that the limit variable W would have had there been just one ancestor of type h.

6.1.2 Continuous time

In Chapter 3 we saw that expectations of branching processes in continuous time grow at an exponential rate (or decrease or converge to a constant). Only the elementary case of Markov branching availed itself to an analysis in terms of basic calculus. Otherwise, we had to rely upon results from (Markov) renewal theory. The number α that gives the rate of exponential increase was named the Malthusian parameter of the process. In deterministic population dynamics it is often denoted by r and referred to as a reproduction rate.

We shall see that virtually all supercritical branching processes themselves (i.e., not only their expectations) grow approximately as $e^{\alpha t}$. Again, the mathematics that underlies the derivation of these results is accessible to experts only, but the results themselves are understandable and useful.

Here, as for mean growth (see the Remark in Section 3.3.1), we have to exclude the fundamentally discrete case of reproduction that occurs only at lattice times. We do not mention this exception usually.

First, consider the case of single-type general branching processes, as introduced in Section 3.3.

Theorem 6.3 *Consider a single-type general branching process Z_t that starts from one newborn ancestor at time zero. As $t \to \infty$,*

$$Z_t \sim We^{\alpha t}(1 - \ell)/\alpha\beta, \tag{6.18}$$

where W is a random variable that equals 0 if the population dies out, is positive otherwise, and satisfies

$$\mathbb{E}[W] = 1. \tag{6.19}$$

(Besides non-latticeness, this assumes a logarithmic moment condition.)

The result can be generalized to the size of a population started by several individuals in the same way as described for discrete-time processes.

Recall that it follows from Section 3.3, Equations (3.39) and (3.45), that $\ell = 1/m$ for Bellman–Harris processes. For these processes also, the mean age at childbearing satisfies

$$\beta = \mathbb{E}\left[Te^{-\alpha T}\right] \tag{6.20}$$

(see Section 3.5).

Often growth is described in terms of the *doubling time* of a population, t_d. We have come across this concept in Examples 3.1 and 3.2 on bacterial growth

(Section 3.2) and Example 5.1 on cell growth (Section 5.3). As the name says, this is the time it takes for a population to grow from a certain level to twice that level. Properly speaking, this time depends upon both starting level and the special form of the growth function. However, as we have seen, supercritical populations tend to grow exponentially, and for exponential growth curves the starting point does not influence the doubling time. Hence, the straightforward general definition is that t_d is determined by

$$e^{\alpha t_d} = 2 . \tag{6.21}$$

Thus, the relation between Malthusian parameter and doubling time is that

$$t_d = \ln 2 / \alpha . \tag{6.22}$$

Example 6.1: Cell growth. In the age-dependent cell cycle model from Section 3.2 disregard the possibilities of quiescence and cell death, so that cells simply divide into two daughter cells after completion of their cell cycle,

$$T = G_1 + S + G_2 + M . \tag{6.23}$$

This is a Bellman–Harris process with $m = 2$ and life span T. Thus, the Malthusian parameter α is determined by (see Section 3.3)

$$2\mathbb{E}[e^{-\alpha T}] = 1 . \tag{6.24}$$

It follows that $\ell = 1/2$, whatever the cell cycle distribution.

Since the exponential function is convex, Jensen's inequality (see the Appendix) implies that

$$1/2 = \mathbb{E}[e^{-\alpha T}] \geq e^{-\alpha \mathbb{E}[T]} , \tag{6.25}$$

with equality if and only if T has zero variance. In other words, if we denote the mean cycle time by

$$t_c = \mathbb{E}[T] , \tag{6.26}$$

then

$$e^{\alpha t_c} \geq 2 = e^{\alpha t_d} , \tag{6.27}$$

and we have proved that the doubling time must be shorter than the mean cycle time, except in the degenerate case for which all cycle times coincide.

Whether the difference between t_c and t_d is of any significance depends upon the distribution of cycle times. This is also needed to determine α and β. To this end assume that T is distributed normally with mean t_c and variance σ_c^2. Since a cell cycle is, in principle, a period during which a large number of given tasks must be performed, and this takes a well-determined time with some variance, this is a reasonable assumption. As the mean cycle time is usually considerably larger than σ_c, we can disregard the theoretical nuisance of negative life spans. They have a negligibly small probability of occurring. Then, for any non-negative number u,

$$\mathbb{E}[e^{-uT}] = \frac{1}{\sigma_c\sqrt{2\pi}} \int_{-\infty}^{+\infty} e^{-ua} e^{-\frac{(a-t_c)^2}{2\sigma_c^2}} \, da = e^{u^2\sigma_c^2/2 - ut_c} . \tag{6.28}$$

Note that this is the Laplace transform of the normal distribution, which is derived in any probability textbook. Equating it to $1/2$ for $u = \alpha$ yields

$$\alpha^2 \sigma_c^2 / 2 - \alpha t_c = -\ln 2 \,, \tag{6.29}$$

with the two solutions

$$\alpha = \frac{(t_c \pm \sqrt{t_c^2 - 2\sigma_c^2 \ln 2})}{\sigma_c^2} \,. \tag{6.30}$$

We now multiply the numerator and denominator by the same factor, namely $\left(t_c \pm \sqrt{t_c^2 - 2\sigma_c^2 \ln 2}\right)$. After some algebra this gives

$$\alpha = \frac{2 \ln 2}{t_c \pm \sqrt{t_c^2 - 2\sigma_c^2 \ln 2}} \,, \tag{6.31}$$

and thus

$$t_d = \frac{\ln 2}{\alpha} = \frac{1}{2} \left(t_c \pm \sqrt{t_c^2 - 2\sigma_c^2 \ln 2} \right) \,. \tag{6.32}$$

The fact that $t_d = t_c$ precisely if $\sigma_c = 0$ shows that the relevant solution is

$$t_d = \frac{1}{2} \left(t_c + \sqrt{t_c^2 - 2\sigma_c^2 \ln 2} \right) \,. \tag{6.33}$$

Equations like this have a twofold importance. First, if all entities in Equation (6.33) can be estimated accurately, the relation can be used to check the assumption of normally distributed life spans. The second situation occurs if the normality has been deemed correct. Population entities like doubling time are measured easily, whereas cycle times can be more difficult to observe. If only a few measurements of cycle times are available, we may have a good estimate of t_c, but not so of σ_c. In such situations the variance in cycle time might be better estimated through Equation (6.33) rather than directly from the observations.

The easiest way to calculate the mean age at childbearing, β, for this example is to differentiate the Laplace transform,

$$\mathbb{E}\left[T e^{-uT}\right] = \frac{d}{du} \mathbb{E}\left[-e^{-uT}\right] \,, \tag{6.34}$$

which, from Equation (6.28), gives

$$\mathbb{E}\left[T e^{-uT}\right] = \left(t_c - u\sigma_c^2\right) e^{u^2 \sigma_c^2 / 2 - u t_c} \,. \tag{6.35}$$

Finally, $u = \alpha$ yields

$$\beta = \mathbb{E}\left[T e^{-\alpha T}\right] = \left(t_c - \frac{\ln 2}{2 t_d} \sigma_c^2\right) / 2 \,, \tag{6.36}$$

since $\mathbb{E}\left[e^{-\alpha T}\right]$ equals $1/2$.

Experimental values for cell cycle lengths in the literature vary depending upon cell type and growth conditions, such as in vitro, in which human cells are often observed to have cycles between, say, 10 and 24 hours, or in vivo, in which life spans can be much longer.

$$\diamond \diamond \diamond$$

Now, we formulate the exponential growth property for supercritical multi-type processes. As in the discrete-time case, this involves eigenvectors u and v, which are defined in Section 3.3. The distribution of reproduction, as well as that of life spans, may now vary between the different types of individuals. Thus, we must discern a set of life-span distributions,

$$\mathbb{P}_j \left(T \leq a \right) = L_j \left(a \right), \ j = 1, \ldots, d . \tag{6.37}$$

For $j = 1, \ldots, d$, further define

$$\ell_j = \mathbb{E}_j \left[e^{-\alpha T} \right], \tag{6.38}$$

which equals

$$\ell_j = \alpha \int_0^\infty e^{-\alpha t} L_j(a) \, da , \tag{6.39}$$

as in the single type case. We use type suffixes to indicate that the probability or expectation is for an individual of the said type, or, when a whole process is under consideration, to indicate that there was just one ancestor, of the type indicated.

Theorem 6.4 *Consider a supercritical, indecomposable multi-type branching process, which starts from one newborn individual of type h. There exists a random variable $W \geq 0$ that (under logarithmic moment conditions) has expectation u_h, and is such that, as $t \to \infty$, the number of type j individuals satisfies*

$$Z_{tj} \sim W e^{\alpha t} u_j (1 - \ell_j) / \alpha \beta , \tag{6.40}$$

and $W = 0$ precisely if the process dies out.

The result can be generalized to the size of a population started by several individuals in the same way as described for discrete-time multi-type processes. It can also be formulated for general type spaces (i.e., situations in which there might be more than finitely many types). The type could be one of countably many possible alleles: it could be body mass at birth or, for that matter, body mass evolution during the individual's whole life. Still, under very general conditions the Malthusian alternative, exponential growth or extinction, continues to hold. For a strict formulation of this we refer to the mathematical literature, such as Jagers (1992) or, for a more reader-friendly version, Jagers (1991).

6.2 Asymptotic Composition and Mass Growth

The asymptotic distribution of types in populations can be derived easily from the results stated in Section 6.1. First, consider discrete-time multi-type processes. As remarked before, from Theorem 6.2 it follows that in the limit the composition of populations that do not die out is fixed: $\lim_{n \to \infty} Z_{ni} / Z_{nk} = u_i / u_k$. Note that this is true regardless of the initial population size. A similar conclusion can be stated for continuous-time multi-type processes. From Theorem 6.4 we see that the proportion of individuals of type i settles down to a limiting value of

$$\lim_{t \to \infty} \frac{Z_{ti}}{|Z_t|} = \frac{u_i (1 - \ell_i)}{\sum_j u_j (1 - \ell_j)}, \tag{6.41}$$

as the population grows. If all life-span distributions are equal this reduces to u_i, since the u_j sum to 1. Hence the name "stable type distribution" for the vector u. As in discrete-time processes, the result does not depend on the initial population size or composition.

These are the initial results on stabilization of population composition. However, other compositional properties, besides types, can be studied, such as, for example, age distribution. (We touch on this topic when indecomposable multi-type Galton–Watson processes are introduced in Chapter 2.) Results on asymptotic age distributions go back to Leonhard Euler (1767), but have been rediscovered over and over again, in demography, biological population dynamics, cell kinetics, and other areas.

First, note that populations can be measured in many different ways. Up to now, we have considered one such measure mainly, namely the total population size. This corresponds to counting the number of individuals alive at a certain time. Populations can also be measured in terms of body mass, DNA content, or volume, and subsets of populations could be measured by counting only individuals with special properties, such as being young, old, or firstborn. Indeed, the problem of measuring a population could be said to consist of two parts, first determining the number of individuals born, and then determining for each individual born whether it be counted or not, and with what weight. Here are some examples of different population measures:

1. The *accumulated population*: each individual born is given weight one, be it alive or not.
2. The *live population*: each individual born is given weight one, provided it is still alive.
3. The population *not older than age a*: each individual born is given weight one, provided it is still alive and its age does not exceed a.
4. The *population mass*: each living individual is weighed, and all the weights added.
5. The number of firstborns.

All of these are counting measures, except population mass (property 4 above): individuals that have some specified property are counted, others are not. We denote such a property by D. In all but the last case D is an *individual* property. Being firstborn, on the contrary, is a relational property; it describes the individual's position in its sibship.

Each of these measures gives rise to different stochastic processes, but they all exhibit the same exponential growth $\sim CWe^{\alpha t}/\beta$. Here, β (defined in Section 3.3.1 for the general case) gives the timescale (generation length), W is a random variable that equals zero if and only if the population dies out, and the only entity affected by the choice of measure is C.

As seen in Theorem 6.3 for the single type case, when the population measure is the number of individuals alive, Z_t (the second example of the list), C equals $(1 - \ell)/\alpha$. From the definitions given in Section 3.3.1 it thus follows that in this case

$$C = \int_0^\infty e^{-\alpha a}(1 - L(a))\, da = \int_0^\infty e^{-\alpha a} p(D, a)\, da \ , \qquad (6.42)$$

where $p(D, a)$ denotes the probability of an a-aged individual being alive.

This holds more generally; that is, if $p(D, a)$ denotes the probability that an individual of age a has property D, then the number Z_t^D of individuals with property D at time $t \to \infty$ satisfies

$$Z_t^D \sim \hat{p}(D) W e^{\alpha t} / \beta \ , \qquad (6.43)$$

where

$$\hat{p}(D) = \int_0^\infty e^{-\alpha a} p(D, a)\, da \ . \qquad (6.44)$$

It follows that the asymptotic proportion of individuals with property D among all alive is

$$\lim_{t \to \infty} \frac{Z_t^D}{Z_t} = \frac{\alpha \hat{p}(D)}{1 - \ell} \ , \qquad (6.45)$$

provided the population does not die out. [This is generally true for individual properties, barring some mathematical subtleties, whereas relational properties should not be dependent upon circumstances all the way back to the dawn of the population. This last proviso should not bother any biologist, but is made more precise in Jagers and Nerman (1996). Suffice it here to note that a property that is determined by all individuals a bounded number of generations back is well qualified.]

The classic special case is the property of being alive and not older than a given age a (property 3 above). Denoting this property by D_a, we obtain

$$p(D_a, u) = \begin{cases} 1 - L(u) & \text{if } u \le a \\ 0 & \text{otherwise .} \end{cases} \qquad (6.46)$$

Hence,

$$\hat{p}(D_a) = \int_0^a e^{-\alpha u}(1 - L(u))\, du \ , \qquad (6.47)$$

and, as $t \to \infty$, the proportion of the population not older than a converges to

$$\frac{\int_0^a e^{-\alpha u}(1 - L(u))\, du}{\int_0^\infty e^{-\alpha u}(1 - L(u))\, du} \ , \qquad (6.48)$$

provided the population does not die out. This is called the *stable age distribution* of the population, already introduced in Chapter 3 [see Equation (3.82)]. There it is shown that this ratio equals the asymptotic ratio of the *expectations* of the number of individuals not older than a and of Z_t. As we see now, it also equals the asymptotic ratio of these *numbers* themselves.

Example 6.2 We continue with the cell kinetics of Example 6.1. A much-studied entity in cell proliferation is the *mitotic index*, or fraction of cells in mitosis. A cell is in mitosis when its age a satisfies

$$G_1 + S + G_2 = T - M \leq a < T . \tag{6.49}$$

Thus, the probability of being in mitosis is $\mathbb{P}(T - M \leq a < T)$ when the cell is of age a. If we denote the property of being in mitosis by M as well, the mitotic index is

$$i_m = \frac{\hat{p}(M)}{\hat{p}(D_\infty)} = \frac{\alpha \hat{p}(M)}{1 - \ell} . \tag{6.50}$$

It turns out that

$$\hat{p}(M) = \mathbb{E}[e^{-\alpha(T-M)} - e^{-\alpha T}]/\alpha . \tag{6.51}$$

(For the calculations that show this and many similar relations, see the very end of this Example.) However, since mitosis is short (around 20 minutes) as compared to the whole cell cycle (of, say, 10 hours)

$$e^{-\alpha(T-M)} - e^{-\alpha T} = e^{-\alpha T}(e^{\alpha M} - 1) \approx e^{-\alpha T}\alpha M . \tag{6.52}$$

Moreover, the length of mitosis is roughly independent of T so that

$$\hat{p}(M) \approx \mathbb{E}[e^{-\alpha T}]\mathbb{E}[M] = \ell \mathbb{E}[M] . \tag{6.53}$$

Since $\ell = 1/2$, it follows from Equations (6.50) and (6.21) (or the formula afterward) that the mitotic index should stabilize around

$$i_m = \frac{\alpha \ell \mathbb{E}[M]}{1 - \ell} = \alpha \mathbb{E}[M] = (\ln 2) t_m/t_d \approx 0.693 t_m/t_d , \tag{6.54}$$

in terms of the mean duration of mitosis, t_m. This value can be compared to the naive relation

$$i_m = \frac{t_m}{t_c} , \tag{6.55}$$

which is valid in non-expanding deterministic populations.

The *proof* of Equation (6.51) involves two ideas:

- the order of expectation and integration can be interchanged; and
- the probability of an event D is the same as the expectation of the indicator function of D (see the Appendix),

$$1_D = \begin{cases} 1 & \text{if } D \text{ occurs} , \\ 0 & \text{otherwise} , \end{cases} \tag{6.56}$$

that is $\mathbb{P}(D) = \mathbb{E}[1_D]$.

We consider the events $\{T - M < u \leq T\}$ for various u and conclude

$$\mathbb{E}[e^{-\alpha(T-M)} - e^{-\alpha T}]/\alpha = \mathbb{E}[\int_{T-M}^{T} e^{-\alpha u} \, du]$$

$$= \mathbb{E}[\int_0^{+\infty} e^{-\alpha u} 1_{\{T-M<u\leq T\}} \, du] = \int_0^{+\infty} e^{-\alpha u} \mathbb{E}[1_{\{T-M<u\leq T\}}] \, du$$

$$= \int_0^{+\infty} e^{-\alpha u} \mathbb{P}(T - M < u \leq T) \, du = \hat{p}(M) , \tag{6.57}$$

as was to be proved. $\diamond \diamond \diamond$

The main result of this section can be stated more generally for arbitrary properties D and indecomposable multi-type processes as follows. Define $p_j(D, a)$ as the probability that an individual of type j and age a has property D, and

$$\hat{p}_j(D) = \int_0^\infty e^{-\alpha a} p_j(D, a)\, da \ . \tag{6.58}$$

Then the number Z_{tj}^D of individuals of type j with property D at time $t \to \infty$ satisfies

$$Z_{tj}^D \sim W e^{\alpha t} u_j \hat{p}_j(D)/\beta \ . \tag{6.59}$$

It follows that the asymptotic proportion of type j individuals with property D, among all type j individuals that are alive, is

$$\lim_{t\to\infty} \frac{Z_{tj}^D}{Z_{tj}} = \frac{\hat{p}_j(D)}{\hat{p}_j(D_\infty)} \ , \tag{6.60}$$

provided the population does not die out. In this, D_a still denotes the event of being alive and not older than a, so that D_∞ simply means being alive and

$$\hat{p}_j(D_\infty) = \int_0^\infty e^{-\alpha u}(1 - L_j(u))\, du \ , \tag{6.61}$$

in terms of the life-span distribution L_j of type j individuals.

If we, finally, consider some continuous scale property, such as the mass of individuals, we would have to replace the probability of the property by the expected size of the property at the age of the individual. Thus, if $m_j(a)$ is the expected volume or mass of a j-type individual of age a, the unit of interest becomes

$$\hat{m}_j = \int_0^\infty e^{-\alpha a} m_j(a)\, da \ , \tag{6.62}$$

and the asymptotic mass of a population will be

$$\sim W e^{\alpha t} \sum_j u_j \hat{m}_j/\beta \ . \tag{6.63}$$

6.3 Reproductive Value

R.A. Fisher – the most prominent figure of both twentieth century quantitative genetics and statistical inference – introduced the important concept of *reproductive value*. The idea is to measure the relative contribution of an individual to the future population. Such a measure should depend upon the genetic setup as well as the age. Once passed its reproductive period in life, the individual does not contribute at all to future populations, and the very young risk not attaining reproductive age. Thus, the reproductive value should be at its maximum at the age the individual enters its most reproductive phase of life.

Fisher considered single-type populations in continuous time and with his strong intuition claimed that the reproductive value of an individual aged a, "as is easily seen," should be given by

$$v(a) = e^{\alpha a} \int_a^\infty e^{-\alpha u} \mu'(u) \, du / (1 - L(a)) . \tag{6.64}$$

Caswell (2001) writes that he has not met anyone who finds this "easily seen," and proceeds to the nowadays frequent establishment of reproductive value in terms of multi-type discrete-time branching processes, in which type = age class. However, Fisher's original concept has a direct interpretation in terms of general, continuous-time branching processes.

Indeed, consider a population at some time, to be called 0 for simplicity, and one individual, alive and of age a at this time. If the individual gives birth to one child at age $u > a$ (i.e., after time $u - a$), at time t ($> u - a$) the expected size of the population that stems from that child is

$$\mathbb{E}[Z_{t-u+a}] \sim e^{\alpha(t-u+a)}(1 - \ell)/\alpha\beta , \tag{6.65}$$

as t becomes large. This we know already from the discussion in Section 3.3 (see also Theorem 6.3). Now, assume that an individual's reproduction in the future is not affected by her having survived up to age a. Then her rate of offspring production, given that she attained this age, is

$$\mu'(u) \, du / (1 - L(a)) . \tag{6.66}$$

It follows that the expected size of all the population that stems from the a-aged individual is

$$\int_a^t \mathbb{E}[Z_{t-u+a}]\mu'(u) \, du / (1 - L(a)) \tag{6.67}$$

at time t. However, as $t \to \infty$ this behaves like

$$\frac{(1 - \ell)}{\alpha\beta \, (1 - L(a))} \int_a^t e^{\alpha(t-u+a)} \mu'(u) \, du \sim$$

$$e^{\alpha t} e^{\alpha a} \frac{(1 - \ell)}{\alpha\beta \, (1 - L(a))} \int_a^\infty e^{-\alpha u} \mu'(u) \, du$$

$$= e^{\alpha t} v(a)(1 - \ell)/\alpha\beta . \tag{6.68}$$

It follows that the relative contribution of an individual aged a to that of a newborn is $v(a)/v(0)$, which confirms Fisher's idea.

Obviously, this can be generalized to the multi-type case, types that mirror genotypes, and even to cases that do not possess the Markov property mentioned (that future reproduction, given the age, is independent of past life and reproduction history), an assumption that is usually tacitly made in demography and matrix-formulated population biology (see Box 6.1).

Box 6.1 Several types

Consider the multi-type case, but keep the Markovian property so that future reproduction is influenced only by type and present age, and ask for the reproductive value of an individual of type h, aged a. Theorem 6.4 and arguments as above lead to expressions in which:

1. Survival probability $1 - L_h(a)$ replaces $1 - L(a)$ in the single-type case;
2. Reproduction rate μ' is also type specific, μ'_{hj}, the rate of offspring of type j born to a type h individual; and
3. Reproductive value of the genotype h enters through $\mathbb{E}_h[W] = u_h$.

In the case of discrete time, the survival probability disappears (all changes occur at the turn of the year, or corresponding entity), and the reproduction rate is replaced by entries of a mean matrix. For the case in which age in discrete time is modeled as type, the u_h themselves, therefore, often are referred to as reproductive values [see Sections 2.3 and 6.1.1, or Caswell (2001) for an extensive discussion].

6.4 Populations Bound for Extinction

The discussion in Section 5.4 shows that subcritical populations usually succumb quickly, but can persist for quite some time, albeit with a small probability. Therefore, it is natural to ask what is the size of old, non-extinct branching processes. If the process is supercritical, we saw that the population is probably quite large. In the subcritical case, however, it can be shown that this is not the case. On the contrary, given that the population does not die out, its size stabilizes. Unfortunately, in general it is difficult to describe the distribution of this stabilized size.

In the geometric case, things are easier, as usual. If the conditional generating function and the probability $1 - Q_n$ of non-extinction are examined, it can be proved that if Z^* denotes the size of an old non-extinct population,

$$\mathbb{P}(Z^* = k) = (1 - m)m^{k-1}, k = 1, 2, \ldots, \text{ and } \mathbb{E}[Z^*] = 1/(1 - m) . \quad (6.69)$$

In other words, in an old persistent branching population with one single ancestor, geometric offspring distribution, and 0.95 children per individual, we should expect a population size of some 20 individuals.

This has a bearing on the interpretation of paleontological data, and generally on evolutionary history. As we have pointed out, most species have died out, but what were their populations like during their glory days? What can be said about their sizes, about the numbers of dinosaurs, trilobites, Neanderthals, or mammoths, at the times when they were around on earth?

You may object that a population that dies out does not have to be subcritical. Indeed, one of the very points of branching process theory is the establishment of the paradoxical possibility of frequent extinction in spite of general growth. Thus, the extinct species population could well have been supercritical, but just suffered from bad luck.

However, as a matter of fact supercritical populations, which are known to die out later, behave as subcritical populations. Contrary to what might be guessed, they do not first grow exponentially at their Malthusian rate (see Sections 2.2 and 6.1), and later drop drastically. Instead, their size stabilizes, as that of a long-lasting subcritical population: if τ is the time of extinction, then

$$\lim_{n \to \infty} \mathbb{P}(Z_n = k | n < \tau < \infty) \tag{6.70}$$

exists for any supercritical population. In particular, for geometric processes with $m > 1$,

$$\lim_{n \to \infty} \mathbb{E}[s^{Z_n} | n < \tau < \infty) = s(m - 1)/(m - s) . \tag{6.71}$$

We thus recover the distribution of Z^* above, but with $1/m$ in place of m:

$$\lim_{n \to \infty} \mathbb{P}(Z_n = k | n < \tau < \infty) = (m - 1)/m^k = (1 - 1/m)/(1/m)^{k-1},$$

$$k = 1, 2, \dots . \tag{6.72}$$

The important message, however, is that both types of models, sub- or supercritical, yield the same type of distributions as those of the size of now-extinct species.

Another useful characteristic of a branching process is the accumulated population size, Y_n, up to generation n,

$$Y_n = Z_0 + Z_1 + \cdots + Z_n . \tag{6.73}$$

If $Z_0 = 1$, then

$$\begin{aligned}
\mathbb{E}[Y_n] &= \mathbb{E}[Z_0 + Z_1 + \cdots + Z_n] \\
&= \mathbb{E}[Z_0] + \mathbb{E}[Z_1] + \cdots + \mathbb{E}[Z_n] \\
&= 1 + m + \cdots + m^n . \tag{6.74}
\end{aligned}$$

If $m < 1$, the process dies out rapidly, and the total number of individuals ever born,

$$Y = Z_0 + Z_1 + \cdots + Z_n + \cdots , \tag{6.75}$$

is finite, and so is its expected value

$$\mathbb{E}[Y] = \sum_{k=0}^{\infty} m^k = \frac{1}{1 - m} . \tag{6.76}$$

This formula was used by Becker (1974) to evaluate the offspring mean in subcritical branching processes.

If $m \geq 1$ then $\mathbb{E}[Y_n] \to \infty$, as $n \to \infty$. However, if we condition on the event that a supercritical process dies out sooner or later (this condition gives nothing new in the critical case) and denote the extinction moment by τ, we obtain

$$\mathbb{E}[Y | \tau < \infty] = \frac{1}{1 - f'(Q)} , \tag{6.77}$$

where $Q = P(\tau < \infty)$ is the extinction probability of the process (observe that $f'(Q) < 1$; see Figure 5.2). To establish Equation (6.77), we write

$$\mathbb{E}[Y \mid \tau < \infty] = \frac{\mathbb{E}[Y ; \tau < \infty]}{P(\tau < \infty)} . \qquad (6.78)$$

Further,

$$\mathbb{E}[Y ; \tau < \infty] = \sum_{n=0}^{\infty} \mathbb{E}[Z_n ; n < \tau < \infty] . \qquad (6.79)$$

Now observe that

$$\mathbb{E}[Z_n ; n < \tau < \infty] = \sum_{k=1}^{\infty} k\mathbb{P}(Z_n = k; n < \tau < \infty)$$

$$= \sum_{k=1}^{\infty} k\mathbb{P}(Z_n = k)Q^k , \qquad (6.80)$$

since each of the populations that stem from the $Z_n = k$ individuals at time n should die out. Thus, we obtain

$$\mathbb{E}[Z_n ; n < \tau < \infty] = Q \sum_{k=1}^{\infty} k\mathbb{P}(Z_n = k)Q^{k-1}$$

$$= Qf_n'(Q) = Q\left(f'(Q)\right)^n . \qquad (6.81)$$

Substituting this sequentially in Equations (6.79) and (6.78) gives Equation (6.77). In particular, in the supercritical geometric case $Q = q/p = m^{-1} < 1$ and

$$f'(Q) = \frac{qp}{(1 - pQ)^2} = \frac{1}{m} . \qquad (6.82)$$

Hence,

$$\mathbb{E}[Y \mid \tau < \infty] = \frac{1}{1 - m^{-1}} = \frac{m}{m - 1} = 1 + \frac{1}{m - 1} . \qquad (6.83)$$

Note that this expectation is monotone, *decreasing* to 1 when m increases. And we obtain an (at first sight) unexpected result: if a *supercritical* process dies out rapidly (and, therefore, the total number of individuals ever born in the population is small), this may indicate that the expected offspring number is actually quite large!

Using Equation (6.72) [or, directly, Equation (6.74)] and applying Equation (A.55) from the Appendix, one can check that

$$Var[Y] = \frac{1}{m - 1} + \frac{1}{(m - 1)^2} \qquad (6.84)$$

and, therefore, the variance of the limiting process decreases (vanishes) as $m \to \infty$. A more detailed discussion can be found in Bruss and Slavtchova-Bojkova (1999).

6.5   Interaction and Dependence

Most of the exposition has concerned classic branching processes, in which each individual is supposed to live and multiply independently of everything else. Here we return to the dynamics of the processes with dependence that were touched upon in Section 1.4, and introduced in Sections 2.6 and 3.5. One fundamental result, which does not require independence between individuals, is the extinction or explosion dichotomy of Section 5.2. That result leads us into investigating what happens to a population in which individual reproductive behavior tends to stabilize in population growth. Typically, the larger a tumor, the slower its growth, to cite just one example.

The independence assumption was defended by a reference to "small" populations that live in circumstances in which a limitation of resources or space is not crucial. However, philosophically, it is also a natural idealization of the fact that in population growth the individual acts as a motor. The population changes its size as a result of individual action, and in this sense the latter is a primary, independent force.

Once properties of such ideal populations of independent individuals have been established, the question arises as to what extent they hold true in real populations, in which interaction abounds. Such interaction may be specific, as that in predator–prey interplay, or in symbiosis or parasitism, and should be modeled thus. Or it could be of a less precise nature, and competitive or collaborative. We focus on the latter general situation, and aim at a broad understanding of the implications of dependence, rather than to penetrate the detailed form of some special types of interaction.

Such general dependences could be *local* in the sense that they concern a few individuals only, such as closest neighbors or siblings. Or else they could be global, and mirror how resources are depleted as the total population grows in a given environment, or for a situation in which resources arrive in a flow, individual reproduction behavior could be dependent upon the size or density of the present population.

As we have learned, such populations still either become extinct or grow beyond all limits. However, two questions arise:

1. How will the extinction probability and time to extinction be affected by dependence?
2. How is the rate of growth affected? For example, is it still exponential, and does it have the same parameters?

6.5.1 Sibling dependence

We first consider the interaction between siblings only. However, later we discuss more general groups of relatives, such as first cousins, second cousins, etc., and even other groups of interacting individuals.

A simple example (Olofsson 1997) demonstrates how the extinction probability can be affected drastically. Consider a binary splitting Galton–Watson process in which the probability of obtaining two children is one-half. This process is critical ($m = 1$), and hence must die out. Now assume, instead, that of two sisters only one may reproduce, say because the parental precinct can sustain one of them only. Since the numbering of sisters is arbitrary, the probability that a given sister reproduces is one-half, and if she so does, she produces two children. The probabilities of begetting two or zero children thus remain one-half, and m therefore still equals 1, but now there is a heavy dependence between sisters: if one has two children, the other has none. Both processes are critical, but the population with dependence *never dies out*.

Here is a more realistic example. Many insects, like various moths or butterflies, lay eggs in large clutches, and their larvae compete for food. Clutch size therefore influences larval growth and survival. Larval size, in its turn, affects pupal survival chance and adult fecundity. In this case, the population cannot be compared easily to a corresponding Galton–Watson process, since not only do the siblings affect each other's survival chances, but also they influence each other's reproduction.

The latter example provides a hint about how to analyze populations with sibling dependence. Different sibships are independent of each other, and their reproduction is determined by their sizes. In other words, if we consider the process as a population of sibships, each sibship being a "macro-individual," as it were, then the process is a multi-type branching process, the type of each macro-individual being its size. And, as such, if it does not die out, it grows exponentially. This idea is illustrated in Example 6.3.

Example 6.3 Consider a butterfly species with clutch sizes of either one or two, so sibships can have two different sizes. This can be modeled as a branching process with two different types. Suppose that clutch size does not affect survival chance, but adult fecundity only. An individual that had no sibs lays two clutches of size two. An individual that grew up with a sib lays one clutch of size two with chance 1/2 and one clutch of size one and one of size two with chance 1/2. This means that a macro-individual (sibship) of size two (a type two macro-individual) has, on average, two offspring of type one and $\frac{1}{4} \times 2 + \frac{1}{2} \times 1 = 1$ offspring of type two. Thus, the mean matrix of the multi-type process is

$$\begin{pmatrix} 0 & 2 \\ 1 & 1 \end{pmatrix}.$$

(6.85)

The dominant eigenvalue of this matrix equals 2 and therefore the process is supercritical.

$$\diamond \ \diamond \ \diamond$$

When we leave the Galton–Watson context to consider more general branching processes with sibling dependences, entire sibships continue to be the suitable "macro-individuals." However, their type description becomes more complex: not only may sibship size play a role, but also properties like times between successive bearings may have a say. Nevertheless, exponential growth remains unaffected. We summarize this as a theorem, for proof of which, see Olofsson (1997).

Theorem 6.5 *Supercritical branching processes with sibling dependence grow exponentially. If the processes are homogeneous in the sense that individual reproduction distributions are unaffected by sibship structure, the growth rate is the same as if individuals were independent with the same individual reproduction law.*

As pointed out, results of this type hold quite widely. In principle, it is enough to establish exponential growth of the expected population, independently of earlier history, plus a bound on the correlation between sizes of the progeny of different individuals. However, for such, relatively complex, general results we refer to Jagers (1997a).

6.5.2   Global dependence

The prototype for global dependence is population size dependence. This is formulated most easily for a discrete-time process with non-overlapping generations. Given the size, Z_n, of the population at time n, its members reproduce independently, as in a Galton–Watson process, but according to a probability distribution that may be influenced by Z_n. To express this dependence on population size we may write $p_k(Z_n)$ for the probability of begetting k children as a function of the population size. As pointed out in Section 2.6, an alternative is to consider a situation in which reproduction is influenced by the total population up to now,

$$Y_n = Z_0 + Z_1 + \ldots + Z_n . \tag{6.86}$$

If it is assumed that the mean reproduction decreases in its argument, then because $Y_{n+1} \geq Y_n$ renders this a special case of monotonically deteriorating environment (i.e., the mean reproduction, though possibly random, is never increasing), we have a set of models of interest in themselves. This is, in a sense, a simpler case than dependence upon the present population size, which might fluctuate wildly, even if it tends to infinity.

Klebaner (1984, 1985) proved Theorem 6.6 below, which shows that if the convergence of reproduction means stabilizes fast enough, the population growth is exponential. For Theorem 6.6, let

$$m(z) = \sum_{k=1}^{\infty} k p_k(z) \quad \text{and} \quad \sigma^2(z) = \sum_{k=0}^{\infty} (k - m(z))^2 p_k(z) \tag{6.87}$$

denote the expectation and variance, respectively, of the number of children per individual in a z-sized population.

Theorem 6.6 *If for some m > 1*

$$\sum_{z=1}^{\infty} |m(z) - m|/z < \infty \quad \text{and} \quad \sum_{z=1}^{\infty} \sigma^2(z)/z^2 < \infty, \tag{6.88}$$

then either the population dies out or it grows as m^n, as $n \to \infty$.

The proof of this is far beyond the reach of the present book, but the essence of the theorem is quite natural: if the stabilization of the mean reproduction [i.e., the convergence of $m(z)$ toward m] is rapid enough as the population size z increases, we have exponential growth at that rate (and otherwise not).

Example 6.4 Imagine a (bird) population in which a couple can raise, on average, m individuals within their own precinct. If they are not disturbed by other population members, they can raise an additional average of a young. The probability of the latter being the case is, of course, a decreasing function of population size. Various forms can be thought of. One case might be that birds disturb each other independently and with the same probability p, so that the probability that this one couple remains undisturbed is $(1 - p)^{z-1}$ (if, for simplicity, population size z is thought of as the number of couples). Then we obtain

$$m(z) = m + a(1 - p)^{z-1}, \tag{6.89}$$

and the theorem is satisfied with a wide margin.

$$\diamond \diamond \diamond$$

The generalization of Theorem 6.6 to more general processes has proceeded along different paths. It holds for multi-type Galton–Watson processes (Klebaner 1989b, 1991), and from that a similar growth result has been deduced for very general discrete-time populations (Jagers and Sagitov 2000).

Indeed, consider a population that evolves in discrete time, say year by year. Assume that what happens to an individual during a year may be influenced by her age and previous reproduction history, but also by population size. To be able to analyze this model in terms of a multi-type Galton–Watson setup, assume that there is a maximal number of children as well as a maximal life span. Then criteria that correspond to those in Theorem 6.6 yield exponential growth.

For processes in continuous time, the situation is more complex. Provided the population has a basic Markov property, namely that it is Markovian in age structure, that exponential growth occurs was established recently, under some extra conditions (Jagers and Klebaner 2000). Markovian age structure is a very common assumption in population dynamics, and means that the future behavior of an individual, given her age, is independent of her past. Classic demographic models, as well as age-structured population dynamics or any model with age-dependent birth and death intensities, build upon this assumption. Nevertheless, it is of course, strictly speaking, untenable: if a mother has just had children she is less probable to have another litter immediately – or, at least, it is not clear that her tendency to give birth is unaffected by her earlier bearing history, to put it mildly.

For completely general processes, exponential growth under population size dependence has been established only under a stronger condition, of the type

(Jagers 1997b)

$$\sum_{z=1}^{\infty} |m(z) - m| < \infty.$$ (6.90)

Thus, we can summarize that exponential growth tends to remain even in cases of dependence, provided the population is supercritical in the sense that the reproduction rate exceeds 1. That this holds true both for local dependences and for dependence upon the whole population indicates that it should be a robust phenomenon, valid in dependent structures of quite different kinds.

Let us underline that we view dependence, be it upon close relatives or upon the population globally, merely phenomenologically, as summarizing the underlying and possibly much more complex causal relations between external environment and internal physiology (cf. Kooijman 1993).

From the discussion in Section 5.2, we know that dependence upon population size cannot produce stationary populations. Now, it seems it cannot even produce growth slower than exponential. This is, however, not true, as we show in the following slightly more technical section. If the reproduction probability distribution approaches criticality as the population grows, and does so at the very same rate as the growth of the population (this basically means that the size of the deviation from criticality of the expected number of children in a population of size z is of the order $1/z$, as $z \to \infty$), then the population actually either dies out or else grows, not exponentially, but at a slower, linear rate – there are also other, polynomial growth rates.

6.5.3   Slowly growing populations

F.C. Klebaner

In this section we look at branching models in which the offspring distribution depends on the population size rather than the population density. In particular, we concentrate on the case in which mean offspring numbers stabilize as the population size grows larger and larger and approaches one, namely near-critical processes. The near-supercritical case is treated in Section 6.5.2.

Again,

$$Z_{n+1} = \sum_{i=1}^{Z_n} \xi_{i,n}(Z_n),$$ (6.91)

with the distribution of $\xi_{i,n}$ dependent on Z_n. If $m(z) = \mathbb{E}[\xi(z)]$ approaches m as $z \to \infty$, the process is called near-critical if $m = 1$ and near-supercritical if $m > 1$. Recall that a classic branching population survives with positive probability if and only if it is supercritical, and when it survives it grows at the rate m^n. Interesting phenomena occur in near-critical populations. Under some conditions, not only can they survive, as shown in Chapter 5, but also they can grow linearly, at the rate n, or more generally polynomially, at the rate n^α (see Klebaner 1983).

Linear growth occurs in the following situation: $m(z)$ should approach 1 as the population size z increases to infinity, and moreover it should approach 1 at the rate $1/z$. The variances $\sigma^2(z) = \mathrm{Var}[\xi(z)]$ should approach some limit σ^2. The condition on third moments of offspring distribution in Theorem 6.7 follows from the moments being bounded, as is certainly biologically reasonable.

Theorem 6.7 *Suppose that $z(m(z) - 1) \to c$, $\sigma^2(z) \to \sigma^2$, $\sigma^2 < 2c$, and that $\mathbb{E}[|\xi(z) - m(z)|^3]/\sqrt{z} \to 0$, all when $z \to \infty$. Then Z_n/n converges in distribution to some random variable, which is 0 with the extinction probability Q. Furthermore, conditioned on non-extinction, it equals c if $\sigma^2 = 0$, and if $\sigma^2 > 0$ it has a Gamma distribution with parameters $2c/\sigma^2$ and $2/\sigma^2$.*

Note that if the condition on the limit variance is violated and $\sigma^2 \ge 2c$, extinction occurs. Therefore, it can be said that such processes "live on the boundary between extinction and survival." For definition of Gamma distributions, see the Appendix.

Under somewhat different conditions, this result was given by Klebaner (1984), Höpfner (1985), and Kersting (1992), who obtained the best possible form.

Example 6.5 Recall the example of binary splitting from Section 5.8, in which the probability of division into two is $p(z) = 1/2 + 1/(2z)$, given that the population size is z. In this case, as shown, there is a positive chance of survival. Further, $m(z) = 2p(z) = 1 + 1/z$, and $z(m(z) - 1) \to 1$, $\sigma^2(z) = 4p(z) - m^2(z) \to 1$. The third moments clearly are bounded by $2^3 = 8$. Hence, if the population does not die out, Z_n/n converges to a $\Gamma(2, 2)$ variable (again, see the Appendix).

$$\diamond \diamond \diamond$$

Example 6.6 A further application is provided by polymerase chain reaction (PCR) models, which have the mean "offspring" number per molecule

$$m(z) = 1 \times (1 - p(z)) + 2 \times p(z) = 1 + p(z) = 1 + \frac{K}{K + z}, \qquad (6.92)$$

when population size is z, provided Michaelis–Menten kinetics can be assumed (see Section 7.5). The variance of the offspring distribution of any single individual is

$$\sigma^2(z) = 4p(z) + 1 - p(z) - m^2(z) = p(z)(1 - p(z)) = \frac{Kz}{(K + z)^2}. \qquad (6.93)$$

Since the offspring number per molecule does not exceed two, the third moment is, clearly, bounded, and thus satisfies the condition of Theorem 6.7. It follows from Theorem 6.7 that

$$Z_n/n \to K \qquad (6.94)$$

in probability as $n \to \infty$. As we show later, convergence actually holds in stronger senses.

$$\diamond \diamond \diamond$$

A more general, polynomial growth may occur in near-critical processes. For different types of growth, including polynomial, see Klebaner (1989a) and Kersting (1992).

Theorem 6.8 *Suppose that there exist constants $0 < c < \sigma^2/2$ and $\alpha < 1$, such that as $z \to \infty$,*

$$z^{1-\alpha}(m(z) - 1) \to c, \sigma^2(z)/z^\alpha \to \sigma^2 > 0, \qquad (6.95)$$

$$\mathbb{E}[|\xi(z) - m(z)|^3]/z^{1/2+\alpha} \to 0. \qquad (6.96)$$

Then $Z_n/n^{1/(1-\alpha)}$ converges in distribution. The limit W equals zero only if the population dies out. If this does not occur, the limit has a generalized Gamma distribution, that is, $W^{(1-\alpha)}$ is Gamma with parameters $(2c - \sigma^2\alpha)/(\sigma^2 - \sigma^2\alpha)$ and $2/(\sigma^2(1 - \alpha)^2)$.

Polynomial growth typically occurs in a situation in which the number of offspring is 0 with a very large probability and very large with a very small probability, as illustrated in the Example 6.7.

Example 6.7 Given that the population size is z, let the number of offspring be $a(z) = [\sigma^2 z^\alpha]$, where the brackets denote the integer part of a number and $\alpha < 1/2$, with probability $p(z) = 1/a(z) + c/(a(z)z^{1-\alpha})$ and zero with the complementary probability. Then $m(z) = a(z)p(z)$ satisfies the assumption, as do the variances and third moment. Therefore, conditioned on non-extinction, $Z_n/n^{1/(1-\alpha)}$ converges to a limit that has a generalized Gamma distribution. This example might be appropriate for modelling insects, in cases where there is a great risk of larvae dying, but otherwise they give birth to a large number of individuals.

$$\diamond \diamond \diamond$$

For near-supercritical processes, $m(z) \to m > 1$, Theorem 5.9 (Section 5.8) gives sharp conditions for geometric growth with rate m^n. A description of growth rates close to exponential in near-supercritical processes is given in Küster (1985) and Keller *et al.* (1987).

These processes are studied in the framework of a more general growth model,

$$X_{n+1} = X_n + g(X_n)(1 + \xi_{n+1}), \qquad (6.97)$$

where $X_0 > 0$, $g(t)$ is strictly positive, $g(t)/t \to 0$ as $t \to \infty$, and (ξ_n) is a so-called zero-mean, square-integrable martingale difference sequence, that is

$$\mathbb{E}[\xi_{n+1}|X_0, \cdots, X_n] = 0 \qquad (6.98)$$

and

$$\mathbb{E}[\xi_{n+1}^2|X_0, \cdots, X_n] = \sigma^2(X_n). \qquad (6.99)$$

Under several regularity conditions, the authors show that extinction or explosion occurs, that is, $\lim_{n\to\infty} X_n = 0$ or $\lim_{n\to\infty} X_n = \infty$ and that $\mathbb{P}(\lim_{n\to\infty} X_n \to \infty) > 0$. The growth rate of the (X_n) is characterized with help of the function $G(t) = \int_1^t ds/g(s)$ (which tends to ∞ as $t \to \infty$). If $A = \{X_n \to \infty\}$, then $G(X_n)/n \to 1$ with probability 1 on A. It is also possible to compare the stochastic recursion Equation (6.97) to the deterministic one. Let $a_0 = 1$ and $a_{n+1} = a_n + g(a_n)$ define the deterministic solution of Equation (6.97), then it is possible to estimate how closely X_n tracks a_n [see Keller *et al.* (1987) for details].

6.6 Growth of Populations with Sexual Reproduction
G. Alsmeyer

In Section 5.9, we studied the effect of sexual reproduction on criticality and extinction risk of branching processes. Here, we consider the ultimately supercritical case ($m_\infty > 1$) and take a look at the question of how such populations grow in the event of survival. Since m_∞ describes the asymptotic growth rate per generation if the population becomes large, it is not unreasonable to believe that Z_n grows as m_∞^n in the event of survival. However, even for the simple Galton–Watson process, the famous Kesten–Stigum theorem has already shown that this is true only under an additional condition on the offspring distribution. Defining the normalized process

$$W_n = \frac{Z_n}{m_\infty^n}, \quad n \geq 0, \tag{6.100}$$

we have that

$$\mathbb{E}[W_{n+1}|W_n = im_\infty^{-n}] = \mathbb{E}[W_{n+1}|Z_n = i]$$
$$= m_\infty^{-(n+1)}\mathbb{E}[Z_{n+1}|Z_n = i] = im_i m_\infty^{-(n+1)} \leq im_\infty^{-n} \tag{6.101}$$

for all i. Thus, $(W_n)_{n\geq 0}$ constitutes a non-negative supermartingale and, therefore, converges to a finite random variable W with expectation $\mathbb{E}[W|Z_0 = i] \leq \mathbb{E}[W_0|Z_0 = i] = i$ for all i. In other words, the long-run population growth rate is at most m_∞. This should not come as a surprise, because the average unit reproduction means are bounded by this value. However, when is m_∞^n also the correct normalization in the sense that the limiting variable W is positive in the event of survival?

The difficulty of this question is best understood when the approximation of Z_n by Wm_∞^n is considered as a two-step result. Writing Z_n as the product $\frac{Z_n}{Z_{n-1}} \cdot \frac{Z_{n-1}}{Z_{n-2}} \cdot \ldots \cdot \frac{Z_1}{Z_0}$, replace first each factor $\frac{Z_k}{Z_{k-1}}$ with its conditional expectation given Z_{k-1}, that is $m_{Z_{k-1}}$, and then the latter with its limit (and upper bound) m_∞ as Z_{k-1} tends to infinity. We thus arrive at the decomposition

$$W_n = V_n \cdot \prod_{k=0}^{n-1} \frac{m_{Z_k}}{m_\infty}, \tag{6.102}$$

where $V_0 = Z_0$ and

$$V_n = \frac{Z_n}{m_{Z_0} \cdot \ldots \cdot m_{Z_{n-1}}}, \quad n \geq 1. \tag{6.103}$$

Now, since

$$V_{n+1} = V_n \frac{Z_{n+1}}{Z_n m_{Z_n}}, \tag{6.104}$$

and the conditional expectation of $\frac{Z_{n+1}}{Z_n}$ given $Z_0, \ldots, Z_n$ equals m_{Z_n}, a similar computation as in Equation (6.101) shows that $(V_n)_{n\geq 0}$ constitutes a non-negative martingale, that is, $\mathbb{E}[V_{n+1}|V_n = v] = v$ for all n and v. Hence V_n converges to a random variable V, which is at least as large as W because $V_n \geq W_n$ for all n. Taking limits in Equation (6.102) now yields

$$W = V \cdot \prod_{k \geq 0} \frac{m_{Z_k}}{m_\infty}. \qquad (6.105)$$

So Z_n indeed grows as m_∞^n in the event of survival, provided that both the martingale limit V and the infinite product $\prod_{k \geq 0} \frac{m_{Z_k}}{m_\infty}$ are positive in that event. The latter obviously holds true if m_j converges to m_∞ sufficiently quickly. However, without a restriction to special mating functions it seems difficult to translate these requirements into conditions on the offspring distribution $(p_{j,k})_{j,k \geq 0}$. González and Molina (1996, 1997) did some related work for general ζ, but circumvented the problem by directly imposing conditions on the *derived* quantities $d_j = m_j - m_\infty$, essential ones being that $d_j \leq g(j)$ for all j and a suitable concave function g that satisfies $\sum_{j \geq 1} j^{-1} g(j) < \infty$. By adding further conditions not reported here they could prove that Z_n grows as m_∞^n (W is positive) in an event of positive probability, but not necessarily the full event of survival.

At least for monogamous populations with mating function $\zeta(x, y) = \min(x, y)$, Bagley (1986) was able to provide a satisfactory answer, stated in Theorem 6.9 below, which is actually the perfect analog of the corresponding result for asexual populations described by simple Galton–Watson processes. Recall that p_j^F and p_k^M denote the probabilities that a couple has exactly j female and k male offspring, respectively, and hence $p_j^F = \sum_{k \geq 0} p_{j,k}$ and $p_k^M = \sum_{j \geq 0} p_{j,k}$.

Theorem 6.9 *Let $(Z_n)_{n \geq 0}$ be an ultimately supercritical Galton–Watson process with a monogamous mating function ζ and an offspring distribution $(p_{j,k})_{j,k \geq 0}$ that satisfies Assumption 5.2 (Section 5.9), then $\sum_{k \geq 1} p_k^F k \ln k < \infty$ implies that W is positive in the event of survival, that is, $\mathbb{P}(W > 0|Z_0 = i) = 1 - Q_i$ for all i, while $\sum_{k \geq 1} p_k^F k \ln k = \infty$ implies $W = 0$.*

As one can verify easily, $\sum_{k \geq 1} p_k^F k \ln k < \infty$ and $\sum_{k \geq 1} p_k^M k \ln k < \infty$ are equivalent conditions under Assumption 5.2 because $0 < \theta < 1$.

Sketch of Proof. We content ourselves with the following very intuitive heuristic argument under the additional assumption $m^F \neq m^M$ or, equivalently, $\theta \neq \frac{1}{2}$. To be specific, suppose $m^F < m^M$. If the population survives and hence grows to infinity, eventually the total number of female offspring produced by a generation is always smaller than the respective number of male offspring. In fact, the Law of Large Numbers even shows that

$$\frac{1}{Z_n}(M_{n+1} - F_{n+1}) = \frac{1}{Z_n} \sum_{k=1}^{Z_n} (Y_{n+1,k} - X_{n+1,k}) \qquad (6.106)$$

tends to $m^M - m^F > 0$, if $Z_n \to \infty$. Consequently, for large n the number of couples that form the $(n + 1)$th generation just equals the number of female offspring of the previous one, whence $(Z_n)_{n \geq 0}$ ultimately behaves as the simple branching process obtained by considering only the females. The assertions of the theorem now follow by invoking the

Kesten–Stigum theorem. It is quite clear that the heuristic just given remains true if the offspring distribution satisfies Assumption 5.3 (Section 5.9). Since $m_\infty = \min(m^F, m^G)$, Theorem 6.9 shows that, even if $m^F = m^M$, a surviving monogamous population grows at the same order of magnitude as one of its associated asexual counterparts in which females and males, respectively, reproduce without mating. Of course, the probability of survival is always smaller in the case with sexual reproduction. The same heuristic becomes exact for (unilateral) promiscuous populations because, in the case of survival, the number of couples precisely equals the number of females in each generation. The male subpopulation enters into the analysis only by causing an increased chance of extinction.

6.7 Immigration in Subcritical Populations

In Section 2.10 we calculated the expected size of populations with recurrent immigration. Here we continue to study the properties of such populations and concentrate on the situation in which the local branching process is subcritical. From Section 2.10 we know that in that case the average number of individuals stabilizes. Here we show that, moreover, the distribution of the population size is stationary.

Let $f(s)$ be the local reproduction generating function and denote the probability generating function of the number of invaders per period by $g(s)$. We denote the probability generating function of the number of individuals in generation n in the resultant branching process with immigration by $F(n, s) = \mathbb{E}[s^{Z_n}]$. Applying the conditional expectation argument (see the Appendix), we obtain

$$F(n, s) = \mathbb{E}[s^{\xi_1 + \xi_2 + \cdots + \xi_{Z_{n-1}} + Y_n}]$$

$$= \mathbb{E}[\mathbb{E}[s^{\xi_1 + \xi_2 + \cdots + \xi_{Z_{n-1}} + Y_n} | Z_{n-1}]]$$

$$= \mathbb{E}[\mathbb{E}[s^{\xi_1} s^{\xi_2} \cdots s^{\xi_{Z_{n-1}}} s^{Y_n} | Z_{n-1}]], \qquad (6.107)$$

where, as before, ξ_i is the number of children of the ith individual in generation $(n-1)$, and Y_n is the number of immigrants in the nth generation. Since Y_n is independent of Z_{n-1}

$$F(n, s) = \mathbb{E}[\mathbb{E}[s^{\xi_1} s^{\xi_2} \cdots s^{\xi_{Z_{n-1}}} | Z_{n-1}]]\mathbb{E}[s^{Y_n}] = \mathbb{E}[f(s)^{Z_{n-1}}]g(s)$$

$$= F(n-1, f(s))g(s) = F(n-2, f_2(s))g(f(s))g(s) = \cdots$$

$$= F(0, f_n(s))g(f_{n-1}(s))g(f_{n-2}(s)) \cdots g(s). \qquad (6.108)$$

Thus, if the population is initiated at $n = 0$ by a random number of invaders distributed according to the probability generating function $F(0, s) = g(s)$,

$$F(n, s) = \prod_{k=0}^{n} g(f_k(s)), \qquad (6.109)$$

and if there were a non-random number z_0 of individuals at $n = 0$,

$$F(n, s) = (f_n(s))^{z_0} \prod_{k=0}^{n-1} g(f_k(s)). \qquad (6.110)$$

We show that a non-trivial stationary distribution exists in both cases if the local process is subcritical, that is, for any $s \in [0, 1)$, $\lim_{n \to \infty} F(n, s)$ exists and is positive.

Since $g(s) < 1$ for any $s \in [0, 1)$, the product $\prod_{k=0}^{n-1} g(f_k(s))$ monotonically decreases in n and, therefore, there exists a finite limit

$$\lim_{n \to \infty} \prod_{k=0}^{n-1} g(f_k(s)) = \prod_{k=0}^{\infty} g(f_k(s)) . \tag{6.111}$$

On the other hand, since the generating function increases with s,

$$0 \le 1 - f_n(s) \le 1 - f_n(0) \le m^n \tag{6.112}$$

and, therefore, $\lim_{n \to \infty} f_n(s) = 1$. Thus, for any fixed z_0,

$$\lim_{n \to \infty} (f_n(s))^{z_0} = 1 . \tag{6.113}$$

Combining this result with Equation (6.111), we see that if $F(0, s) = f^{z_0}(s)$ then

$$\lim_{n \to \infty} F(n, s) = \lim_{n \to \infty} (f_n(s))^{z_0} \prod_{k=0}^{n-1} g(f_k(s)) = \prod_{k=0}^{\infty} g(f_k(s)) . \tag{6.114}$$

The same limiting expression appears if $F(0, s) = g(s)$. To show that the limiting quantity is positive we apply Jensen's inequality (see Appendix) to obtain

$$F(n, s) = \mathbb{E}[s^{Z_n}] \ge s^{\mathbb{E}[Z_n]} . \tag{6.115}$$

Hence, recalling Equation (2.106), we conclude that in both cases

$$\lim_{n \to \infty} F(n, s) \ge s^{\lim_{n \to \infty} \mathbb{E}[Z_n]} = s^{\lambda/(1-m)} > 0 \tag{6.116}$$

as desired (where λ is the expected number of immigrants per period, see Section 2.10).

Unfortunately, the right-hand side of Equation (6.114) can rarely be calculated explicitly. However, using Equation (6.114) we can derive estimates for the average length of a life period of the process. Recall that a life period is the interval between the moments when the first invader (or invaders) came to an empty site until the first moment when the site becomes empty again. Information on the length of such periods may be used, for example, in epidemiology and ecology. In the context of epidemics, such periods correspond to the durations of outbreaks of diseases that do not lead to full epidemics (Section 7.6). Further, they correspond to periods of occupancy of sites in metapopulations (Section 7.7).

To calculate the average length of a life period it is convenient to select 0 as the starting point of the period. That is, we suppose that $Z_{-1} = 0$ and $Z_0 > 0$ and calculate the average value of the quantity

$$\tau = \min\{k > 0 : Z_k = 0\} . \tag{6.117}$$

(This does not impose any restrictions, it is just for mathematical convenience.) It can be shown (see Box 6.2) that Theorem 6.10 is true.

Box 6.2 Proof of Theorem 6.10

Since Z_0 is distributed as Y_0, the number of invaders at moment 0, provided that $Y_0 > 0$, is

$$F(0, s) = \frac{\mathbb{E}[s^{Y_0}; Y_0 > 0]}{\mathbb{P}[Y_0 > 0]} = \frac{\mathbb{E}[s^{Y_0}] - \mathbb{E}[s^{Y_0}; Y_0 = 0]}{1 - \mathbb{P}[Y_0 = 0]} = \frac{g(s) - g(0)}{1 - g(0)} \,. \tag{a}$$

Now we consider the development of this process up to time n. Taking into account that $\tau \geq 1$, we have

$$F(n, s) = \sum_{k=0}^{n-1} \mathbb{P}[\tau = n-k] \mathbb{E}[s^{Z_n} \mid \tau = n-k] + \mathbb{P}[\tau > n] \mathbb{E}[s^{Z_n} \mid \tau > n] \,. \tag{b}$$

Observe that if $\tau = n-k$ for some $k < n$, the site is empty at time $n-k$ and, starting from this moment, the population develops through invaders and their offspring only. In other words,

$$\mathbb{E}[s^{Z_n} \mid \tau = n - k] = \prod_{j=0}^{k} g(f_j(s)) \,. \tag{c}$$

However, on account of Equation (a) and Equation (6.108)

$$F(n, s) = F(0, f_n(s)) \prod_{k=0}^{n-1} g(f_k(s)) = \frac{g(f_n(s)) - g(0)}{1 - g(0)} \prod_{k=0}^{n-1} g(f_k(s)) \,. \tag{d}$$

Thus, from Equations (a) and (b)

$$\frac{g(f_n(s)) - g(0)}{1 - g(0)} \prod_{k=0}^{n-1} g(f_k(s)) = \sum_{k=0}^{n-1} \mathbb{P}[\tau = n - k] \prod_{j=0}^{k} g(f_j(s))$$
$$+ \mathbb{E}[s^{Z_n} \mid \tau > n] \mathbb{P}[\tau > n] \,. \tag{e}$$

Define $b_j = \mathbb{P}[\tau \geq j]$ and recall that

$$r_k = \prod_{j=0}^{k} g(f_j(0)), \ k \geq 0 \,. \tag{f}$$

Observe that $\mathbb{P}[\tau = n - k] = b_{n-k} - b_{n-k+1}$. Since, given that $\tau > n$ and $Z_n \geq 1$, we have

$$\mathbb{E}[s^{Z_n} \mid \tau > n] \Big|_{s=0} = 0 \,, \tag{g}$$

substituting $s = 0$ in Equation (e) gives

$$\frac{g(f_n(0)) - g(0)}{1 - g(0)} r_{n-1} = \sum_{k=0}^{n-1} (b_{n-k} - b_{n-k+1}) r_k \,. \tag{h}$$

It is an easy exercise to check that for $n \geq 1$

$$\sum_{k=0}^{n-1} (b_{n-k} - b_{n-k+1}) r_k = r_{n-1} b_1 + \sum_{k=1}^{n-1} b_{n-k+1} (r_{k-1} - r_k) - r_0 b_{n+1} \,. \tag{i}$$

continued

Box 6.2 *continued*

Substituting this in (e), recalling that $r_0 = g(0)$ and $b_1 = 1$ and rearranging, we see that

$$g(0)b_{n+1} = r_{n-1}(1 - \frac{g(f_n(0)) - g(0)}{1 - g(0)}) + \sum_{k=1}^{n-1} b_{n-k+1}(r_{k-1} - r_k) . \tag{j}$$

Since $g(f_n(0)) = \frac{r_n}{r_{n-1}}$,

$$g(0)b_{n+1} = \frac{1}{1 - g(0)}(r_{n-1} - r_n) + \sum_{k=1}^{n-1} b_{n-k+1}(r_{k-1} - r_k) . \tag{k}$$

Since τ is a positive integer-valued random variable, $\mathbb{E}[\tau] = \sum_{n=1}^{\infty} b_n$ (see the Appendix). Using this relation and summing the left-hand side of Equation (k) over $n \geq 1$, we obtain

$$g(0) \sum_{n=1}^{\infty} b_{n+1} = g(0) \sum_{n=2}^{\infty} b_n = g(0)(\mathbb{E}[\tau] - 1) , \tag{l}$$

whereas summing the right-hand side of Equation (k) gives

$$\frac{1}{1 - g(0)} \sum_{n=1}^{\infty} (r_{n-1} - r_n) + \sum_{n=1}^{\infty} \sum_{k=1}^{n-1} b_{n-k+1}(r_{k-1} - r_k)$$

$$= \frac{1}{1 - g(0)} (r_0 - \lim_{n \to \infty} r_n) + \sum_{k=1}^{\infty} (r_{k-1} - r_k) \sum_{n=k+1}^{\infty} b_{n-k+1}$$

$$= \frac{1}{1 - g(0)} (r_0 - \lim_{n \to \infty} r_n) + (r_0 - \lim_{n \to \infty} r_n)(\mathbb{E}[\tau] - 1) . \tag{m}$$

Combining the equalities above and recalling that $r_0 = g(0)$ and $\lim_{n \to \infty} r_n = r$, we obtain from Equations (l) and (m)

$$(\mathbb{E}[\tau] - 1)g(0) = (g(0) - r) \left(\frac{1}{1 - g(0)} + (\mathbb{E}[\tau] - 1) \right) , \tag{n}$$

which leads to the expression in Theorem 6.10.

Theorem 6.10 *The average length of a life period is given by*

$$\mathbb{E}[\tau] = \frac{(1 - r)g(0)}{r(1 - g(0))} , \tag{6.118}$$

where $r = \prod_{k=0}^{\infty} g(f_k(0))$.

The proof is given in Box 6.2. One may think that the calculation of r is difficult. However, since $r_n = \prod_{k=0}^{n} g(f_k(0))$ converges to r rather quickly, r_n provides a good approximation. Indeed,

$$r_n - r = \prod_{k=0}^{n} g(f_k(0)) - \prod_{k=0}^{\infty} g(f_k(0)) = \prod_{k=0}^{n} g(f_k(0))(1 - \prod_{k=n+1}^{\infty} g(f_k(0))) . \quad (6.119)$$

Further,

$$1 - \prod_{k=n+1}^{\infty} g(f_k(0)) \leq \sum_{k=n+1}^{\infty} (1 - g(f_k(0))) \leq \sum_{k=n+1}^{\infty} \lambda(1 - f_k(0))$$

$$\leq \sum_{k=n+1}^{\infty} \lambda m^k = \frac{\lambda m^{n+1}}{1 - m} , \quad (6.120)$$

and therefore,

$$\prod_{k=n+1}^{\infty} g(f_k(0)) \geq 1 - \frac{\lambda m^{n+1}}{1 - m} = \frac{1 - m - \lambda m^{n+1}}{1 - m} . \quad (6.121)$$

Thus, for the absolute error in the approximation of r by r_n,

$$0 \leq r_n - r \leq \frac{\lambda m^{n+1}}{1 - m} , \quad (6.122)$$

and if $\lambda m^{n+1} < 1 - m$, then we find for the relative error

$$0 \leq \frac{r_n - r}{r} = \frac{1}{\prod_{k=n+1}^{\infty} g(f_k(0))} - 1 \leq \frac{1 - m}{1 - m - \lambda m^{n+1}} - 1$$

$$= \frac{\lambda m^{n+1}}{1 - m - \lambda m^{n+1}} . \quad (6.123)$$

6.8 Quasi-stationarity: General Remarks
A.D. Barbour

The branching process is a good model for describing the growth of a population, as long as the assumption of a fixed life-history distribution is reasonable, irrespective of population size or density. This is a plausible approximation to the truth in the early stages of population growth, and makes branching processes excellent models for answering questions about the survival or extinction of populations, and in particular about the life-history strategies to be adopted by colonizing species. For instance, their reproductive strategy should ideally be chosen to maximize the Malthusian growth rate, rather than the lifetime mean number of offspring per individual. However, if a population successfully establishes itself, it cannot maintain exponential growth for ever; limitations on resources, such as space, light, and food, or increases in predation and disease lead to changes in fecundity and mortality, and the assumption of a fixed life-history distribution has to be abandoned.

The simplest scenario is that in which increasing population density leads to a gradual reduction in the lifetime mean number of successful offspring m per individual. Initially, the growing population has mean offspring number $m = m_0 > 1$.

As pressure on resources grows, because of increasing population density, the value of m sinks, as does the Malthusian rate of growth. When the value of m has fallen as low as 1, the rate of growth is reduced to zero, and no further population growth takes place, at least on average. If random variation increases the population density, living conditions become even worse, the mean number of offspring per individual falls below 1, and the population density tends to decline, until it again reaches the level at which $m = 1$. If random variation reduces the population density, living conditions improve, m rises above 1, and the population density tends to increase until $m = 1$ once more. Thus the actual population density fluctuates at values close to the "carrying capacity," the density K at which $m = 1$. It varies because of random effects, or "shocks": the environment may fluctuate in a way that can best be described as random, and even in a constant environment, the randomness inherent in individual reproductive success influences the population density. However, whenever the system is disturbed from its natural equilibrium, the resultant changes in m have the effect of directing the population density back toward equilibrium once more. A snapshot of an established population thus reveals a population density that is not exactly K, but takes a random value in the neighborhood of K (see Barbour 1976).

An example of this, first for simplicity without the randomness, is afforded by the logistic growth model

$$\frac{dn(t)}{dt} = r_0 n(t)\{1 - n(t)/K\}, \qquad n(0) = n_0 , \tag{6.124}$$

which describes the way in which the population density $n(t)$ varies as a function of real time, starting from some initial density n_0. At low densities, n increases exponentially at the basic reproductive rate r_0, which corresponds to the exponential growth of the branching process in continuous time; for the population to be successful in growing, it is necessary for r_0 to be positive, which is equivalent to having $m_0 > 1$. At higher densities, the growth rate is tempered by the factor $(1 - n/K)$, which reduces the actual net reproductive rate as n increases, in such a way that $dn/dt > 0$ if $n < K$ and $dn/dt < 0$ if $n > K$; the population density tends toward an equilibrium at $n = K$, as long as $n_0 > 0$. If randomness is added, the population density n, when observed in "equilibrium," becomes a random variable with a mean close to K. Its variance depends principally on two competing effects: the strength and intensity of the random shocks, which disturb n away from K, and the strength of the drift back toward K implicit in the differential equation (6.124).

The drift back to equilibrium is relatively easy to describe. In the deterministic model, Equation (6.124), one can re-express the differential equation in terms of the difference y between the population density n and its equilibrium value K, $n(t) = K + y(t)$, where, near equilibrium, $|y(t)| \ll K$. This then leads to the differential equation $dy(t)/dt = -r_0 y(t)$ with solution

$$y(t) = y(0)e^{-r_0 t} , \tag{6.125}$$

which is *approximately* satisfied by y. The form of this solution describes an exponential decline in the size of the departure from equilibrium, with "relaxation rate" r_0 characterizing the rate at which the decline takes place. This exponential rate can also be expressed in terms of its "half-life" τ, the amount of time that elapses before the disturbance $y(t)$ drops to half its original value $y(0)$, such that $y(\tau) = \frac{1}{2}y(0)$. Then it is easy to see from Equation (6.125) that the half-life τ is given by $\tau = (1/r_0)\ln_e 2$. Half-life and relaxation rate are thus different, but equivalent, ways to describe the strength of the drift back to equilibrium. We use the latter because, like drift, it is a rate.

The relaxation rate in the Model (6.124) takes the value r_0, but this is actually coincidental. There is a more general model,

$$\frac{dn}{dt} = r_0 n f(n/K), \qquad n(0) = n_0 , \qquad (6.126)$$

in which $f(x)$ represents the fraction of the basic reproductive rate achieved by individuals at a population density of xK, and can be chosen more flexibly than the choice $f(x) = 1 - x$ implicit in Equation (6.124); f is typically a decreasing function, with $f(0) = 1$ (so that r_0 indeed becomes the reproductive rate at population density zero) and with $f(1) = 0$ (so that the carrying capacity K is indeed the equilibrium of the system). In this more general model, it can be shown that the drift back toward K has relaxation rate

$$\gamma = -r_0 f'(1) . \qquad (6.127)$$

In Model (6.124), $f'(1) = -1$, and Equation (6.127) gives the relaxation rate r_0, as previously calculated. Note that, in general, $f'(1) < 0$ because f is decreasing, and so the relaxation rate is, in fact, always positive. Thus, as might be expected, the scale of the variation about K depends, in part, upon the way in which the reproductive success of the individuals reacts to changes in population density in the neighborhood of the carrying capacity, through $f'(1)$, and is not determined by the initial net reproductive rate r_0 alone.

Many stochastic models are broadly approximated by Equation (6.124) or (6.126). Perhaps the simplest is a "Markov" model, in which individuals are exposed to mortality at a density-dependent rate $\mu(n/K)$ per unit time, and give birth at a rate $\lambda(n/K)$ per unit time, irrespective of their age and experience. This is formally understood in the sense that, in a population of size $N = nA$ in a habitat of area A, an "event" takes place after a random time that has a negative exponential distribution with mean $1/[nA\{\lambda(n/K) + \mu(n/K)\}]$; the event is a birth, with probability $\lambda(n/K)/\{\lambda(n/K) + \mu(n/K)\}$, in which case N is increased by 1, or a death, with probability $\mu(n/K)/\{\lambda(n/K) + \mu(n/K)\}$, in which case N is decreased by 1, and the process then begins again with the new value of N. None of these assumptions seems particularly attractive at first sight; fortunately, the pattern of behavior of the model mimics that of many more realistic models that incorporate "demographic randomness."

In this Markov model, if A and therefore N are large, it can be shown that the behavior of $\bar{n}(t) := \mathbb{E}n(t)$, the expected population density, is close to that of the solution to the differential equation

$$\frac{d\bar{n}}{dt} = \bar{n}\{\lambda(\bar{n}/K) - \mu(\bar{n}/K)\}, \qquad \bar{n}(0) = n_0 . \qquad (6.128)$$

Thus, if λ is a decreasing function (fecundity declines as population density increases) and μ is an increasing function (mortality increases with increasing population density), and if $\lambda(1) = \mu(1)$ (fixing the carrying capacity as K), then Equation (6.128) is the same as Equation (6.126), with $r_0 = \lambda(0) - \mu(0) > 0$ and $f(x) = (\lambda(x) - \mu(x))/(\lambda(0) - \mu(0))$, and with relaxation rate $\gamma = -r_0 f'(1) = \mu'(1) - \lambda'(1)$. It can further be shown that, in "equilibrium," when the random shocks occasioned by demographic randomness are balanced by the inherent drift back toward K, and the fluctuations about the equilibrium density K have a distribution that remains constant over time, the population density n is distributed approximately normally, with mean K and variance V given by

$$V = \frac{1}{2A}K\{\lambda(1) + \mu(1)\}/\gamma = \frac{KA\nu_1}{A^2\gamma} , \qquad (6.129)$$

where ν_1 denotes the common value of $\lambda(1)$ and $\mu(1)$. In this formula, the relaxation rate γ of the differential equation (6.128), which describes the behavior of the mean drift, appears in the denominator. To interpret the remainder, $2KA\nu_1$ is the rate at which random shocks – in this model, birth and death events – occur in equilibrium, and $1/A^2$ is the squared magnitude of these shocks. When the number of individuals in the population changes by 1, the population density n changes by $1/A$. Thus, the equilibrium variance is given by half the rate of incidence of squared shocks, divided by the relaxation rate, a somewhat quaint formula (except that a variance is an average of squares, which explains the appearance of *squared* shocks; the dimensions are $\{\text{area}\}^{-2}$).

The variance V is inversely proportional to the typical size of the population, as reflected in A. Thus, the larger the value of A, the smaller the value of V, and the more reasonable it becomes to approximate the population density by the constant K. Indeed, such an approximation is plausible provided the fluctuations are small compared to the mean density; that is, provided that $\sqrt{V} \ll K$ or, even more explicitly, provided that

$$KA \gg \nu_1/\gamma . \qquad (6.130)$$

Note that KA is approximately the expected total population size in equilibrium.

An alternative random variant of Equation (6.126) is obtained by supposing that randomness arises only through variations in the environment. For instance, the constant carrying capacity K in Equation (6.126) could be replaced by a randomly varying function $K(t)$, generated much as $n(t)$ in the previous model by a combination of random shocks, which tend to disturb $K(t)$ away from its natural equilibrium value $\overline{K}$, and a natural exponential relaxation rate of return toward $\overline{K}$. The intensity of squared shocks, the product of the rate at which random shocks occur and their average squared magnitude, and thus the analog of $2KA\nu_1/A^2$, could be denoted by σ^2, and the relaxation rate by β, for instance. Then, provided that $V = \sigma^2/2\beta$ is small compared to 1, the random fluctuations in $n(t)$, the solutions of

$$\frac{dn}{dt} = r_0 n f(n/K(t)), \qquad n(0) = n_0 , \qquad (6.131)$$

are small compared to the typical population density $\overline{K}$, and the approximation of $K(t)$ by the fixed value $\overline{K}$ is not at all bad. If, in addition, the relaxation rate γ of the underlying differential equation (6.126) is small compared to β, the population density reacts relatively slowly to the random changes in the environment, and is correspondingly affected less by them, as a result of "averaging"; this further improves the approximation that results by assuming $K(t)$ to be fixed at $\overline{K}$.

In the above examples, and many other, more complicated variants, the behavior around equilibrium can be described reasonably by such combinations of random shocks and relaxation toward equilibrium. In a model with "equilibrium" variance V and relaxation rate β, with $\sqrt{V}$ small compared to the equilibrium density K, a reasonable approximation near equilibrium is usually provided by an Ornstein–Uhlenbeck diffusion. This can be understood by thinking of the random shocks as driving a Brownian motion, which is, however, constrained by the relaxation effect to remain near equilibrium. Over very short time intervals, the randomness looks exactly like that of Brownian motion, but the further the process wanders from equilibrium, the stronger is the force of the drift back toward it. If the infinitesimal variance (volatility) of the Brownian motion is denoted by σ^2 – as in the intensity of squared shocks in the previous example – then the equilibrium variance V is again given by $\sigma^2/2\beta$; and, much as in the usual Central Limit Theorem, the equilibrium distribution of the random population density is approximately normal, with mean K and variance V. Looking more closely at the values of the population density in (stochastic) equilibrium, the correlation between $n(0)$ and $n(t)$ falls off exponentially, being given by $e^{-\beta t}$ (and they are jointly normally distributed); thus, it is possible to estimate both σ^2 and β from measurements of the population density at a number of different times, $t_1 < t_2 < \cdots < t_n$, by considering the sample variance and the sample autocorrelations.

To illustrate this last point, suppose that population densities $n_1, n_2, \ldots, n_r$ have been measured at equally spaced times, $t, t + h, \ldots, t + (r - 1)h$. Then the sample variance $v(0)$ of $n_1, n_2, \ldots, n_r$ is an estimate of the equilibrium variance $V = \sigma^2/2\beta$, and each of the sample autocovariances at lags $l = 1, 2, \ldots, L$, defined by

$$v(l) := \frac{1}{r - l} \sum_{j=1}^{r-l} (n_j - \overline{n})(n_{j+l} - \overline{n}) , \qquad (6.132)$$

where $\overline{n}$ denotes the sample mean of $n_1, n_2, \ldots, n_r$, is an estimate of $Ve^{-l\beta h}$. Here, h and L should be chosen carefully in such a way that $v(l)$ declines visibly between t and $t + Lh$; however, systematic effects, such as seasonal variation, also have to be accounted for, so this procedure is not particularly easy to realize in practice. Nonetheless, if this has been achieved, and if all the $v(l)$ are substantially positive, a linear regression of $\ln_e v(l)$ against l, for $0 \leq l \leq L$, should display an intercept $\ln_e(\sigma^2/2\beta)$ and slope $-\beta h$; for values of $v(l)$ near or below zero, a more sophisticated fitting procedure should be used.

Since the whole of the random behavior of the Ornstein–Uhlenbeck process is determined by the values of the two parameters β and σ^2, it provides a simple and practical description of behavior near equilibrium. In addition, it has a number of other useful features. For instance, if $n(0)$ is known, the conditional distribution of $n(t)$ is normal with mean $n(0)e^{-\beta t}$ and variance $\sigma^2(1 - e^{-2\beta t})/2\beta$, which shows in very concrete terms how the predictive value of the current value of the population density decreases with time. What is more, the model is Markovian; extra information about $n(s)$ at times $s < 0$ has no extra predictive value.

Of course, many other possible combinations of environmental and demographic randomness could be envisaged, and many more complicated interactions between species that may combine to maintain an equilibrium; indeed, the equilibrium may not consist of a static, fixed state, but may take the form of an (almost) steady periodic orbit, a more complicated "strange attractor," or even "chaotic" behavior. Thus, the Ornstein–Uhlenbeck model is no more than one possibility, but it can easily occur, and it can be verified easily in practice, because it depends on only two parameters.

So far, in this discussion of randomness at equilibrium, the emphasis has been on fluctuations about equilibrium that are small when compared to the average population density. What happens if this is not the case? The answer is (biologically) both simple and disturbing: extinction. Obviously, a population cannot regenerate itself from a value of zero, and any chance fluctuation that reduces the population to zero results in extinction. For populations in which the equilibrium density is not large compared to the size of the typical fluctuations about equilibrium, such a chance event occurs relatively quickly, and the population becomes extinct thereafter – indeed, one cannot in such circumstances reasonably talk about "equilibrium"; the natural limit is extinction, even though r_0 may be greater than 0 (or, equivalently, $m_0 > 1$).

Technically speaking, the same is true even when the random fluctuations about the equilibrium density are relatively small. Let the random process run indefinitely, and at some time a fluctuation will occur that is large enough to bring the population to extinction. Indeed, all individuals simultaneously having poor reproductive success for many generations or a long period of adverse environmental conditions might be enough for this. However, as long as the typical fluctuations in the population density are small enough, the expected length of time until such an event occurs is very long; no-one seriously expects the human race to die out in the foreseeable future just as a result of no-one choosing to have children, although theoretically this could happen. Until such a chance event occurs, the fluctuations in population density are as if this theoretical possibility did not exist, at least if the branching process approximation in small populations has $m_0 > 1$. Such behavior is called quasi-stationary. It is indistinguishable from a true stochastic equilibrium over long time periods, and exhibits just the same sort of stability in the distribution

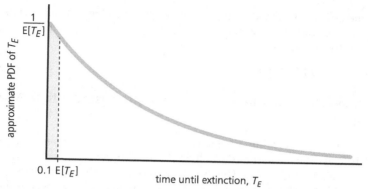

Figure 6.2 Approximate probability density function (PDF) of the time until extinction T_E. The shaded area represents the probability of the event $\{T_E \leq \frac{1}{10}\mathbb{E}T_E\}$.

of fluctuations about the equilibrium density, even though the theoretical possibility of extinction exists.

A quantity of critical interest is therefore the expected length of time that a quasi-equilibrium is likely to persist. Is the system effectively stable, in the sense that the expected time to extinction is longer than the expected lifetime of the planet, or is extinction likely to occur in the next few years? Endangered populations are those for which the latter is the case; the longer the expected time to extinction, the less threatened the population. Unfortunately, the expected time to extinction is a quantity that depends very much on the detail of the random processes that govern the population size. Simple rules of thumb are unreliable, except perhaps as qualitative guides, and should always be supplemented by (computer) calculations within the actual model. This said, in a model with natural carrying capacity $\overline{K}$, "equilibrium" variance V, and relaxation rate β, the time until a fluctuation below a level $\overline{K} - k\sqrt{V}$ occurs, for k "large, but not too large," has an expectation that is, roughly,

$$(\sqrt{2\pi}/k\beta)\exp\{\tfrac{1}{2}k^2\} ; \tag{6.133}$$

k needs to be large because the formula is an asymptotic formula for a particular, Ornstein–Uhlenbeck diffusion model, yet k must not be too large, so that this diffusion model remains a reasonable approximation to the true model. And, as observed above, if an Ornstein–Uhlenbeck approximation is reasonable, both V and β can, in principle, be estimated from the observed autocovariances of the population density.

The main interest in Formula (6.133) is in the dependence on k, the number of standard deviations represented by the magnitude of the fluctuation being considered; the expected time until such a random fluctuation occurs increases as the exponential of the *square* of k. This helps to explain why quasi-equilibria are very much a fact of life; they can really persist for very long times, if the conditions are suitable. However, a warning is also necessary here; the approximate distribution

190   *Branching Processes: Variation, Growth, and Extinction of Populations*

of the time until extinction T_E is typically exponential, so that an actual extinction time of one-tenth of the expected time to extinction occurs with a probability of about $1/10$ (see Figure 6.2),

$$\mathbb{P}[T_E \leq \tfrac{1}{10}\mathbb{E}T_E] \approx 1/10 . \qquad (6.134)$$

6.9 Quasi-stationary Behavior in a Simple Discrete-time Model
G. Högnäs

In many population-modeling situations we start by assuming certain properties or rules of behavior on the part of the individual and extrapolate these to formulate a model for, say, the population density. While this approach has many undeniable advantages – it may be the only practicable way of analysis – we lose something in the process as well.

In this section we look at some aspects of modeling individuals versus population densities. As they should, the two formulations yield results that are similar in many respects, but some of the analogies are not so clear-cut and need a subtler mathematical treatment. An introduction to, as well as a motivation for, the study is given in Section 1.4. Klebaner (1993), Klebaner and Nerman (1994), and Klebaner *et al.* (1998) found similar results as those presented here, on related models.

Consider a homogeneous population that consists of a single species. We are interested in modeling the total number of individuals at discrete time-points $t = 0, 1, 2, \ldots$ as a simple random process. Assume that the population dynamics is governed by the size of the total population only. We do not, for example, differentiate between individuals by age; one could think of certain insect populations in which all individuals, in a sense, belong to the same generation. (Another convenient way to think of is that we model the number of individuals in generation $t + 1$, which depends on the size of the preceding tth generation only.) This influence on the population is summarized in a *constant* environmental parameter b. If the total population in generation t is small, almost all the individuals produce offspring. We assume that the actual offspring sizes of the individuals are independent random variables. When the total population Z_t grows larger, individual reproduction declines through competition or cannibalism, so that the probability of an individual not reaching maturity (and thus producing no offspring) is $1 - \exp(-bZ_t)$, as argued in Section 1.4. We assume, however, that the individual offspring distribution *on condition of survival until maturity* remains one and the same $p_0, p_1, p_2, \ldots$.

Thus, the population grows by a *size-dependent branching* mechanism, as follows. If the size of the population at time t is $n > 0$, then each individual, independently of the others, does not survive to produce offspring (or is otherwise unable to produce offspring) at time $t + 1$ with probability $1 - \exp(-bn)$, and produces k offspring with probability $e^{-bn}p_k$, where $p_0, p_1, p_2, \ldots$ is the given offspring distribution. If the size of the population ever hits 0, the population is extinct.

Note. The sole source of randomness is *demographic stochasticity*, the variation in the actual offspring of individuals who are assumed identical in all other respects.

◇ ◇ ◇

Denote by Z_t the total number of individuals in the tth generation. We can write our model in the form

$$Z_{t+1} = \begin{cases} \sum_{j=1}^{Z_t} \xi_{j,t}, & \text{if } Z_t > 0 \\ 0, & \text{if } Z_t = 0 \end{cases} \quad t = 0, 1, 2, \ldots, \tag{6.135}$$

where $\xi_{1,t}, \xi_{2,t}, \ldots, \xi_{Z_t,t}$ are mutually independent identically distributed random variables. The common distribution depends on Z_t, however: $\mathbb{P}(\xi_j = k \mid Z_t = n) = \exp(-bn)p_k$ for $k = 1, 2, \ldots$, and $\mathbb{P}(\xi_j = 0 \mid Z_t = n) = 1 - \exp(-bn)(1 - p_0)$. (Here $n > 0$ and $j = 1, 2, \ldots, n$.)

We see immediately that the process in Equation (6.135) has two salient features (which are characteristic of a good population model according to Hassell 1974):

1. The population has the potential to increase exponentially for small populations;
2. but there is a density-dependent feedback that reduces the actual rate of increase as the population grows.

To avoid non-essential technical difficulties we take $0 < p_1 < 1$. In addition, we assume throughout that the offspring distribution has an exponential moment, that is, the moment generating function (MGF) $\mathbb{E}[e^{s\xi}]$ is finite for some positive s. (This implies, for example, that all moments $\mathbb{E}[\xi^k]$ are finite.) The mean value $\sum_{k=0}^{\infty} k p_k$ of the offspring distribution is denoted by m. We make the natural, but important, assumption that $m > 1$.

The process in (6.135) is a Markov chain on the non-negative integers. 0 is absorbing and the positive integers $\{1, 2, 3, \ldots\}$ form one communicating class (because of our assumptions $m > 1$ and $0 < p_1 < 1$). This means that all positive integer values are, in principle, possible for Z_t, no matter what initial value $Z_0 > 0$ is chosen.

For comparison purposes we prefer to work with the normed chain bZ_t, with state space the set of non-negative multiples of b. Denote with E the set of positive multiples $\{b, 2b, \ldots\}$. Again, regardless of starting value in E, the process bZ_t may visit any other state in E. The transition probability matrix restricted to $E \times E$ is denoted by Q. The matrix Q depends on b, of course, but we usually suppress the explicit reference to b. Q is substochastic, that is, the probability of staying inside E is < 1. This is the same as saying that there is a positive probability of extinction at each time step.

We use the notation $\mathbb{P}_x[A]$ for the conditional probability of the event A given that $bZ_0 = x$ (i.e., the Markov chain is started at the value x).

We have, for $x, y \in E$,

$$Q_{xy} = \mathbb{P}[b(\xi_{1,t} + \xi_{2,t} + \cdots + \xi_{Z_t,t}) = y \mid bZ_t = x] = \mathbb{P}_x[bZ_1 = y]. \tag{6.136}$$

The conditional mean of bZ_{t+1}, given that $bZ_t = x$, explains why it is appropriate to call the model a stochastic Ricker model.

$$\mathbb{E}[bZ_{t+1} \mid bZ_t = x] = b \cdot \frac{x}{b}\mathbb{E}[\xi_{1,t} \mid bZ_t = x] = xe^{-x}m \equiv f(x). \quad (6.137)$$

It turns out that for an arbitrary, but fixed, time t, bZ_t approaches the deterministic process of the form

$$Z_{t+1} = mZ_t \exp(-Z_t) \equiv f(Z_t), \quad (6.138)$$

as the environmental influence decreases (b here goes to 0). More precisely, the limit of bZ_t (with $bZ_0 = x$) is $f^t(x)$, the tth iterate of $f(x) = mxe^{-x}$ evaluated at x.

This is very natural indeed: the conditional variance of bZ_1 given $bZ_0 = x$ is of the order of $bx \exp(-x)$ so, for small $b > 0$, bZ_1 is likely to lie close to its mean $f(x)$ [and, similarly, bZ_2 is close to the second iterate $f^2(x) = f(f(x))$, etc.].

Remark. When the environmental parameter b is small, the process should be almost a Galton–Watson branching process. This is true, but only as long as Z_t remains small (and the normed process bZ_t is near zero). When bZ_t becomes moderately large the inhibitive influence asserts itself and the process no longer looks like a Galton–Watson process.

◇ ◇ ◇

The row sums of Q are always strictly less than 1. In fact, the sum s_x of the xth row of Q is given by

$$s_x \equiv \sum_{y\in E} Q_{xy} = 1 - (1 - e^{-x}(1 - p_0))^{\frac{x}{b}}, \quad x \in E \quad (6.139)$$

(the conditional probability of $bZ_{t+1} \neq 0$, given that $bZ_t = x$). For large x, $s_x \approx \frac{x}{b}e^{-x}(1 - p_0)$. Hence the row sums stay at a positive distance from 1. This implies that the chain is absorbed at 0 regardless of where it started. Clearly, the only stationary probability distribution for the chain bZ_t is δ_0, the point mass at 0. For a Markov chain, a *stationary distribution*, call it ρ, is sustainable or steady state in the sense that if Z_0 is distributed according to ρ, so are $Z_1, Z_2, Z_3, \dots$. In other words, the whole Markov chain has distribution ρ.

While the short-term dynamics of bZ_t are fairly well approximated by the deterministic process of (6.138), the (very) long-term asymptotics are widely different: the deterministic process never dies out and there is a large number of different possible scenarios dependent only on the choice of initial value (see Box 6.3).

There is, however, a kind of medium-term description of the Markov chain bZ_t that fits neatly here. A remarkable feature of the deterministic dynamic system (with $m \ll m_c$, see Box 6.3) is that orbits from most starting points x eventually end up in (an arbitrarily small neighborhood of) *the unique attracting periodic cycle*. For small b the so-called *quasi-stationary distribution*, which may be viewed as the limiting conditional distribution of bZ_t *under the condition that it is not yet extinct*, is a good approximation of the uniform distribution on the attracting periodic cycle of Equation (6.138).

Note. For population density models, properties 1 and 2 above lead to models of the form $Z_{t+1} = f(Z_t)$ for unimodular functions f ["one-humped" functions, increasing from the origin up to some point c and decreasing on (c, ∞)]. In fisheries management, f is called a *stock-recruitment function*. Examples of such models are the Ricker (1954) model, $Z_{t+1} = m Z_t \exp(-b X_t)$, and the Hassell (1974) model,

$$Z_{t+1} = \frac{r Z_t}{(1 + Z_t)^b} , \tag{6.140}$$

where m and r represent the intrinsic growth rate for small population densities and b represents the inhibitive density-dependent feedback, usually attributed to the environment. A somewhat similar model is the much-studied *logistic* one

$$Z_{t+1} = r Z_t (1 - \frac{Z_t}{K}), 0 < r \leq 4, 0 < Z_t < K , \tag{6.141}$$

where the maximal value K is the *carrying capacity*.

The dynamic systems generated by unimodular functions f, as above, exhibit a rich variety in their behavior (see Box 6.3). The short-term evolutions of these models are rather similar, but unfortunately the asymptotic behavior depends strongly on detailed knowledge of f.

$$\diamond \diamond \diamond$$

A probability measure π on E is a *quasi-stationary distribution* for the Markov chain $b Z_t$ absorbed at 0 if

$$\pi_y = \frac{\sum_{x \in E} \pi_x Q_{xy}}{\sum_{z \in E} \sum_{x \in E} \pi_x Q_{xz}}, \quad y \in E . \tag{6.142}$$

As indicated above, a probabilistic interpretation of the quasi-stationary distribution π can also be given. If the initial distribution (the distribution of $b Z_0$) is π, then the conditional probability distribution after the first step, given that it is non-zero, is also π,

$$\mathbb{P}_\pi[b Z_1 = x \mid b Z_1 \neq 0] = \pi_x, \quad x \in E , \tag{6.143}$$

where we use the notation $\mathbb{P}_\pi$ to indicate that the initial distribution is π (see Ferrari *et al.* 1992). Thus, the quasi-stationary distribution has a probabilistically (and biologically!) appealing interpretation: it is an *equilibrium distribution conditional on non-extinction*.

Suppose that m is fixed and such that the deterministic Ricker model in Equation (6.138) admits an attracting periodic cycle (e.g., take m small, less than $m_c \approx 14.76$). Then there exists a quasi-stationary distribution $\pi(b)$ for the Process (6.135), the notation indicating that the quasi-stationary distribution, of course, depends on b.

Then, as $b \to 0$, the quasi-stationary distribution $\pi(b)$ approaches the uniform distribution on the unique attracting periodic cycle of the deterministic process of Equation (6.138). Furthermore, the mean time to extinction (the lifetime of the process) is exponential in $\frac{1}{b}$.

Box 6.3 Asymptotic behavior

We give here a description of the asymptotic behavior of the Ricker model, Equation (6.138); the behavior of the Hassell model is basically the same.

We first rewrite f in the form $f(x) = x \exp(r - x)$, commonly used in the dynamic systems literature. We observe [see May (1976) and Devaney (1989) for the terminology] that the asymptotics of the process given by Equation (6.138) is then governed by the parameter $r = \ln(m)$. If $1 < r < 2$, there is a stable or attracting fixed point, an *attractor* p ($= r$), and all trajectories of the process in Equation (6.135) converge geometrically fast to the fixed point. At $r_1 = 2$ there is a *bifurcation* into an attracting periodic cycle of length 2. For $r = 2$ the trajectories continue to converge to the fixed point p, but slower than geometrically; p is said to be *weakly attracting*. For $r_1 < r < r_2 \approx 2.526$ there is an attracting two-cycle, which attracts all orbits *except* for those that lead in a finite number of steps to the now repelling fixed point p. When $r = r_2$ we still have convergence, but slower, of the "typical" trajectories to the two-cycle.

At r_2, the two-cycle $\{P_1, P_2\}$ becomes repelling and an attracting cycle of period four appears. Again the trajectories converge to the attracting four-cycle, *except* those that lead to the repelling cycles of periods 1 and 2 in a finite number of steps.

Numerical studies suggest (see May 1976) that the bifurcations continue and produce, successively, periodic cycles of periods 8, 16, 32, ... until at $r_c \approx 2.692$ the system in (6.135) admits *all* periods of length 2^n, with n a non-negative integer. [The corresponding $m_c = \exp(r_c)$ is about 14.76.]

If $r \ll r_c$ we note that:

■ There is exactly one attracting (or weakly attracting, if r happens to be exactly at a bifurcation point) periodic cycle;
■ There are finitely many periodic cycles the lengths of which are 2^k, $k = 0, 1, 2, \ldots, n$.

The trajectories all converge to the (weakly) attracting periodic cycle, except for those that start from the inverse images of the repelling periodic points.

For $r > r_c$ there appear to be infinitely many periodic cycles, all of which may be repelling. There are certain parameter intervals (*periodic windows*) for which the system admits attracting periods. In these cases the prime period is not restricted to multiples of 2.

Finally, we note that the arithmetic mean of all the periods, stable or unstable, equals r, the fixed point for the mapping $x \exp(r - x)$.

If $T = T(b)$ is the time to extinction, the latter statement means that there are some positive constants c, C independent of b, so that

$$\exp(c\frac{1}{b}) < \mathbb{E}[T] < \exp(C\frac{1}{b}) . \tag{6.144}$$

Thus, even for moderately small b, the lifetime is very long, indeed.

Let us first examine the time to extinction of bZ_t. By Equation (6.139) we see that the probability of extinction from x in one step is

$$1 - s_x = (1 - e^{-x}(1 - p_0))^{\frac{x}{b}} . \tag{6.145}$$

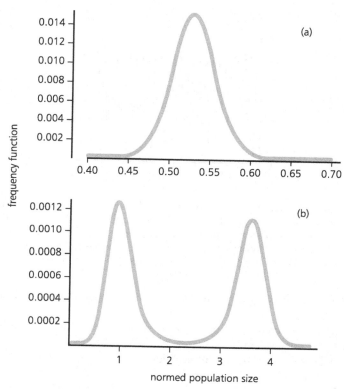

Figure 6.3 The quasi-stationary distributions of bZ_t with (a) $m = 1.7$ ($r = 0.5306$) and (b) $m = 10.0242$ ($r = 2.3050$). The offspring distribution is concentrated on three points (a) and an approximate geometric distribution (b). The values of b are 0.001 and 0.0016, respectively. For $r = 0.5306$ the deterministic dynamic system, Equation (6.137), has a stable fixed point at r. For $r = 2.3050$ the deterministic system has a stable two-periodic cycle {0.9314, 3.6786}.

To obtain an intuitive understanding of it, imagine that the process stays in a small set, a small neighborhood of the attracting cycle, until extinction. Let $d \equiv (1 - e^{-c}(1 - p_0))^c$ be the maximum of the function $(1 - e^{-x}(1 - p_0))^x$ over this set. Then we see that the time T to extinction is bounded above by a geometrically distributed time with mean $d^{\frac{1}{b}}$. Likewise, we obtain a lower bound of the same form.

It is remarkable that the process bZ_t approaches the attracting periodic cycle, even if it happens to start exactly on an unstable periodic point. The deterministic process that starts there is, of course, stuck in the unstable period, but the demographic stochasticity "shakes bZ_t loose" from the unstable period. The time it remains close to the unstable period is of the order of $\ln(\frac{1}{b})$ which, for small b, is a very short instance compared to the time to extinction (Högnäs 1997). The

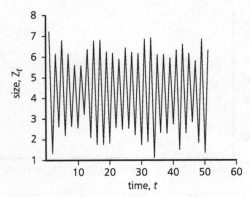

Figure 6.4 A small perturbation of a two-periodic deterministic system. Simulation of $Z_{t+1} = Z_t \exp(r - b_{t+1}Z_t)$, where $\ln m = r = 2.3050$ and b_t is independent and uniform on the interval [0.5,0.6]. The theoretical mean $\mathbb{E}[Z_t] = \frac{r}{\mathbb{E}[b_t]} = 4.1909$, Gyllenberg *et al.* (1994). The deterministic system $Z_{t+1} = m Z_t \exp(-b Z_t)$ has the attracting two-period $P_1 = \frac{0.9314}{b}$, $P_2 = \frac{3.6786}{b}$.

distribution of bZ_t is often very quickly close to the "conditional equilibrium distribution" $\pi(b)$.

In this sense, the stochastic process bZ_t is much smoother and more robust than its deterministic counterpart.

Let us elaborate on Figure 6.3 further. For $r = 0.5306$ the deterministic model of Equation (6.138) has an attracting fixed point at 0.53. If we observe the successive values of Z_t, they all eventually fall very close to 0.53. If we carry out long-term statistics on Z_t, we obtain a delta spike at 0.53. The quasi-stationary distribution of the stochastic model in Equation (6.135) is much broader than a point mass, but, when b decreases, the quasi-stationary distribution becomes higher and thinner, and increasingly concentrated around 0.53.

If $r = 2.3050$, the deterministic model of Equation (6.138) has an attracting two-cycle, so asymptotically Z_t oscillates between the two values 0.9314 and 3.6786. The stationary distribution is concentrated around these two points, which gives them the mass $\frac{1}{2}$ each. The corresponding frequency function diagram consists of two delta spikes in 0.9314 and 3.6786. The stochastic model with $r = 2.3050$ has a quasi-stationary distribution that may be seen as an approximation of this two-point distribution. If b is very small the two humps around 0.9314 and 3.6786 are very thin and very high. The heights of the two humps are approximately the same and the probability mass under the two humps is very close to $\frac{1}{2}$ each.

Note. For *fixed*, moderate $b > 0$, the time to extinction decreases dramatically with increasing m. In these cases the deterministic model oscillates wildly between very large and near-zero densities. But as $b \to 0$ we nevertheless obtain the same asymptotics [on condition that m lies in a periodic window (Ramanan and Zeitouni 1999), see Box 6.3].

◇ ◇ ◇

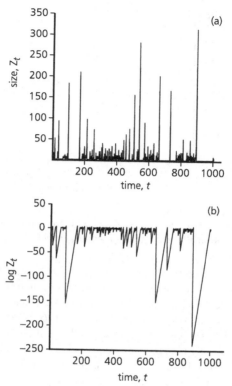

Figure 6.5 Growth–catastrophe behavior. Simulation of $Z_{t+1} = Z_t \exp(r - b_{t+1}Z_t)$, where $\ln m = r = 2.3050$ and b_t is independent and uniform on the unit interval. (a) Natural scale and (b) logarithmic scale.

The next modification is to introduce environmental stochasticity into our branching model. The discussion above indicates that it might be worthwhile first to look at what happens in the long run in density models with a random environment.

Models for population density with a random environmental parameter b were investigated by Gyllenberg *et al.* (1994) and Vellekoop and Högnäs (1997), where

$$Z_{t+1} = mZ_t \exp(-b_{t+1}Z_t) , \tag{6.146}$$

and the sequences b_t are independent and uniform on $[0, \infty)$.

If b is allowed to vary in a small closed interval that does not contain 0, the behavior of the process in Equation (6.146) differs very little from that of the deterministic system in (6.138) (see Figure 6.4).

The periodic behavior of the deterministic system is destroyed easily by moderate variations in b, however. When the parameter varies widely (by orders of magnitude) the *growth–catastrophe* phenomenon appears. This behavior is characterized by rapid growth, followed by a huge decline in one step to near-zero

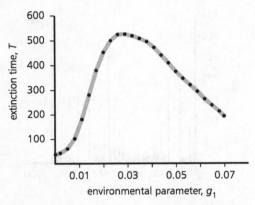

Figure 6.6 Observed extinction times for simulated Z_t as in Equation (6.147), with the environmental parameter taking two values g_1, g_2 with equal probability. The horizontal axis gives the value of the smaller parameter g_1 while g_2 is held fixed at 0.07.

density. From this very low level the process recovers almost deterministically as the inhibitive term $b_{t+1}X_t$ in the exponent is much less than $\ln m$ (see Figure 6.5). It was shown in Gyllenberg *et al.* (1994) and Vellekoop and Högnäs (1997) that if b_t is strictly positive, the stochastic process of (6.146) is positive recurrent (i.e., there exists an invariant probability measure on the positive real line). This result remains true even if b_t may take the value 0 provided the probability is small (Gyllenberg *et al.* 1994). Recall that $b = 0$ means that the environment is neutral, and so exerts no inhibitive influence on the growth of the population density.

Let us return to the size-dependent branching model with the additional feature of independent environmental stochasticity expressed through an independent sequence of parameters b_t.

The full model is thus

$$Z_{t+1} = \begin{cases} \sum_{j=1}^{Z_t} \xi_{j,t}, & \text{if } Z_t > 0 \\ 0, & \text{if } Z_t = 0 \end{cases} \quad t = 0, 1, 2, \ldots, \tag{6.147}$$

where $\xi_{1,t}, \xi_{2,t}, \ldots, \xi_{Z_t,t}$ are mutually independent identically distributed random variables. The common distribution depends on both Z_t and b_{t+1}: $\mathbb{P}[\xi_j = k \mid Z_t = n, b_{t+1} = g] = \exp(-gn)p_k$ for $k = 1, 2, \ldots$ and $\mathbb{P}[\xi_j = 0 \mid Z_t = n, b_{t+1} = g] = 1 - \exp(-gn)(1 - p_0)$, $n > 0$ and $j = 1, 2, \ldots, n$. b_{t+1} is assumed independent of $Z_0, Z_1, \ldots, Z_t$.

When the random variable b is confined to a closed interval that does not include 0, the qualitative results are similar to the case of fixed b. The process (6.147) admits a quasi-stationary distribution and the extinction time is roughly exponential in $\frac{1}{\mathbb{E}[b]}$. (A very large relative variation in b shortens the time, though; see Figure 6.6).

At the other extreme we have the case in which b may take the value 0 with a positive probability p, say. The process of Equation (6.147) exhibits a growth–catastrophe behavior that leads to short extinction times, as a long period of growth

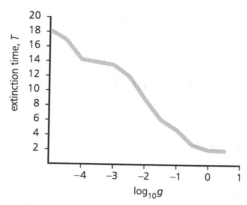

Figure 6.7 Observed times of extinction T for simulated Z_t as in Equation (6.147), with the environmental parameter taking the values 0 and g with probability $\frac{1}{2}$ each. T is plotted against $\log_{10} g$.

makes the population too large, which results in extinction when the environmental variable b happens to be a little larger. In such cases the time of extinction increases very slowly, polynomially in $\frac{1}{\mathbb{E}[b]}$, as $\mathbb{E}[b] \to 0$ (see Figure 6.7). It is conjectured that if p is large enough ($> \frac{1}{2}$?), then the process fails to admit a quasi-stationary distribution altogether.

7

Specific Models

7.1 Coalescent Processes: Reversed Branching

P. Jagers and S. Sagitov

7.1.1 Coalescence

Branching viewed backward is *coalescence*, the process of merging or clumping. It arises naturally in the structuring or formation of dispersed matter of various kinds and in various scales, from that of molecule aggregates in colloids (so-called *micelles*) to galaxies. It has been studied by physicists through computer simulation (e.g., Nilsson *et al.* 2000) and in a series of interesting mathematical articles by Aldous (1999).

Evolution can be viewed as a grand multi-type branching process, with new species that arise through mutation (see Jagers 1991; Jagers *et al.* 1992; Taib 1992). The study of the origin of species is then time-reversed branching (i.e., coalescence). In genetics, the latter is also used to trace the roots of the genetic composition of populations and its development. It was within this area that the first pure coalescence model, the Kingman coalescent (1982a), was formulated as a reverse counterpart to the diffusion approximation of the renowned Wright–Fisher model (Fisher 1930; Wright 1931; see also Ewens 1979).

The object of genetic models is thus population composition rather than size. Indeed, most population genetics even assumes that population size is completely constant over generations. As we show later, the Wright–Fisher model can be obtained as Galton–Watson branching with Poisson reproduction, conditioned at a constant population size. In the same vein, most population genetics simplifies the flow of time into generation counting. Instead, it is lineage that counts. What are the relations among n individuals sampled out of a total population of size N?

For (diploid) genetics mating also cannot be disregarded. It is usually modeled as random, that is, the two genes of an individual are selected independently at random from any individual in the preceding generation. (The latter are thus, implicitly or explicitly, thought of as hermaphroditic.) Most of the time genes are passed unchanged, but sometimes new alleles appear through mutation. We consider only *neutral* mutations, so that the reproduction law remains one and the same throughout the process. In such cases, what is the distribution of alleles at a locus?

If mating is thus assumed to be random and population size is thought to be given, gene branching might be studied and simulated directly, with individual reproduction behavior bypassed by means of a general exchangeable population model (to be further discussed in Section 7.1.4). A vector of random variables is said to be *exchangeable* if any permutation of its components has the same distribution. In symbols, if $i_1, i_2, \ldots, i_N$ is any reordering (permutation) of the numbers $1, 2, \ldots, N$, then the vector $(v_{i_1}, v_{i_2}, \ldots, v_{i_N})$ has the same joint distribution function as has $(v_1, v_2, \ldots, v_N)$. In an exchangeable population model N is the size of the reproducing generation and $v_1 + v_2 + \cdots + v_N$, which is usually also required to equal N, is the size of the following generation.

7.1.2 The Wright–Fisher population model

The best-known exchangeable reproduction model is the *Wright–Fisher* population. It depicts a population in (an imagined) equilibrium in which the population size N (the number of genes) is constant over generations, time thus being discrete. It is assumed that gene number i in the current generation is passed on to the next generation in v_i copies so that the joint distribution of gene offspring sizes $(v_1, \ldots, v_N)$ is symmetrically multinomial. In line with earlier notation, for binomial, geometric, and Poisson random variables this can be denoted by the vector being multinomial $(N, \frac{1}{N}, \ldots, \frac{1}{N})$ (see the Appendix).

A remarkable feature of the Wright–Fisher model, which emphasizes its concentration upon lineage at the expense of size fluctuations, is the mentioned interpretation of it as a Galton–Watson process with Poisson reproduction law, conditioned to have constant population sizes. Indeed, if the offspring distribution is Poisson(m) and $\xi_1, \xi_2, \ldots$ are the numbers of offspring of the individuals in the nth generation (somehow numbered), then provided $i_1 + i_2 + \cdots + i_N = N$,

$$
\begin{aligned}
&\mathbb{P}(\xi_1 = i_1, \xi_2 = i_2, \ldots, \xi_N = i_N) \\
&= \mathbb{P}(\xi_1 = i_1, \xi_2 = i_2, \ldots, \xi_N = i_N \mid \xi_1 + \cdots + \xi_N = N) \\
&= \frac{\frac{m^{i_1}}{i_1!}e^{-mi_1}\frac{m^{i_2}}{i_2!}e^{-mi_2} \cdots \frac{m^{i_N}}{i_N!}e^{-mi_N}}{\frac{(mN)^N}{N!}e^{-mN}} = \frac{N!}{i_1!i_2!\ldots i_N!}\left(\frac{1}{N}\right)^N .
\end{aligned} \tag{7.1}
$$

Otherwise the probability is obviously equal to zero.

The reverse description of how Wright–Fisher populations evolved is particularly simple: the N children are partitioned into sibships with multinomially distributed sizes $(v_1, \ldots, v_N)$ by letting them select their parents independently at random from among the N individuals that constitute the preceding generation. This leads to a simple algorithm to simulate n alleles sampled from a Wright–Fisher population:

1. Build an ancestral tree for n sampled genes that result from n coalescing random walks over the set of N points;
2. Impose mutations on the ancestral tree.

The reverse approach renders simulation faster because it deals only with the fate of direct ancestors of the sampled genes. Figure 7.1 depicts a possible outcome of the reverse simulation algorithm with $n = 6$. Mutations are numbered in time sequence and the resultant six alleles are labeled according to the mutations that

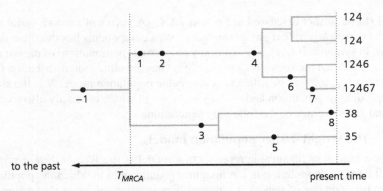

Figure 7.1 An example of simulating six sequences using the Kingman coalescent. Dots stand for mutations numbered in their order of occurrence.

created them. Mutations prior to the *most recent common ancestor* (MRCA) do not contribute to the sample polymorphism.

As $N \to \infty$, the forward picture of the Wright–Fisher model can be simplified by a diffusion approximation. However, there is a reverse counterpart to this diffusion approximation, called the Kingman coalescent (Kingman 1982a). It is an asymptotic ancestral process and the subject of Section 7.1.3. It is a basic coalescent process with several extensions that reflect such evolutionary and demographic features as recombination, population subdivision, variable population size, and selection. A reader interested in the theory of coalescent processes and its applications in population genetics is referred to Chapters 6–8 of Balding *et al.* (2001). A recent review by Rosenberg and Nordborg (2002) contains further references. Important earlier surveys are Tavaré (1984), Hudson (1991), and Donnelly and Tavaré (1995).

7.1.3 The Kingman coalescent

Consider the genealogical tree of n genes sampled at random from a Wright–Fisher population with constant population size N. Let R_k^N be the number of branches in the tree k generations before the generation sampled. For a fixed N the sequence $(R_k^N)_{k=0,1,\ldots}$, the *ancestral process*, has monotone trajectories that decrease from $R_0^N = n$ down to $R_T^N = 1$ at the random time $T = T_{MRCA}$, when the tree reaches the MRCA of the sampled genes. It is a Markov chain and it has been shown (Watterson 1975) that its transition probabilities have the form

$$\mathbb{P}(R_{k+1}^N = j \mid R_k^N = i) = c_{i,j} N(N-1) \cdots (N-j+1)/N^i , \qquad (7.2)$$

where $c_{i,j}$ are combinatorial constants independent of N (actually, so-called Stirling numbers of the second kind). In particular, $c_{i,i} = 1$ and $c_{i,i-1} = \binom{i}{2} = i(i-1)/2$, so

$$\mathbb{P}(R_{k+1}^N = i \mid R_k^N = i) = 1(1 - 1/N) \cdots (1 - (i-1)/N)$$
$$= 1 - (1 + 2 + \cdots + i - 1)/N + o(1/N)$$
$$= 1 - i(i-1)/(2N) + o(1/N), \tag{7.3}$$

as $N \to \infty$. Similarly,

$$\mathbb{P}(R_{k+1}^N = i - 1 \mid R_k^N = i) = i(i-1)/(2N) + o(1/N), \tag{7.4}$$

and the remaining probabilities

$$\mathbb{P}(R_{k+1}^N = j \mid R_k^N = i) = o(1/N), \, j \le i - 2. \tag{7.5}$$

If N is large (as compared with n^2), then Equation (7.3) means that the chain R_k^N stays in the same state for many generations, until it jumps to a new state. The number of generations between two jumps should be comparable with the population size N. If such a jump takes place, in view of Equation (7.5), the probability is high that the number of branches in the gene tree reduces by just one. This corresponds to a pairwise coalescent. Equations (7.3)–(7.5) imply the following convergence result (cf. Kingman 1982a), which suggests a remarkably convenient way to simulate the gene tree of a large Wright–Fisher population.

Theorem 7.1 *If R_k^N is the number of branches in the gene tree of a Wright–Fisher population with population size N, then as $N \to \infty$ the process*

$$R_{[Nt]}^N \to R_t, \, t \ge 0, \, N \to \infty. \tag{7.6}$$

Here $[Nt]$ means the integral part of Nt (i.e., the largest integer that does not exceed Nt), and the limit process R_t, to be described below, is the *Kingman coalescent*. The arrow stands for so-called weak convergence, which is actually a fairly strong convergence concept in stochastic processes. For an introduction to this, see more advanced probability literature (Billingsley 1968). Suffice it here to say that it, usually, implies convergence of the distributions at a fixed t (problems possibly arise at values of t where the limit has a jump with positive probability). However, it also implies convergence of distributions of vectors,

$$(R_{[Nt_1]}^N, R_{[Nt_2]}^N, \ldots, R_{[Nt_r]}^N) \to (R_{t_1}, R_{t_2}, \ldots, R_{t_r}), \tag{7.7}$$

and even of functionals like the supremum of the process.

The Kingman coalescent is a Markov process that describes the asymptotic distribution of a gene tree for large Wright–Fisher populations. The process R_t yields the number of branches in the Kingman coalescent. The theorem shows that time t in the limit tree corresponds to $[Nt]$ generations in the Wright–Fisher population, counted backward in time. Equations (7.3)–(7.5) for the ancestral process are transferred into the defining properties of the Kingman coalescent,

$$\mathbb{P}(R_{t+\Delta} = j \mid R_t = i) = q_{ij} \cdot \Delta + o(\Delta), \, j \ne i, \tag{7.8}$$

and

$$\mathbb{P}(R_{t+\Delta} = i \mid R_t = i) = 1 - \lambda_i \cdot \Delta + o(\Delta), \ \Delta \to 0, \tag{7.9}$$

where

$$q_{ij} = \begin{cases} \binom{i}{2} & \text{if } j = i - 1, \\ 0 & \text{otherwise}, \end{cases} \tag{7.10}$$

and $\lambda_i = \sum_{j \neq i} q_{ij}$ is equal to

$$\lambda_i = \binom{i}{2} = i(i-1)/2. \tag{7.11}$$

These relations lead to a simple wait-and-jump description of the limit process R_t. First, according to Equation (7.9) the waiting time T_i spent by the limit process at the state i is distributed exponentially with parameter λ_i, so that

$$\mathbb{E}[T_i] = \frac{2}{i(i-1)}, \ \text{Var}[T_i] = \frac{4}{i^2(i-1)^2}, \ i \geq 2. \tag{7.12}$$

Second, in view of Equations (7.10) and (7.11), the jump probabilities $p_{ij} = q_{ij}/\lambda_i$ satisfy

$$p_{ij} = \begin{cases} 1 & \text{if } j = i - 1, \\ 0 & \text{otherwise}. \end{cases} \tag{7.13}$$

This means that the number of branches will be $i - 1$ immediately after it was i. In other words, the limit process of Theorem 7.1 is what is known as a *pure death process*, which consecutively visits all the states $\{n, n - 1, \dots, 2, 1\}$ with waiting times as described.

7.1.4 Exchangeable population models

The Kingman coalescent was derived as a limit of a very special, albeit classic, genetic model, the Wright–Fisher model. The convergence result of Theorem 7.1 can, however, be extended to a broad class of population models, which thus widens the applicability of the coalescent far beyond Wright and Fisher's framework.

Exchangeable population models (cf. Cannings 1974) generalize the Wright–Fisher setup by allowing the offspring sizes $(\nu_1, \dots, \nu_N)$ to have any symmetric (exchangeable) joint distribution. Exchangeability, as pointed out, means that the joint distribution does not change if the family indices are permuted. This implies that all offspring sizes $\nu_1, \dots, \nu_N$ have the same distribution and therefore the same expectation and variance. Since

$$N = \mathbb{E}[\nu_1 + \cdots + \nu_N] = \mathbb{E}[\nu_1] + \cdots + \mathbb{E}[\nu_N], \tag{7.14}$$

the common expected offspring number must be $\mathbb{E}[\nu_i] = 1$. We denote the variance by σ_N^2.

Exchangeability of gene inheritance is natural under the assumption of only neutral mutations. This class of population models also encompasses all Galton–

Watson processes conditioned to have a constant population size N in all generations.

Kingman (1982b) presents two simple conditions, namely

$$\sigma_N^2 \to \sigma^2, \ N \to \infty, \ 0 < \sigma^2 < \infty, \tag{7.15}$$

and

$$\sup_N \mathbb{E}[\nu_1^k] < \infty, k \geq 3, \tag{7.16}$$

which ensure a convergence to the coalescent,

$$R_{[Nt/\sigma^2]}^N \to R_t, \ N \to \infty, \tag{7.17}$$

for the ancestral process R_k^N of a population with exchangeable reproduction. The limit process is again the Kingman ancestral process, while the ancestral process of the exchangeable population is accelerated by a factor of σ^2, which reflects that the ancestral lines merge faster in a population with greater variation in offspring size.

Minimal conditions for convergence to the Kingman coalescent were found by Möhle and Sagitov (1999). A simplified version of their result is given in Theorem 7.2.

Theorem 7.2 *Let* $R_k^N, k = 0, 1, 2, \ldots$ *be the number of branches in the gene tree of an exchangeable population with constant population size N that satisfies Condition (7.15). The convergence (7.17) to a Kingman coalescent holds if and only if*

$$\mathbb{E}[\nu_1^3]/N \to 0, \ N \to \infty. \tag{7.18}$$

Biologically, Theorem 7.2 says that in a large population of N individuals, the ancestral process is of the Kingman type, which does not allow mergers of more than two lines simultaneously, if and only if a single sibship attains a very large size with a diminutive probability only (but see Box 7.1).

There are two types of natural follow-up questions:

1. How large does N need to be to ensure validity of the approximation (7.17) (a million, a billion, or only 100 or 1000)?
2. In general, when the restriction (7.15) is removed, what are the reproduction laws that lead to coalescent processes that allow triple mergers?

The answer to the first question is given in Möhle (2000), in which the distributions of two random vectors connected by

$$(R_{[Nt_1/\sigma^2]}^N, R_{[Nt_2/\sigma^2]}^N, \ldots, R_{[Nt_r/\sigma^2]}^N) \to (R_{t_1}, R_{t_2}, \ldots, R_{t_r}) \tag{7.19}$$

are compared. It was found that the corresponding probabilities differ at most by the value

$$c_1 \mathbb{E}[\nu_1^3]/N + c_2 \mathbb{E}[\nu_1^2 \nu_2^2]/N + c_3/N. \tag{7.20}$$

Here, the constants c_1, c_2, and c_3 depend, in a complicated way, on the sample size n, the dimension k, and the variance σ^2. This implies, in particular, that under the Kingman moment conditions the convergence rate in Equation (7.17) is c/N. In general, the dominating term in the estimate is $c_1 \mathbb{E}[v_1^3]/N$, in accordance with Theorem 7.2.

The second item was analyzed recently by Sagitov (1999, 2003). It turns out that the ancestral process allows mergers of more than two lines simultaneously if the largest sibship $v_{(1)}$ in the generation exceeds Nx with a probability that satisfies a smallness condition. We obtain convergence to the Kingman coalescent with a more flexible scaling of time,

$$(R^N_{[Nt/\sigma_N^2]})_{t \geq 0} \to (R_t)_{t \geq 0}, \ N \to \infty, \tag{7.21}$$

if for any fixed number $x > 0$,

$$N\sigma_N^{-2} P(v_{(1)} > Nx) \to 0. \tag{7.22}$$

If there is a non-zero limit, we could easily obtain multiple mergers with three or more branches that join at one node. If, furthermore, the second-largest sibship $v_{(2)}$ satisfies

$$N\sigma_N^{-2} P(v_{(2)} > Nx) \to 0, \tag{7.23}$$

for all $x > 0$, it is impossible to encounter two large sibling groups in the same generation among $[N/\sigma_N^2]$ consecutive generations, which effectively prohibits simultaneous mergers in a coalescence with multiple mergers.

Example 7.1 However, even though situations in which Equation (7.22) is not satisfied seem impossible in the realm of mammals, they could certainly be thought of among e.g., reptiles, fish, or insects. As an illustration, assume that $N = 100$ is large enough for Theory 7.2 to apply. In such a small population of, say, turtles, it could certainly happen once in a century that one of the turtles is the mother of, maybe, half the population. In other terms, the probability of begetting 50 offspring or more does not seem to be less than 0.0001.

As a consequence, the probability that the maximal sibship exceeds 50 individuals may well be larger than 0.01.

$$\diamond \ \diamond \ \diamond$$

Schweinsberg (2003) considers an important family of the exchangeable reproduction models with fixed population size N that arise from supercritical Galton–Watson processes starting with N individuals. If $X_1, \ldots, X_N$ are the offspring numbers in the initial generation with $E(X_1) = \mu$, then assuming $\mu > 1$ we expect the total offspring number $Y = X_1 + \cdots + X_N$ to be much larger than N, with a high probability.

The offspring numbers $v_1, \ldots, v_N$ of the corresponding fixed-size model are defined in a two-step procedure. The first step yields the Galton–Watson offspring numbers $X_1, \ldots, X_N$. The second step defines v_i as the number of survivors among X_i siblings after N individuals have been sampled randomly to survive from the Y offspring produced at the first step.

Box 7.1 Non-negligible sibships

The more mathematical reader might like to know what happens if the restrictions (7.15) and (7.18) are removed.

Theorem 7.3 *Assume Equation (7.23). If for all $0 < x < 1$*

$$N\sigma_N^{-2}P(\nu_{(1)} > Nx) \to \int_x^1 y^{-2}\,dF(y),\ N \to \infty, \tag{a}$$

F being an arbitrary probability distribution function on the unit interval not concentrated at zero, the weak convergence (7.21) holds toward a coalescent with multiple mergers.

Given (a), the limit process R_t has the transition rates

$$q_{ij} = \binom{i}{j-1}\int_0^1 x^{i-j-1}(1-x)^{j-1}\,dF(x) \tag{b}$$

and

$$\lambda_i = \int_0^1 \frac{1 - (1-x)^i - ix(1-x)^{i-1}}{x^2}\,dF(x). \tag{c}$$

In particular, if $F(x)$ is concentrated at zero, the limit process R_t corresponds to the Kingman coalescent with the transition rates given by Equations (7.10) and (7.11).

Furthermore, the expectation that the intensity of merger slows down as the number of branches in the gene tree decreases turns out to be correct, $\lambda_n > \lambda_{n-1} > \cdots > \lambda_2 = 1$, and $\lambda_i \leq \binom{i}{2}$. Thus, the Kingman coalescent brings about the fastest merger for any given number of ancestral lines.

In the special case of the so-called *Beta-coalescent*, with $F(x)$ being a Beta(a, b) distribution function, the key quantities q_{ij} and λ_i are easy to compute using simple recursion relations that involve parameters $a > 0$ and $b > 0$. As an example of a coalescent pattern dramatically different from the Kingman coalescent, consider the Beta$(1, 1)$-coalescent with $F(x) = x$ that corresponds to the uniform distribution over the (0,1) interval (Bolthausen and Sznitman 1998). In this case, we have a linear relationship $\lambda_i = i - 1$ instead of the quadratic (7.11).

Theorem 7.3 suggests a simple way to model genealogies with rare multiple mergers based on the Beta$(a, 1)$-coalescent with a small positive parameter $a \approx 0$: if $F(x)$ has density $f(x) = ax^{a-1}$, the transition rates become

$$q_{i(i-1)} \approx \binom{i}{2}\left(1 - \sum_{k=1}^{i-2}\frac{a}{k}\right),\ q_{i(i-j)} \approx \binom{i}{2}\frac{2a}{(j-1)j(j+1)}\ \text{for } j \geq 2, \tag{d}$$

so that $\lambda_i \approx \binom{i}{2}(1 - a\delta)$, where $\delta = \sum_{k=1}^{i-2}\frac{1}{k} - \frac{1}{2} + \frac{1}{i(i-1)}$.

For a thorough mathematical treatment of the coalescent with the transition rates as in Equation (b) consult Pitman (1999). A full classification of the coalescent processes that arise in exchangeable populations is discussed in Möhle and Sagitov (2001), which covers the possibility of simultaneous mergers of ancestral lines that correspond to a non-zero limit in Equation (7.23). The coalescent with simultaneous mergers is further analyzed by Schweinsberg (2000).

According to Theorem 4 in Schweinsberg (2003), if $\mu > 1$ and

$$P(X_1 > k) \sim Ck^{-a}, \ k \to \infty \tag{7.24}$$

for some finite constants $C > 0$ and $a > 0$, then the weak convergence (7.21) holds with the coalescent limit R_t, depending on the parameter value a:

- If $a \geq 2$, the limit R_t is the Kingman coalescent.
- If $1 \leq a < 2$, the limit is the Beta$(2 - a, a)$-coalescent and, in particular, if $a = 1$, it is the Bolthausen–Sznitman coalescent already mentioned. If $1 < a < 2$, the timescale N/σ_N^2 is proportional to N^{a-1}, and $N/\sigma_N^2 \sim \ln N$ in the case $a = 1$ (see Box 7.1).
- If $0 < a < 1$, the limit process belongs to a certain one-parameter class of coalescent processes with simultaneous multiple mergers.

7.2   Ancestral Inference in Branching Processes

S. Tavaré

7.2.1   Introduction

The topic of inference for branching processes is classic and many articles and books have been devoted to it. Common themes include estimation of the offspring mean, the offspring distribution, and the age of the process (cf. Stigler 1970; Guttorp 1991, 1995). In this subsection we illustrate some computational approaches to ancestral inference for branching processes when the effects of mutations among individuals in the population are taken into account. Our examples are from population genetics (in which the timescale is of the order of thousands of years) and from cancer biology (in which the timescale is of the order of years). The techniques illustrated here are but the tip of the inferential iceberg, but they serve to illustrate the crucial interplay between the simulation of a stochastic model and any inference about its parameters.

7.2.2   Inference in the coalescent

Coalescent trees.   In Section 7.1 the *coalescent* was introduced as a model for ancestral relationships among a set of chromosomal segments sampled from an evolving population. In the case of a population that has a constant but large number N of chromosomal segments, we showed that when time is measured in units of N generations, the coalescent tree of a sample of n segments can be described as follows. We begin with n tips and wait for an amount of time T_n that has an exponential distribution with mean $2/n(n-1)$ time units before choosing at random two of the tips to coalesce. The coalescent tree now has $n-1$ nodes (which corresponds to $n-1$ ancestors of the sample), and we then wait a further time T_{n-1} that has an exponential distribution with mean $2/(n-1)(n-2)$ time units until, once again, choosing at random two of the nodes to coalesce. We can continue this description using mutually independent exponential random variables, the waiting time while there are j ancestors of the sample having a mean of $2/j(j-1)$ time units. Eventually, the segments in the sample can be traced back to a common

(a)

(b)

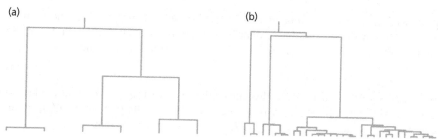

Figure 7.2 Coalescent trees for samples of size (a) 6 and (b) 32 from a population of constant size.

ancestor. Figure 7.2 shows two simulated coalescent trees for samples of size $n = 6$ and $n = 32$.

The height of the coalescent tree, which is the time to the most recent common ancestor (T_{MRCA}) of the sampled segments, is

$$W_n = T_n + T_{n-1} + \cdots + T_2 \,, \tag{7.25}$$

and the length of the tree is

$$L_n = nT_n + (n-1)T_{n-1} + \cdots + 2T_2 \,, \tag{7.26}$$

the sum of the lengths of all the branches in the tree. The means of W_n and L_n are given by

$$\mathbb{E}[W_n] = 2\left(1 - \frac{1}{n}\right), \quad \mathbb{E}[L_n] = 2\sum_{j=1}^{n-1} \frac{1}{j} \,, \tag{7.27}$$

these being multiplied by N to convert coalescent time into generations.

Mutation in the coalescent. The variation observed in the chromosomal segments in the sample is a consequence of mutation in the ancestry of the sample. There are many models for the effects of such mutations, depending on the type of data under consideration. In this subsection we use the so-called *infinitely-many-sites model*, the simplest description of variation in a set of DNA sequences. We suppose that mutations occur only at locations in the DNA segment at which mutations have not occurred before. The sequences in the sample then exhibit a number of *segregating sites*, positions in the DNA at which the members of the sample are not identical. In modern parlance, such locations are called single nucleotide polymorphisms (SNPs). A consequence of this description is that each mutation which occurs in the ancestry of the sample results in a SNP.

The rate at which mutations occur in the region is determined by the compound parameter θ, defined by

$$\theta = 2Nu \,, \tag{7.28}$$

where u is the mutation probability in the region per segment per generation. Mutations are superimposed on the coalescent tree of the sample according to Poisson processes of rate $\theta/2$, independently in each branch. It follows that the number S_n of SNPs in the sample has a distribution determined by the length L_n of the coalescent tree; given $L_n = l$, S_n has a Poisson distribution with mean

$$\mathbb{E}[S_n \mid L_n = l] = \theta l/2 . \tag{7.29}$$

Inference about θ and W_n.   In this subsection we illustrate a computational technique to simulate observations from the posterior distribution of (θ, W_n), given that $S_n = k$. To do this, set $T = (T_n, T_{n-1}, \ldots, T_2)$ and note that

$$\begin{aligned}
f(\theta, T \mid S_n = k) &\propto \mathbb{P}(S_n = k \mid \theta, T)\pi(\theta, T) \\
&= \text{Po}(\theta L_n/2)\{k\} \, \pi(\theta, T) ,
\end{aligned} \tag{7.30}$$

where we define

$$\text{Po}(\lambda)\{k\} = e^{-\lambda}\frac{\lambda^k}{k!} \tag{7.31}$$

with $\text{Po}(0)\{0\} = 1$. In Equation (7.30), $\pi(\theta, T)$ denotes the prior distribution of (θ, T), which is typically the product of the prior π for θ and the "prior" for T, determined by the coalescent model. The prior for θ can be used to incorporate known information about θ. For example, in many problems the size of the mutation rate u in Equation (7.28) is known, at least approximately, as is that of N. This information can be used to design the prior π. A common alternative is to use an uninformative prior for θ, in the form of a density uniform over an interval.

In practice, the density implicit in Equation (7.30) is hard to evaluate in a useful form and it is much simpler to simulate observations from the distribution instead. This is achieved readily by the rejection algorithm:

■ A1. Simulate θ from $\pi(\theta)$ and $t = (t_n, \ldots, t_2)$ from the coalescent model, and calculate $l = nt_n + \cdots + 2t_2$;

■ A2. Accept (θ, t) with probability

$$h = \text{Po}(\theta l/2)\{k\} , \tag{7.32}$$

and return to A1.

Accepted observations clearly have the required density, as can be seen by simple calculation. We make three observations about this approach.

First, it is more efficient to replace h in A2 by h/c where

$$c = \max_{\theta, l} \text{Po}(\theta l/2)\{k\} = \text{Po}(k)\{k\} , \tag{7.33}$$

which can result in considerable gains of speed.

Second, it is not necessary to compute the probability h in A2. Instead, the number of mutations k' on the tree of length l can be simulated, and A2 replaced with

■ A2′. Accept (θ, t) if $k' = k$.

In this example, h can be computed easily so this simulation-based approach is not necessary. However, the alternative approach is far more general than the first

Table 7.1 Inference about θ and W for Yakima data based on 5000 simulated values.

	$T_{MRCA} W$	Mutation rate θ
First quartile	1.05	0.019
Mean	1.68	0.024
Median	1.46	0.023
Third quartile	2.07	0.029

because the likelihood h does not need to be known (in theory or computationally) to use the method. Note, though, that the gains in speed mentioned in the first observation do not seem to be available in this approach.

Third, there is no need to restrict the algorithm to a coalescent with constant size. All that is required to handle the case of deterministic fluctuations in population size is to simulate from the appropriate coalescent distribution for T. In a similar way, we can simulate observations from the posterior distribution of the coalescent topology (and not just the branch lengths); all that is required is to simulate a coalescent tree and proceed as before. Many other applications of these and related algorithms can be found in Tavaré *et al.* (1997).

Example 7.2 To illustrate these ideas, we use some molecular data obtained as part of a larger study on mitochondrial variation observed in Amerindian populations in the USA (Ward *et al.* 1991; Shields *et al.* 1993). Among the aims of this study was the development of methods to infer population history from DNA sequence variation, and in particular to gain an understanding of the way in which the Americas were settled. A convenient place to read more about this field of research is the 2 March 2001 issue of *Science*.

The particular data we use for illustration here comprise a set of $n = 42$ Yakima mitochondrial DNA sequences, each of length 360 base pairs, given in Shields *et al.* (1993; see also Markovtsova *et al.* 2000a, p. 404). The observed base frequencies in the sequences are

$$(\pi_A, \pi_G, \pi_C, \pi_T) = (0.328, 0.113, 0.342, 0.217), \tag{7.34}$$

and there are 20 distinct sequences and 31 SNPs in the sample.

In the absence of other information, we chose a wide uniform prior for θ, and used a constant population size coalescent to model T. The results of 5000 accepted runs of the algorithm are given in Table 7.1 and Figure 7.3. The posterior distribution of W does not differ enormously from its prior determined by the coalescent model. The parameter N is approximately 600, so if we assume a generation time of 20 years, the mean height of the coalescent tree is about 20 000 years.

7.2.3 Approximate Bayesian computation

The Yakima data used in Example 7.2 have been discussed in a coalescent framework by Markovtsova *et al.* (2000a, 2000b), where the posterior distributions of θ and W_n were found by Markov chain Monte Carlo methods using the full sequence data rather than the summary statistic $S_n = 31$. One reason for basing our inference on statistics such as S_n, rather than the full data, is a practical one: we

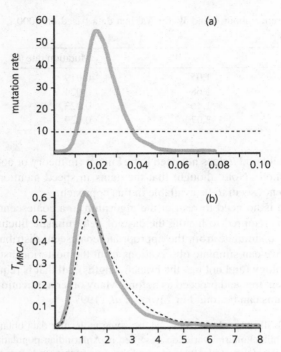

Figure 7.3 Posterior density of θ (a) and W_n (b). Dotted lines show prior density.

hope to generate observations much more quickly than when using other stochastic computation methods. In exchange, we are left with observations from the density $f(\theta, W \mid S_n = k)$ as opposed to the full density $f(\theta, W_n \mid \mathcal{D})$, where $\mathcal{D}$ denotes the complete sequence data. Approaches that use summary statistics for inference are called approximate Bayesian computation (ABC). The consequences of such reductions can be complicated and unexpected; see Beaumont *et al.* (2002) for a number of related examples and other approaches, as well as historical references on ABC.

7.2.4 Inference for tumor histories

In the next example we adapt the same type of approach to a discrete setting that involves inferences about the history of a tumor.

The data and the problem. It is difficult to infer tumor histories by using direct observation of a patient. Adenomas, thought to be precursors of cancer, are removed if they are detected, and the amount of time required to observe the entire progression of a cancer may be many decades. To overcome the limitations of direct observation it is possible to exploit the pattern of mutations observed in an adenoma or a cancer (Tsao *et al.* 2000). These mutations can be used to estimate the age of the adenoma or cancer, in much the same way as we used variation in mitochondrial sequences to infer aspects of the history of the Yakima. The timescale of the cancer example is of the order of years, in contrast to the Yakima example, which is of the order of tens of thousands of years.

In this example, we study a class of colon cancers known as mutator phenotype cancers. These colorectal cancers have lost DNA mismatch repair (MMR), so they are less able to repair errors during DNA replication. These cancers also have greatly elevated mutation rates. The consequences can be observed most easily in microsatellite (MS) loci. These loci, which may be thought of as runs of a short motif such as CA, show dramatic expansions and contractions in size over small numbers of cell divisions.

It is these mutations that we use to track the history of a cancer. We are able to measure the length variation in a series of such MS loci sampled from cells in a tumor (Tsao *et al.* 2000). The problem is to estimate the time since MMR was knocked out; that is, to estimate the age of the tumor.

Once MMR is lost in a parent cell, the descendant cells derived from it by mitotic division eventually form a final clonal expansion that originates from a single cell and results in a detectable tumor (which we assume has an average size of about 1 cm^3, or about 10^9 cells). Using the MS variation, we estimate the number of divisions Y_0 between loss of MMR and the initiation of the final clonal expansion, and the number of generations Y_1 from that event until the tumor is observed at biopsy.

Once more, this is an "ancestral inference" problem, in which the desired posterior is $f(Y_0, Y_1 \mid \mathcal{D})$. The data $\mathcal{D}$ come from L MS loci, the first of which is measured in n_1 tumor chromosomes, the second from n_2 tumor chromosomes, $\ldots$, and the Lth from n_L chromosomes. The total number of chromosomes used is then

$$n = n_1 + \cdots + n_L . \tag{7.35}$$

In the studies reported below, we sampled X chromosome MS loci from male patients. As males have a single copy of their X chromosome, we can identify each sampled X chromosome with a single cell. This simplifies the required analysis.

We know the somatic size of each locus (i.e., the number of repeats at each locus prior to loss of MMR). All repeat lengths are measured relative to this baseline size. For each MS locus, we are able to estimate the mean MS lengths, $m_1, m_2, \ldots, m_L$, and the variances of these lengths, $s_1^2, s_2^2, \ldots, s_L^2$. These data are, in turn, summarized by the two statistics

$$S_{\text{alleles}}^2 = \text{average of } s_1^2, \ldots, s_L^2 ;$$
$$S_{\text{loci}}^2 = \text{variance of } m_1, \ldots, m_L . \tag{7.36}$$

A model for tumor evolution. One question that has to be addressed is the model used to describe the evolution of the tumor from loss of MMR until detection. The relative sizes of S_{alleles}^2 and S_{loci}^2 give a hint.

In Figure 7.4, taken from Figure 1 in Tsao *et al.* (2000), the results of 1000 simulations of $L = 20$ MS loci measured in $n_i = 25$ chromosomes are summarized (the method used to perform the simulations is given in Section 7.2.5). The

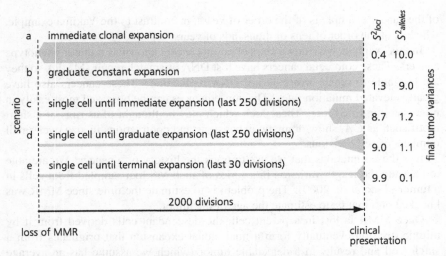

Figure 7.4 Simulations of MS mutation. Different patterns of MS mutations are summarized by the values of S^2_{alleles} and S^2_{loci}. All simulations use 2000 divisions, but different tumor histories. Further details are given in the text. Modified from Tsao *et al.* (2000).

simulations assume a symmetric stepwise mutation model with the chance of the addition and of the loss of one repeat being 0.0025, to give a total mutation rate of 0.005 per division. In each scenario a total of 2000 divisions is assumed, and the final tumor size is, on average, one billion cells.

The results show that it is possible to infer, in broad terms, the form of the tumor history by measuring its MS alleles and estimating S^2_{alleles} and S^2_{loci}. In the analysis that follows, we use the model that corresponds to scenarios (d) and (e) in Figure 7.4: a single progenitor cell lineage that lasts for Y_0 divisions, and a terminal expansion described by symmetric binary splitting for Y_1 generations with parameters chosen to make the average size of the tumor a billion cells. Some experimental justification for this model is given in Figure 2 of Tsao *et al.* (2000).

The genealogy of a sample from a branching process. The data we collect come from a few hundred cells sampled from a tumor that contains about a billion cells. To simulate observations on the MS loci observed in the tumors we could simulate the entire tumor history and then subsample these cells, or we could generate the history of the sample only. The latter approach is the one we used to describe the coalescent: we generate the genealogical history of the cells in the sample, then simulate the effects of mutations at the MS loci in this shared ancestry. This results in a sample of MS loci in the cells in the sample.

Methods to generate the genealogical history of a sample from a branching process are described in Weiss and von Haeseler (1997) in the context of the polymerase chain reaction (PCR), and in Tsao *et al.* (2000) in the present context. Suppose we want to simulate the history of a sample taken after g generations, going back to time 0. The basic idea is to use three passes to generate the MS sample: in the first phase, the numbers of cells that have 0, 1, and 2 descendant cells in generation 1, generation 2, ..., generation g are simulated. This results in a collection of family-size statistics (M_{j0}, M_{j1}, M_{j2}), $j = 0, 1, \ldots, g - 1$, where M_{jl} is the number of families of size l born to cells in generation j. In a given generation, $j + 1$ say, there are $M_{j+1} = M_{j1} + 2M_{j2}$ cells. To generate M_{jl} requires knowledge of the offspring distribution in each generation (which may differ across generations). The branching property means that if the total number of cells in generation j is M_j, the number of cells that have 0, 1, and 2 descendants in the next generation is multinomially distributed with parameters M_j and p_0, p_1, p_2, these being the probabilities of 0, 1, or 2 descendants, respectively, from a given cell in generation j.

The second stage reconstructs the genealogy of the sample taken from generation g using the family sizes (M_{j0}, M_{j1}, M_{j2}) in the order $j = g - 1, g - 2, \ldots, 0$. If the sample has n cells at time g, we assign the n cells at random to ancestors in generation $g - 1$, in accordance with the numbers $M_{g-1,1}$ and $M_{g-1,2}$. Using a "balls in urns" analogy, this is equivalent to choosing without replacement n balls from $M_{g-1,1} + M_{g-1,2}$ urns, $M_{g-1,1}$ of which contain one ball, and $M_{g-1,2}$ of which contain two balls. This done, we count the number n_{g-1} of distinct ancestors (i.e., the number of different urns sampled) in generation $g - 1$, and repeat the assignment of these cells to their parental cells using the counts $M_{g-2,1}$ and $M_{g-2,2}$. Continuing back in this way to time 1 produces a genealogical tree of the sample.

The third stage starts from the top of this genealogical tree by assigning MS lengths to each of the ancestral cells at time 0, and then runs the mutation process down the branches of the tree until arriving at the n cells in the sample at time g. The mutation mechanism we use here is the simplest of a large number of models that have been used in the literature: a MS locus inherits the same length as its parent, plus the addition or deletion of a single motif caused by errors in MMR.

Before exploiting this approach to infer the age of a tumor, we note that the algorithm used to generate the history of a sample of cells can be adapted to arbitrary branching processes. The branching property is the only key assumption: given the history of the process up to time j, the individuals in generation j produce offspring independently and with identical distributions (which may depend on the history up to time j). In particular, the resultant branching process need not even be Markovian. This provides plenty of flexibility to analyze samples from extraordinarily complicated processes about which theoretical results are few and far between. The approach can also be modified to generate genealogical histories of samples from multi-type branching processes.

7.2.5 Example

This example comes from data at 23 loci measured in an adenoma. The sample sizes at each locus varied between 10 and 33, and the observed summary statistics were $S^2_{\text{alleles, obs}} = 0.828$, $S^2_{\text{loci, obs}} = 6.229$.

As described above, we assumed a simple symmetric step-wise mutation model for each MS, with an overall mutation rate of 0.005 per replication. We used uniform priors for Y_0 and Y_1, with ranges $(100, 2100)$ and $(25, 400)$, respectively.

For the ABC approach we simulated observations from the priors for Y_0 and Y_1, and then simulated the history of the n cells that were sampled. Given this genealogy we simulated L MS loci using the given mutation model, and calculated the simulated values $S^2_{\text{alleles, sim}}$ and $S^2_{\text{loci, sim}}$ of the statistics in Equation (7.36). The values of Y_0 and Y_1 were accepted if

$$\left| \frac{S^2_{\text{loci, sim}}}{S^2_{\text{loci, obs}}} - 1 \right| + \left| \frac{S^2_{\text{alleles, sim}}}{S^2_{\text{alleles, obs}}} - 1 \right| < \epsilon, \qquad (7.37)$$

where ϵ is a tuning parameter. Large values of ϵ accept most values and so reconstruct the prior, whereas as $\epsilon \to 0$ only those values of Y_0 and Y_1 that reproduce the data S^2_{alleles} and S^2_{loci} exactly are accepted. The trade-off is in picking values of ϵ that lead to a reasonable number of accepted values in a given time, as well as a reasonable approximation to the required posterior. In the example below ϵ is set to 0.1, which corresponds to an acceptance rate of about 0.6%.

In Table 7.2 summary statistics for the posterior distributions of Y_0, Y_1, and the age $Y = Y_0 + Y_1$ of the tumor are given. The corresponding posterior densities, based on 1000 simulated observations, are given in Figure 7.5. The posterior density of Y is shown in Figure 7.6. A 95% credible interval for Y is $(895, 2197)$ divisions. Assuming one division per day, this translates into an interval of $(2.5, 6.0)$ years, and a mean posterior age of 4.2 years.

In Tsao *et al.* (2000) a different statistical approach was used to assess variability in the estimate of Y. Using data combined from two regions of the adenoma, they found an estimated age of 1300 divisions (3.6 years), with an estimated 95% confidence interval of $(1.3, 5.2)$ years. Despite the different statistical approaches these results are consistent with each other.

We remark that the approach outlined here can also be used to investigate the robustness of the modeling assumptions. All that needs to be changed are the details of the branching process being used and the mutation model; some of this is described in Tsao *et al.* (2000). Different statistical approaches can also be explored in this way, such as by using different metrics in Equation (7.37) and different summaries of the data.

Table 7.2 Inference about Y_0 and Y_1 for adenoma data based on 1000 accepted values. $Y = Y_0 + Y_1$ is the age of the tumor.

	Y_0	Y_1	Y
First quartile	1077.0	170.0	1255.0
Mean	1343.9	186.0	1529.8
Median	1325.0	184.0	1514.0
Third quartile	1614.3	200.0	1790.0

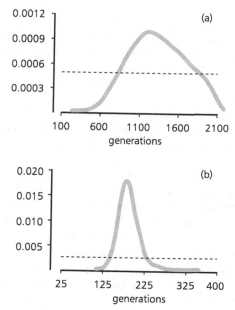

Figure 7.5 Posterior density of Y_0 (a) and Y_1 (b). Dotted lines show prior density.

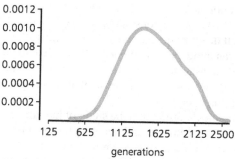

Figure 7.6 Posterior density at age Y of the tumor.

7.3   The Cell Cycle

M. Alexandersson

7.3.1   Introduction

A cell is born, it grows for a while, and then it divides. Even though this is an extremely simplified description, it captures the basic structure of the cell cycle, and serves as a skeleton or a scaffold for more detailed models.

The cell cycle is a complex process with streams of signals and an elaborate succession of events through each stage. Also, because of an extensive control system within and outside the cell, the cycle might be terminated prematurely (apoptosis), or sometimes a cell may not even cycle (quiescence). All these details are important in a biological model, and essential to the understanding of the cell cycle. However, when constructing a mathematical model of such a complex system, it is important to bear in mind what biological questions we want to investigate, and what is observed and measured in real experiments.

One purpose of mathematical modeling of the cell cycle is to provide a way to compare theories of cell kinetics and cell growth at the individual level, with observations at the population level. Another is to predict the development and future of the population. As we show later, the theory of multi-type branching processes is unusually well suited to models and analysis of cell populations. The rules of cell proliferation are based clearly on local processes in and around the cells, and branching process models are constructed by defining the reproduction pattern for a typical individual in the population. With an informal Law of Large Numbers argument, we relate individual properties to the development of the population.

In a general branching process every individual in the population is characterized by a property called its *type*. This is a label inherited at birth that determines the range of choices the individual has during its life, more specifically what probability laws its life career will follow. For instance, the type can be the age of the individual at a given time, its size at birth, or its growth rate. Just as birth types characterize individuals, so the birth-type distribution characterizes the entire population; all other population properties, such as age and size distributions, can be derived from this. In the model considered we assume that all cells proliferate, and that there is no cell death or quiescence. Consequently, the population size grows to infinity, and branching process theory tells us that its composition stabilizes (see Chapter 6), whatever the initial state of the population. This is an important property in cell populations, since it means that it is possible to generate a large population from a single cell.

Population dynamics originates in the study of human populations, in which the fertility is clearly age dependent, and age structure is a natural choice in demography. However, when considering cell populations it seems more appropriate to take the *size* (e.g., mass, volume, or DNA content) of the individuals into account. We assume our cell population to be size structured, which means that the life of a cell depends upon its size rather than on, for instance, its age. In other words, the type of a cell is its size. One of the most influential articles on size-structured cell populations is Bell and Anderson (1967), which introduced what is now called the Bell–Anderson model. In a branching process context this is a single-type process with binary splitting, and clearly follows the basic structure described above. The model we concentrate on in this section is an extension of the Bell–Anderson model, and similar to a class of models sometimes called *transition probability models*. Transition probability models use the idea that the cell cycle consists of a completely deterministic B-phase, and an indeterminate or stochastic A-phase. We assume that there is a *critical* size m_0, such that a newborn cell always has size less than m_0, and then it grows beyond m_0 before it divides. The critical size divides the cell cycle into two independent phases, A and B say, and when the B-phase is deterministic our model coincides with a transition probability model. We also introduce a microheterogeneity in growth: the growth control is supramitotic in the sense that a newborn cell continues to grow with the same growth rate as its mother until it reaches the critical size, at which it chooses a new growth function.

Size-structured models are sometimes criticized because they lead to negative mother–daughter life length correlations, whereas this correlation is often observed to be zero or positive, especially in mammalian cells. In our model this problem does not occur. By dividing the cell cycle into two independent phases, the only dependence between mothers and daughters is through their common growth function, that is through the second (B) phase of the mother and the first (A) phase of the daughter. The size structure means that slow growth leads to longer life lengths and fast growth leads to shorter life lengths, for both the mother and her daughters, and their life length correlation is non-negative.

Our aim here is to give an informal description of a general cell population model, one complex enough to capture many important biological features without having too many parameters. We show some of the ways in which general branching process theory helps us derive facts about our model and conclude with a fuller description of the mother–daughter correlation issue mentioned above. As it turns out, we obtain explicit analytical expressions of the quantities we wish to investigate, and we obtain numerical plots of the composition of the population.

7.3.2 A biological model

The cell model considered here is a version of the two-subcycle model, described and analyzed in Sennerstam and Strömberg (1995), and based on a cell line of multipotent embryonal carcinoma cells (PCC3). In this section we define the model biologically, and in Section 7.3.3 we use the branching process framework to define and analyze it mathematically.

The cell cycle is usually defined to be the period between two consecutive mitotic events (i.e., between two cell divisions). During its life length the cell is thought to progress through four, sometimes five, different phases, G_1, S, G_2, and M, where S is the DNA synthesis phase in which the genome is duplicated, M is the mitotic phase, and G_1 and G_2 are preparation phases or "gaps." The cell cycle is sometimes expanded with a fifth stage, G_0, which is a resting phase between mitosis and G_1, and from which cells are recruited randomly into G_1. In transition

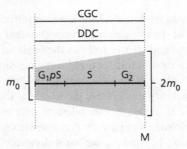

Figure 7.7 The DNA division cycle (DDC) and cell growth cycle (CGC) span the same time interval.

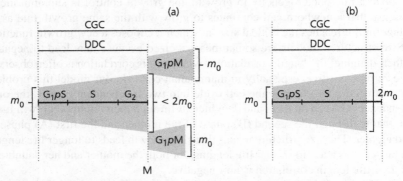

Figure 7.8 (a) When the growth rate is slow, the CGC extends past mitosis. (b) When the growth rate is fast, the cell reaches size $2m_0$ before mitosis.

probability models the A-phase contains G_0, when introduced, and the first part of G_1, and the B-phase consists of the remainder of G_1, as well as S, G_2 and M.

The idea in the two-subcycle model is that the cell cycle consists of two mutually dissociated, simultaneously running subcycles: the DNA division cycle (DDC), which causes a doubling of the genome, and the cell growth cycle (CGC), in which the cell doubles its size. The DDC is assumed to have a fairly constant duration and covers the S, G_2, and M phases and a pre-S phase. The pre-S phase is postulated to be a temporally constant (cf. Sennerstam and Strömberg 1995) late G_1 period (G_1pS) when a cell is committed to enter the S phase. In the simplest case the CGC spans over the same time interval as the DDC, beginning at some cell size m_0, and cells divide equally (Figure 7.7). This is the Bell–Anderson model and is treated in Taib (1999) and Jagers (2001).

However, the cell may vary considerably in growth rate during its CGC, the most common situation being when the cell grows rather slowly and the CGC extends past mitosis (Figure 7.8a). The cell then divides unequally, and each daughter cell continues to grow to complete its mother's CGC in what is called the post-M phase (G_1pM). When the cell reaches the critical size m_0, whenever that occurs

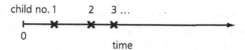

Figure 7.9 Each individual gives birth to a number of children, and the births occur as random points in time.

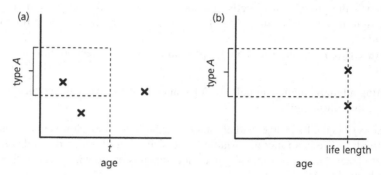

Figure 7.10 (a) A general birth process, in which children are born with different types at different ages of the mother. (b) For cells the number of children is always two, and the daughters are born at the same time, but are assigned different types. *A* denotes an arbitrary set of types.

in the G_1-period, a "start" event is triggered, and the cell is committed to enter its own DDC and its pre-S phase begins.

A third case is when the cell grows rapidly from "start" and the CGC is completed both before the DDC is and before the mitosis is activated (Figure 7.8b). The cell is then thought to enter a new S phase, which generates a tetraploid cell (i.e., a cell with four sets of chromosomes instead of the normal two).

A cell grows with the same growth rate as its mother until it reaches the critical size, where it chooses a new growth function. Two daughter cells are assumed to have similar, but not identical, growth functions through some inheritance from the mother. We give an example of a mathematical representation of this inheritance in the next subsection.

7.3.3 The branching process model

A branching process model is based on the reproduction of a typical individual in the population (Figure 7.9). Births occur at random points in time, and each newly born individual is assigned a type.

Of particular interest is the reproduction of a typical cell, given as the average number of children of certain types of an individual with a given type and age (Figure 7.10).

Once the reproduction pattern of the population has been specified, we can use it to determine the stable composition. The proportions of cells that have certain properties, such as being alive, with certain division sizes, or certain ages, can be determined by counting the number of cells that have the property, and dividing by the total number of cells. Branching process theory tells us that these proportions converge to limits when time goes to infinity (and hence the composition has stabilized). In cell populations what are known as the α- and β-curves are especially important. These are defined as follows:

- $\alpha(a)$ is the proportion of cells that survives past age a, and $\beta_1(a)$ is the proportion of sister cell pairs whose life lengths differ by more than a time units, while
- $\beta_2(a)$, $\beta_3(a)$, ... are the same proportions for cousin, second cousin, ... cell pairs.

Omitting all mathematical details (which can be found in Alexandersson 2000) we assume the following:

1. The type is the birth size, and all other distributions (see below) depend on this.
2. Length of life is a random variable, the distribution of which depends on birth size. Hence, a cell's propensity to divide depends on how large it is rather than on how old it is.
3. The cell divides unequally into two fractions, δ and $1 - \delta$, where $0 < \delta < 1$ is a random variable with distribution symmetric around $\frac{1}{2}$ (i.e., the cell divides in two equal parts on average).
4. There is a critical size m_0 such that all cells are born smaller than m_0, and then grow past m_0 before division.
5. Individual growth is exponential with a growth rate that consists of two parts: a latent factor handed down by the mother, which represents inheritance, and an individual contribution. Hence, the growth rate for a cell is of the form

   growth rate = inherited component + individual component

   where the inherited component is common for both daughters, and the inherited and the individual components are independent.
6. The DDC is constant.

With these assumptions, branching process theory enables us to form an analytical expression for the stable birth-type distribution, and hence also for quantities such as the α- and β-curves. When the model is simulated, as in Sennerstam and Strömberg (1995), we obtain explicit mathematical functions that can be plotted numerically.

We use the parameter values defined in Sennerstam and Strömberg (1995):

Critical cell size (m_0)	7 (relative size units)
DDC	8 (hours)
Inherited component	$N(0.06, 0.005^2)$
Individual component	$N(0, 0.015^2)$

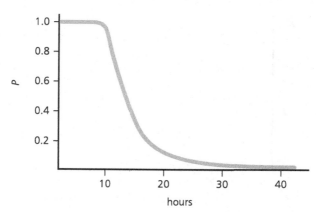

Figure 7.11 The α-curve.

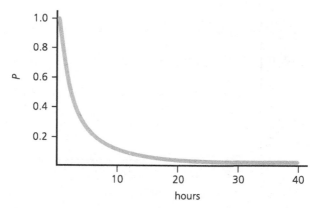

Figure 7.12 The β_1-curve.

where $N(m, v)$ stands for the normal distribution with mean m and variance v. Plots of the α- and β_1-curves, the densities of the stable birth-size distribution and the stable division-size distribution are shown in Figures 7.11–7.14.

7.3.4 Mother–daughter cell cycle time correlation

As mentioned in Section 7.3.1, size-structured models are sometimes criticized because they predict a negative mother–daughter life-length correlation. That the correlation becomes negative is easy to see since, for a cell population in which all the cells have the same growth rate, a long cell cycle leads to a larger division size, and therefore larger daughters. Since life length depends on size, the daughters' life lengths become shorter, and hence the negative correlation. In mammalian cells, however, this correlation is observed to be zero or positive.

In the two-subcycle model, the cells change growth rate, as pointed out, at the critical size, and a new growth rate is chosen independently of the old one. This

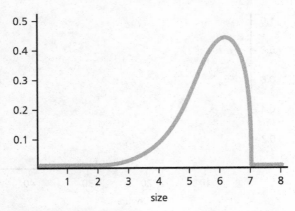

Figure 7.13 Density of the stable birth size.

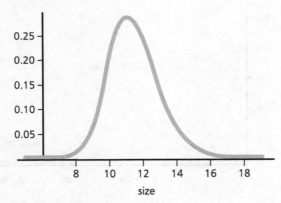

Figure 7.14 Density of the stable division size.

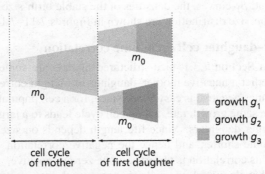

Figure 7.15 The life lengths of the mother and the daughters are dependent through their common growth rate g_2.

means that the length of the G_1 phase of the mother is independent of the length of the G_2 phase of the daughter. By dividing the life length of a cell into two independent parts, separated by the critical size m_0, as shown in Figure 7.15, we see that the mother and daughter are dependent only through the period when they share a growth function (g_2 in Figure 7.15). Slow growth or fast growth affects both mother and daughters in the same way, which leads to a non-negative correlation. For the special case in which the DDC is considered constant, we obtain that the life lengths of the daughters are independent of the mother's life, and the correlation becomes zero. Thus, in this case the mother–daughter correlation is zero as a consequence of constant DDC rather than because of the microheterogeneity in growth. Note that in this case the correlation is zero even if all cells have the same growth rate. A proof of this can be found in Alexandersson (2000).

7.4 Telomere Shortening: An Overview

P. Olofsson and M. Kimmel

7.4.1 Introduction

Shortening of telomeres is one of the supposed mechanisms of cellular aging and death. The hypothesis is that each time a cell divides it loses pieces of its chromosome ends. These ends are called telomeres and consist of repeated sequences of nucleotides (telomere units). When a critical number of telomere units are lost, the cell stops dividing. There is an extensive biological literature on the subject; a good reference for the basic facts and an intuitively appealing explanation of the deletion process is Levy *et al.* (1992). This also seems to be the first article to formulate the problem mathematically, and its essentially deterministic model serves as the basis for all subsequent mathematical models in the literature.

Telomeres are assumed to consist of telomere units, repeated sequences of nucleotides. When a chromosome replicates, each newly synthesized strand loses one telomere unit at one of its ends. This means that the pair of daughter chromosomes each has one old, unchanged strand and one new strand, one unit shorter (see Figure 7.16). Once a critical number of telomere units are lost, a so-called Hayflick checkpoint is reached and the cell stops dividing. Under this assumption, only the length of the shortest telomere matters and thus a chromosome is said to be of type j if its shortest telomere has j remaining units. This leads to a model in which a type j chromosome has two offspring, one of type j and one of type $j - 1$. Cells of type 0 do not divide.

In this section, the attempts to model and analyze this process with mathematical methods are reviewed. The emphasis is put on multi-type branching process models, in which the main features turn out to be reducibility and polynomial growth.

7.4.2 Branching process model

We now describe the telomere loss process as a multi-type branching process. For the generalities of such processes, see, for example, Mode (1971). The type of a

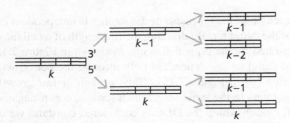

Figure 7.16 Two replication rounds in the simple model for telomere shortening. Note how 5′ ends are replicated completely, whereas 3′ ends have overhangs because of the incomplete replication. The number under the chromosome is the number of remaining telomere units.

chromosome is, as described above, the number of remaining telomere units on its shortest end. Fix one particular chromosome and let the type of the cell be the type of that chromosome. A cell of type j divides into two daughter cells of types j and $j - 1$ if $j \geq 1$; a cell of type 0 does not divide, but stays in the population forever (is immortal). Dividing cells have life lengths that are independent and uniform random variables. Start with one cell of type k and let $M_j(t)$ denote the expected number of cells of type j at time t. The main feature of the resultant multi-type branching process is reducibility, meaning that types do not communicate. As a result of this, the asymptotics are polynomial rather than exponential. The main results of Arino *et al.* (1995) are given in Theorems 7.4 and 7.5.

Theorem 7.4 *Assume that life lengths are independent exponential with mean* $1/\alpha$. *Then*

$$M_j(t) = \frac{(\alpha t)^{k-j}}{(k-j)!} \tag{7.38}$$

for $j = 0, 1, \ldots, k$.

and

Theorem 7.5 *Assume that life lengths are bounded and continuous independent random variables with mean* μ. *Then, as* $t \to \infty$,

$$M_j(t) \sim \frac{t^{k-j}}{(k-j)!\mu^{k-j}} \tag{7.39}$$

for $j = 0, 1, \ldots, k$.

In Olofsson and Kimmel (1999) and Olofsson (2000), probabilistic methods are used to prove these and more general results, which are presented in Section 7.4.3.

7.4.3 Incorporating cell death

In Olofsson and Kimmel (1999), new probabilistic methods of proof are introduced. The model is the same as that above, but the methods of proof are simpler, clearer, and easier to generalize.

Let the ancestor be of type k throughout. The basic idea is to relate the real-time process to the process counted by generation. Recall $M_j(t)$, the expected number of j-type cells at time t, and let $m_j^{(n)}$ be the expected number of j-type cells in the nth generation. The two relate in the following way:

$$M_j(t) = \sum_{n=k-j}^{\infty} m_j^{(n)} (G^{*n}(t) - G^{*(n+1)}(t)) , \tag{7.40}$$

where G is the distribution function of the life length and G^{*n} its n-fold convolution, the distribution of the sum of n independent identically distributed life lengths. To understand this formula, note that at time t cells from any generation may be present. Since there are $m_j^{(n)}$ individuals in the nth generation and each of these is alive at time t with probability $G^{*n}(t) - G^{*(n+1)}(t)$ (born before t, but not yet dead at t), the expected number of cells from the nth generation present at time t is simply $m_j^{(n)}(G^{*n}(t) - G^{*(n+1)}(t))$. Summing over all the generations gives us $M_j(t)$. The identity may be derived formally as the unique solution to a certain renewal equation; for a more general version of this, see Jagers (1989).

In particular, if life lengths are $\exp(\alpha)$, it can be shown that

$$G^{*n}(t) = e^{-\alpha t} \sum_{i=0}^{n-1} \frac{(\alpha t)^i}{i!} \tag{7.41}$$

is, indeed, the Gamma distribution with parameters n and α, and hence

$$M_j(t) = e^{-\alpha t} \sum_{n=k-j}^{\infty} m_j^{(n)} \frac{(\alpha t)^n}{n!} . \tag{7.42}$$

If there is no cell death, $m_j^{(n)} = \binom{n}{k-j}$ and hence

$$M_j(t) = e^{-\alpha t} \sum_{n=k-j}^{\infty} \binom{n}{k-j} \frac{(\alpha t)^n}{n!} = \frac{(\alpha t)^{k-j}}{(k-j)!} e^{-\alpha t}$$

$$= \sum_{n=k-j}^{\infty} \frac{(\alpha t)^{n-(k-j)}}{(n-(k-j))!} = \frac{(\alpha t)^{k-j}}{(k-j)!} \tag{7.43}$$

which is Theorem 7.4.

The first generalization is to introduce possible cell death. As before, let life lengths of proliferating cells be independent exponential random variables with mean $1/\alpha$. Suppose that cells of type 0 live for a time that is exponential with mean $1/\tau$ where $\tau \leq \alpha$. Another way to think of this is that 0-cells live for a time that is exponential with mean $1/\alpha$, just like all other cells, and then "try to divide" and either die or live on. If the probability of survival is p, we have the relation $\tau = \alpha(1 - p)$.

The formula for $M_j(t)$ remains unchanged for $1 \leq j \leq k$, and for $j = 0$ we obtain the asymptotic result in Theorem 7.6.

Theorem 7.6 *As* $t \to \infty$,

$$M_0(t) \sim \frac{1}{1-p} \frac{(\alpha t)^k}{k!} .$$ (7.44)

In the model with immortal cells, the 0-cells dominate, since their growth rate is the fastest. In the model with cell death, the 1-cells have the same growth rate, so asymptotically they dominate together with the 0-cells in the proportions 1 to $1/(1-p)$. Already, this is a qualitative difference between the two models.

Next, we introduce possible death also for the non-zero cells. Thus, let p be the probability of survival for 0-cells and q the corresponding probability for non-zero cells. The asymptotics now depend on the relation between p and q, as the result in Theorem 7.7 shows.

Theorem 7.7 *For* $1 \le j \le k$,

$$M_j(t) = \frac{(\alpha q t)^{k-j}}{(k-j)!} e^{-\alpha(1-q)t} .$$ (7.45)

For $j = 0$ *there are three cases. If* $p = q$, *then*

$$M_0(t) = \frac{(\alpha q t)^k}{k!} e^{-\alpha(1-q)t} .$$ (7.46)

If $p < q$ *then*

$$M_0(t) \sim \frac{q}{q-p} \frac{(\alpha q t)^{k-1}}{(k-1)!} e^{-\alpha(1-q)t} .$$ (7.47)

Finally, if $p > q$, *then*

$$M_0(t) \sim \left(\frac{q}{q-p} \right)^k e^{-\alpha(1-p)t} .$$ (7.48)

Note that the polynomial growth is now asymptotically killed by an exponential decay factor, determined by the larger of the two survival probabilities p and q.

A second important generalization of the model that results from Arino *et al.* (1995) is to relax the exponential assumption and let the life-length distribution be arbitrary. Section 7.4.4 is devoted to this.

7.4.4   General life-length distributions

In Olofsson and Kimmel (1999), the results from Arino *et al.* (1995) were also extended to general life-length distributions. We state Theorem 7.8 with proof.

Theorem 7.8 *Suppose the life lengths are independent with common distribution function* G *and common mean* μ. *Then, as* $t \to \infty$,

$$M_j(t) \sim \frac{t^{k-j}}{\mu^{k-j}(k-j)!}$$ (7.49)

for $0 \le j \le k$.

The proof is based on the following so-called Tauberian theorem (Feller 1971).

Theorem 7.9 *Let H be a non-decreasing function such that the Laplace–Stieltjes transform $\widehat{H}(s) = \int_0^\infty e^{-su} H\,(du) < \infty$ for $s > 0$. Then for any $r > 0$*

$$\widehat{H}(s) \sim \frac{A}{s^r} \quad \text{as } s \to 0 \tag{7.50}$$

if and only if

$$H(t) \sim \frac{At^r}{\Gamma(r+1)} \quad \text{as } t \to \infty . \tag{7.51}$$

Proof of Theorem 7.8 First let $k = 1$. Then

$$M_1(t) = \sum_{n=1}^\infty n(1 - G) * G^{*n}(t) = \sum_{n=1}^\infty n(G^{*n}(t) - G^{*(n+1)}(t))$$

$$= \sum_{n=1}^\infty \sum_{j=1}^n (G^{*n}(t) - G^{*(n+1)}(t)) = \sum_{j=1}^\infty \sum_{n=j}^\infty (G^{*n}(t) - G^{*(n+1)}(t))$$

$$= \sum_{j=1}^\infty G^{*j}(t) \sim \frac{t}{\mu} \tag{7.52}$$

as $t \to \infty$, by the renewal theorem (Asmussen 1987). That this theorem applies can be realized directly: the k-type cells split according to a renewal process and at each renewal an infinite line of $(k-1)$-type cells is initiated, since there is no death. Hence, the number of $(k-1)$-type cells at time t is exactly the number of such renewals up to time t.

In the absence of cell death, clearly $M_1(t)$ is non-decreasing and

$$\widehat{M}_1(s) = \int_0^\infty e^{-su} M_1(du) = \sum_{n=1}^\infty \int_0^\infty e^{-su} G^{*n}(du)$$

$$= \sum_{n=1}^\infty \widehat{G}(s)^n = \frac{\widehat{G}(s)}{1 - \widehat{G}(s)} < \infty \tag{7.53}$$

for $s > 0$ unless $G(0) = 1$. Hence, Theorem 7.8 above applies and

$$\widehat{M}_1(s) \sim \frac{1}{\mu s} \tag{7.54}$$

as $s \to 0$.

Now, consider $M_k(t)$ for an arbitrary k. We obtain

$$\widehat{M}_k(s) = \int_0^\infty e^{-su} M_k(du) = \sum_{n=k}^\infty \binom{n}{k} \int_0^\infty e^{-su}(1 - G) * G^{*n}(du)$$

$$= (1 - \widehat{G}(s)) \sum_{n=k}^\infty \binom{n}{k} \widehat{G}(s)^n = (1 - \widehat{G}(s)) \frac{\widehat{G}(s)^k}{(1 - \widehat{G}(s))^{k+1}} = \widehat{M}_1(s)^k . \tag{7.55}$$

Hence,

$$\widehat{M}_k(s) \sim \frac{1}{\mu^k s^k} \tag{7.56}$$

as $s \to 0$ and applying Theorem 7.8 again yields

$$M_k(t) \sim \frac{t^k}{\mu^k k!} \tag{7.57}$$

as $t \to \infty$. Note that Theorem 7.5 follows as a special case.

The models and main results in Arino *et al.* (1997) and Olofsson (2000) are essentially the same and extend the previous model by allowing for loss of a random number of telomere units in each division. The methods in the first article are deterministic, mainly using differential equations. The second article uses variants of the probabilistic methods introduced above and involves standard results from renewal theory. We state the main result from Olofsson (2000).

$$\diamond \ \diamond \ \diamond$$

Theorem 7.10 *Assume that a cell of type j gives birth to a cell of type j and a cell of type i with probability $p_{j,i}$ for $i = 0, \ldots, j - 1$. Let a cell of type j have the life-length distribution function G_j. Start from a cell of type k and let $M_j^{(a)}(t)$ be the expected number of j-type cells of age less than a at time t. Then, as $t \to \infty$,*

$$M_j^{(a)}(t) \sim \frac{C_{k,j}}{\mu_j} \int_0^a (1 - G_j(y)) \, dy \cdot \frac{t^{k-j}}{(k-j)!} \tag{7.58}$$

$$M_0^{(a)}(t) \sim C_{k,0} \cdot a \cdot \frac{t^{k-1}}{(k-1)!}. \tag{7.59}$$

The constant $C_{k,j}$ is

$$C_{k,j} = \prod_{i=j+1}^{k} \frac{p_{i,i-1}}{\mu_i}. \tag{7.60}$$

7.4.5 Discussion

As pointed out above, the type of a cell is the length of the telomere of one single chromosome. This may be reasonable if one particular chromosome is responsible for stopping cell division [or if there only is one chromosome, such as is the case for the protozoan *Tetrahymena thermophila* (see Larson *et al.* 1987)]. If, however, cell division is stopped whenever any of several chromosomes reaches the critical telomere length, the situation is different. We then consider a minimum of several processes and the dynamics is not the same.

For a simple illustration of this, assume a cell has two chromosomes and that the type of the cell is (j, k), where these are the remaining numbers of telomere units on the respective chromosomes. As soon as one of the chromosomes hits 0, the cell stops dividing.

Now suppose the cell is of type $(1, 1)$. Then the two type 1 chromosomes each produce two daughter chromosomes, one of type 0 and one of type 1. If chromosomes are allocated randomly to daughter cells, the two daughter cells may thus be either of types $(0, 0)$ and $(1, 1)$ or of types $(0, 1)$ and $(1, 0)$, with equal probabilities. In the second case, both cells stop dividing, and in the first case the $(1, 1)$-type continues to divide. Clearly, there is only a finite number of divisions in this model (the total number of cells is a geometric random variable) and the population no longer exhibits polynomial growth.

In humans, theoretically any number between 1 and 46 chromosomes could be involved in stopping cell division. There are, so far, three articles that address these problems, two of which use computer-simulation approaches.

In Arino *et al.* (1995), the number of chromosomes involved is taken to be 40 and the model is compared to the data from Levy *et al.* (1992). The 40 chromosomes are assumed to evolve according to independent branching processes, an assumption that simplifies the analysis and gives a good fit, but that does not address the problems mentioned above with observing the minimum.

In Tan (1999), the number of chromosomes that best fits the data of Jones *et al.* (1985) is investigated. The model allows for random lengths of the deletions and the conclusion is that only a few, most likely only two, chromosomes are responsible for sending the initial signal to stop cell division.

In contrast, Rubelj and Vondraček (1999) find a good fit to the same data using all 46 chromosomes and a model that includes what they call "abrupt telomere shortening." This means that the telomeres decrease in unit steps, but once they have reached a certain length n_0, they may lose anything from 1 to n_0 units according to some specified probability distribution on $\{1, 2, \ldots, n_0\}$.

A comparison of the simulations shows that Tan (1999) mimics the cloning of the laboratory by letting cells reproduce to a certain number, then harvesting half the cells at random, letting them reproduce again, and so on until all the cells have stopped dividing. The procedure in Rubelj and Vondraček (1999) is to start from one cell, choose one of the two daughter cells at random, and continue in this fashion a specified number D_0 of times. This gives a good fit to the bimodal distributions observed in the experiments of Jones *et al.* (1985).

7.5 The Polymerase Chain Reaction

P. Jagers

One of the most powerful tools of modern molecular biology is the polymerase chain reaction, commonly referred to as PCR, a technology that can generate a test sample out of a minuscule amount of DNA. In *quantitative* (also called real-time or kinetic) PCR the successive growth of the amount of DNA is followed from a threshold of observation up to large molecule numbers. In ordinary PCR, the focus is on the end product, which is sufficiently large to detect viruses or mutations, and also for many other uses, such as in paternity testing or forensic matters.

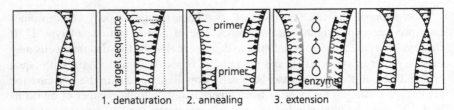

1. denaturation   2. annealing   3. extension

Figure 7.17 The polymerase chain reaction.

The first PCR step is *denaturation* (i.e., heating up to 90°C so that DNA strands separate). In the subsequent *annealing* phase, at around 50°C, short synthetic so-called *primers* of single-stranded DNA bind to the separated PCR sequences. Temperature is raised and an enzyme, the *polymerase*, promotes the synthesis process at the region marked by the primers, the primer extending into a complementary DNA string. Thus, a double-stranded DNA string is obtained, and the number of DNA molecules would be doubled if it were not for stochasticity: primers may fail to attach or other things may go awry. In short, the molecule replicates with a probability p, which is usually called the *efficiency* of the reaction in the present context. The procedure is repeated over and over again, and in a short time a substantial amount can be obtained from just a few DNA strings. The process is illustrated in Figure 7.17. (This is a simplified description, but adequate for our purposes.)

After some possible hesitation in the very first iterations, because of impurities, the amount of DNA amplifies at a seemingly exponential rate. This later slows down into linear growth, but during the most important growth phase, efficiency (i.e., the probability of successful replication) is usually viewed as constant, around 0.95 or even higher.

7.5.1   PCR as a Galton–Watson process

The most natural, albeit simplified, model of PCR describes the exponential growth phase. Thus, the numbers of molecules after iteration $n = 0, 1, 2, \ldots$ constitute a binary Galton–Watson process with the reproduction (or rather, in these circumstances, replication) law

$$\mathbb{P}(\xi = 2) = p, \quad \mathbb{P}(\xi = 1) = 1 - p . \tag{7.61}$$

This means that $m = 1 + p$ and $\sigma^2 = p(1 - p)$. The starting number $Z_0 = z$ may be an unknown parameter to be estimated or a random variable (e.g., with a Poisson distribution arising out of dilution). In such cases it may be of interest to estimate the Poisson parameter.

The process transition probability is

$$\mathbb{P}(Z_{n+1} = j \mid Z_n = i) = \binom{i}{j - i} p^{j-i}(1 - p)^{2i-j} , \tag{7.62}$$

since if $k \leq i$ molecules replicate successfully, the result is $j = i - k + 2k$ molecules in the next round, and $k = j - i$. It follows that the likelihood of a whole quantitative PCR observation from the threshold c of observation up to the last measurement at the nth iteration is

$$
\mathbb{P}(Z_{k-1} < c \le Z_k = z_k | Z_0 = z) \binom{z_k}{z_{k+1} - z_k} p^{z_{k+1}-z_k} (1-p)^{2z_k - z_{k+1}} \times \cdots
$$

$$
\times \binom{z_{n-1}}{z_n - z_{n-1}} p^{z_n - z_{n-1}} (1-p)^{2z_{n-1}-z_n} = \mathbb{P}(Z_{k-1} < c \le Z_k = z_k | Z_0 = z)
$$

$$
\prod_{j=k}^{n-1} \binom{z_j}{z_{j+1}-z_j} \left(\frac{p}{1-p}\right)^{z_n - z_k} (1-p)^{y_{n-1} - y_{k-1}}, \tag{7.63}
$$

where k is the (unobserved) iteration number until the threshold is reached, $z_k, \ldots, z_n$ are the observed molecule numbers, and $y_j = z_0 + z_1 + \cdots + z_j, j = 1, 2, 3, \ldots$ denotes the accumulated number of molecules. Note that $y_n - y_{k-1}$, depends only upon the observable molecule numbers $z_k, z_{k-1}, \ldots, z_n$.

This has a number of interesting consequences. First, the number of iterations until the threshold is passed, usually denoted C_T or CT in PCR literature, together with the first observation form a sufficient statistic (see any basic statistics text-book) to infer the starting number of molecules. Given $C_T = k, Z_k = z_k$, the pair

$$
(Z_n - Z_k, Y_{n-1} - Y_{k-1}) \tag{7.64}
$$

is sufficient and yields the maximum likelihood estimator

$$
\hat{p} = \frac{Z_n - Z_k}{Y_{n-1} - Y_{k-1}} \tag{7.65}
$$

of the efficiency. This is also a consequence of the maximum likelihood estimator of the reproduction mean in a Galton–Watson process generally being given by (see Jagers 1975, p. 47)

$$
\hat{m} = \frac{Y_n - Z_0}{Y_{n-1}}. \tag{7.66}
$$

The starting number of molecules is then estimated naively by

$$
c/(1 + \hat{p})^{C_T}, \tag{7.67}
$$

and it turns out that a more refined analysis does not alter much. Dilution by a factor of, say, $d = 10$, so that the starting number of cells changes from z to z/d, results in a logarithmic increase of C_T, since

$$
zm^{C_T} \approx (z/d)m^{C_T'}, \tag{7.68}
$$

where C_T' denotes the number of iterations until the threshold is reached, if the start was from z/d molecules. However, variance will increase more drastically, actually by d times (*op. cit.*).

7.5.2   Variable efficiency

Actually, efficiency is influenced by a whole range of factors, such as the amount of target DNA, of primers, polymerase, deoxynucleic triphosphate, magnesium chloride, etc. However, many enzymatic reactions are described well by the simplified so-called Michaelis–Menten kinetics (e.g., Cornish-Bowden 1979), and so are the DNA polymerase reactions, as claimed by Schnell and Mendoza (1997a, 1997b); however, also see Lalam *et al.* (2004).

The Michaelis–Menten formula says that reaction velocity is proportional to the free substrate concentration divided by the same plus the (very large) Michaelis–Menten constant K. In a PCR round, the crucial substrates are the target molecules (the other building blocks being present in abundance), so that the rate should be proportional to

$$\frac{s}{K + s} , \tag{7.69}$$

if the number of molecules equals s. If the rate is assumed constant during one PCR iteration, and the durations have the same length, the (expected) number of new molecules produced during one iteration must be proportional to the rate. However, it also equals $sp(s)$, the efficiency p now taken as a function of s. Solving yields

$$p(s) = \frac{A}{K + s} \tag{7.70}$$

for some constant A. Since $p(s) \approx 1$, for little s, A should equal K.

In this manner we arrive at a near critical Galton–Watson process dependent on population size. Such processes are discussed in Sections 2.6, 4.3.2, and 5.8, and PCR is mentioned as an application: the "offspring" number per molecule is

$$m(s) = 1 \times (1 - p(s)) + 2 \times p(s) = 1 + p(s) = 1 + \frac{K}{K + s} . \tag{7.71}$$

Hence, at the beginning of the reaction

$$\mathbb{E}[Z_n | Z_{n-1}] = Z_{n-1} + \frac{K Z_{n-1}}{K + Z_{n-1}} \approx 2 Z_{n-1} , \tag{7.72}$$

or at least $\geq 1.95 Z_{n-1}$, which gives the well-known exponential style growth of $\{Z_n\}$. However, as $n \to \infty$ so does Z_n, and therefore

$$\mathbb{E}[Z_n | Z_{n-1}] \approx Z_{n-1} + K . \tag{7.73}$$

Iterations give

$$\mathbb{E}[Z_n] = \mathbb{E}[Z_{n-1}] + \mathbb{E}\left[\frac{K Z_{n-1}}{K + Z_{n-1}} \right] \cdots$$

$$= \mathbb{E}[Z_0] + \sum_{k=0}^{n-1} \mathbb{E}\left[\frac{K Z_k}{K + Z_k} \right] \sim K n , \tag{7.74}$$

as $n \to \infty$. This shows that ultimately we should expect linear growth, but even more holds true.

Indeed, when the total molecule number is s, the variance of the offspring distribution of any single individual is

$$\sigma^2(s) = 4p(s) + 1 - p(s) - m^2(s) = p(s)(1 - p(s)) = \frac{Ks}{(K + s)^2} . \qquad (7.75)$$

Thus, the variance of the number Z_n of molecules at the nth cycle (see the variance decomposition formula of the Appendix) turns into

$$\text{Var}[Z_n] = \mathbb{E}[\text{Var}[Z_n | Z_{n-1}]] + \text{Var}[\mathbb{E}[Z_n | Z_{n-1}]]$$

$$= \mathbb{E}\left[\frac{K Z_{n-1}^2}{(K + Z_{n-1})^2} \right] + \text{Var}\left[Z_{n-1} + \frac{K Z_{n-1}}{K + Z_{n-1}} \right] . \qquad (7.76)$$

In this, the first term converges toward K from below, as n passes, since then $Z_n \to \infty$. The second expression is $\sim \text{Var}[Z_{n-1}]$, for basically the same reason,

$$\frac{K Z_{n-1}}{K + Z_{n-1}} \to K . \qquad (7.77)$$

It follows that

$$\text{Var}[Z_n] \sim Kn, n \to \infty , \qquad (7.78)$$

which yields the mean square convergence of Z_n/n toward K,

$$\mathbb{E}[(Z_n/n - K)^2] = \text{Var}[Z_n]/n^2 + (\mathbb{E}[Z_n]/n - K)^2 \to 0 . \qquad (7.79)$$

However, there is also almost certain convergence, that is, with probability one the sequence $Z_n/n \to K$ (see Jagers and Klebaner 2003). This is important, as it means that we can follow the successive molecule numbers of any one experiment, and be assured about the convergence. Thus, in spite of their crudeness, Michaelis–Menten kinetics yield branching processes that manage to reproduce the qualitative behavior of PCR: an exponential increase followed by linear growth in molecule numbers. However, this cannot be more than a first step toward a more detailed analysis of how probabilities are determined by circumstances; the Michaelis–Menten approach in itself is not only approximate, but also it is macroscopic rather than probabilistic in essence. In one model (see Lalam *et al.* 2004), a purely exponential growth phase is supposed to be followed by one governed by Michaelis–Menten kinetics, as above. Another generalization is to regard the Michaelis–Menten constant as summarizing the experimental conditions, and thus subject to random variation. This is pursued in Jagers and Klebaner (2004).

7.6   Modeling Measles Outbreaks

V.A.A. Jansen and N. Stollenwerk

7.6.1   Introduction

Epidemiology is one of the areas in biology to which mathematical modeling has been applied most successfully. It was through the mathematical concept of basic reproductive number that Ronald Ross gained the insight that malaria could be controlled if the mosquito population was sufficiently suppressed (Heesterbeek 2002). Later, similar concepts were applied to the eradication of smallpox and the control of many viral diseases through vaccination campaigns (Anderson and May 1991). To date, mathematical models remain an important tool.

Epidemiology studies the distribution of a disease within a host population. Most models used for infectious diseases are based on partitioning the host population into different classes that correspond to the different stages of the infectious process (e.g., susceptible, infectious, and recovered) and the transitions between these classes. Such models work particularly well for diseases caused by microparasites, a class of pathogens that comprises viruses and bacteria and that are characterized by an infection normally accomplished with a single dose that consists of a relatively small number of infective particles (Anderson and May 1991).

Mathematical epidemiology has concentrated on cases for which the disease is endemic or in which large epidemics occur. In this context the reproductive number is an important parameter. It is defined as the average number of secondary infections caused by an infected host over the lifetime of the infection. In a completely susceptible population, this number is known as the basic reproductive number, R_0. If the reproductive number is larger than one, a single infection can lead to a chain reaction of infections and, eventually, to an epidemic or an endemic state. If the reproductive number is smaller than one, large epidemics do not occur and the pathogen is bound to disappear from the population.

If the basic reproductive number is less than one, the numbers of infected individuals hardly ever become truly large and stochastic effects prevail. If this is the case the disease manifests itself in the form of outbreaks that follow the introduction of the disease into the population. The size and duration of an outbreak can vary enormously through chance events. To capture these dynamics a stochastic formalism is needed. In this section we illustrate how the theory of branching processes can be used to describe the epidemiology of pathogens with a reproductive number smaller than one, using the epidemiology of measles as an example. We use this to explain the distribution of disease outbreaks in small island populations, and the effect of a recent decline in vaccination in the UK after a scare about vaccine safety.

7.6.2   The epidemiology of measles

Measles is caused by the measles virus. It is transmitted on close contact via airborne propagules. Infection leads to the development of a typical rash. The infectious period is in the order of a week, after which the hosts recover and develop lifelong immunity. Hosts, therefore, are normally infected only once in their lifetime and, if the force of infection is sufficiently large, this happens at a young age, and hence measles is a childhood disease. Although in unvaccinated populations measles is a common disease, infection is not without danger. In developed countries infection with measles leads to complications in one out of seven cases and is fatal in about one in 5000 cases (Ramsay *et al.* 1994; Carabin *et al.* 2002).

The basic reproductive number of measles lies between 10 and 18 (Anderson and May 1991). In large, unvaccinated populations measles is endemic, and a course of measles is a normal part of childhood. Before mass vaccination was introduced, measles used to follow a cyclic pattern, with a period of about 2 years in Europe and North America. Mathematical models have shown that the annual variation in the transmission together with the disease dynamics can result in a 2-year cycle or more complex dynamics (Bolker and Grenfell 1993; Drepper *et al.* 1994). Since mass vaccination was introduced in the UK in the 1970s, measles has lost its periodic character. Over the past decade the vaccine coverage in most parts of Europe was above 90% (de Melker *et al.* 2001) and measles currently only occurs following introduction of the disease.

In small, isolated populations an outbreak of measles can immunize a large part of the population, so the disease disappears even if the reproductive number was larger than one initially. Therefore, measles cannot persist in small populations (Bartlett 1957). In small island populations this is, indeed, the case and measles is not endemic, but comes in the form of outbreaks after importation of the disease (Rhodes and Anderson 1996; Rhodes *et al.* 1997). In a number of small islands these outbreaks have been documented meticulously, which provides an unparalleled record of outbreak patterns.

In the UK, the safety of the combined measles, mumps, and rubella (MMR) vaccine recently became the focus of a heated debate following concerns over the safety of the vaccine (Wakefield *et al.* 1998). None of the claims regarding the safety of the vaccine have been confirmed (Donald and Muhtu 2002; Consumer's Association 2003), but nevertheless this scare resulted in a decreased uptake of the MMR vaccine. As a result, measles outbreaks have increased in size (Ramsay 2003). We use a branching process to describe the epidemiology of measles in a vaccinated population and demonstrate how this model can be used to estimate the reproductive number in the UK population. Such information is of vital importance in public health policy.

7.6.3 A general model for measles

A basic model for the epidemiology of measles outbreaks is founded on a subdivision of the host population into classes. The allocation of individuals to classes is not static: hosts can move from one class to another. Whenever this happens we speak of a transition. Examples of transitions are infection, which moves a host from the susceptible to the infected class, and recovery, which moves a host from the infected to the recovered class.

Table 7.3 Transition rates for the continuous-time epidemic process.

Event	Type of transition	Rate
Infection	$S \rightarrow S - 1, I \rightarrow I + 1$	$\beta S I / N$
Recovery	$I \rightarrow I - 1, R \rightarrow R + 1$	γI
Death of infected	$I \rightarrow I - 1, S \rightarrow S + 1$	μI
Death of recovered	$R \rightarrow R - 1, S \rightarrow S + 1$	μR
Vaccination	$S \rightarrow S - 1, R \rightarrow R + 1$	νS

We make one important assumption, that the transition rate (i.e., the probability of a host moving from one class to another per unit of time) depends only on the current state of the system and not on the system's history. The state of the system is defined by the numbers of hosts in the different classes (the Markov property discussed in Section 2.3). This assumption is less restrictive than it sounds. It effectively requires us to choose classes in such a way that they contain all the information necessary to predict changes at the current point in time.

We consider three classes: susceptible hosts, infected, and recovered. Susceptible hosts have never been in contact with the virus and have not been vaccinated. The number of susceptible hosts is denoted by S. Upon infection susceptible hosts enter the infected class. The number of infected individuals is given by I. Hosts that acquire immunity to measles, either through exposure to the virus or vaccination, move into the recovered class. The number of recovered hosts is given by R. Models of this type are called SIR models.

The average length of the infectious period for measles is about a week. As this is very short compared to the average lifetime of the human host we assume that the total host population is constant on the timescale of a measles outbreak. The size of the host population is $N = S + I + R$. Susceptible hosts become infected with the virus upon contact with an infected individual. The force of infection (the probability that an individual host acquires the infection per unit of time) is given by $\beta I / N$, where β is the transmission parameter. The force of infection is proportional to the fraction of infected people, I / N, because the number of potential infectious contacts tends to be independent of the population size. The rate at which the number of infected individuals grows is $\beta I S / N$.

An infected host has a probability γ of recovering from the infection per unit of time. The rate at which the number of infected individuals decreases through recovery is thus γI. In addition, birth, death, and vaccination are taken into account. All individuals have a probability μ of dying per time unit. As we wish to keep the population size constant, we replace dead individuals by newborn susceptible individuals. Finally, we model vaccination as a fixed probability per unit of time, ν, of a susceptible host being moved to the recovered class. These transitions are summarized in Table 7.3.

The average duration of infection in this model is $1/\gamma$. The average number of secondary infections is $\frac{\beta S}{\gamma N}$. In a completely susceptible population $S = N$, and hence the basic reproductive number is $R_0 = \frac{\beta}{\gamma}$. In a population in which a

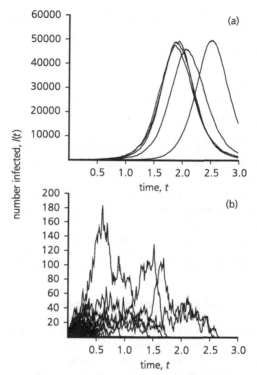

Figure 7.18 Stochastic simulations of the SIR system above and below criticality. For these simulations we used $\beta = 500.1333$, $\mu = 1/75$, and $\gamma = 50$ (corresponding to an average recovery time of about 1 week). (a) For $\nu = 0.108$: if the vaccination dynamics has equilibrated these values correspond to $R_c = 1.099$ and the system is supercritical. After an initial random phase the epidemic looks essentially deterministic (i.e., the graph is smooth) and is well described by the ordinary differential equation system in Section 7.6. In this figure 100 simulations are shown. Most epidemics have faded out and are therefore not visible. (b) For $\nu = 0.1215$: the system has $R_c = 0.989$ after equilibration and the system is subcritical. As can be seen, the outbreaks have a strong stochastic component. In this figure 500 simulations are shown.

fraction c of the individuals is vaccinated and in which the virus is not present the number of susceptible hosts is $S = (1 - c)N$ and the reproductive number is $R_c = \frac{\beta(1-c)}{\gamma} = R_0(1 - c)$.

7.6.4 A deterministic model for endemic measles

If the reproductive number exceeds one an epidemic can start after the introduction of the disease in the population, when it frequently becomes endemic, and affects many people. Once the infection rate has become sufficiently large, stochastic effects play a minor role and changes in the number of hosts in the different classes are largely deterministic (see Figure 7.18a). If this is the case, the dynamics of a well-mixed population can be described by the differential equations

$$\frac{dS}{dt} = \mu N - \frac{\beta I S}{N} - (\mu + \nu) S \tag{7.80}$$

$$\frac{dI}{dt} = \frac{\beta I S}{N} - (\gamma + \mu) I \tag{7.81}$$

$$\frac{dR}{dt} = \gamma I + \nu S - \mu R, \tag{7.82}$$

where the densities S, I, and R are differentiable real-valued functions of time. This is the classic deterministic SIR differential equation model for the dynamics of endemic measles.

Figure 7.18b shows the number of infected individuals over time after introduction into a vaccinated population with a reproductive number smaller than one. These simulations differ from those in Figure 7.18a in that the disease always disappears from the population and that the outbreak size does not become truly large (note the difference in the scales), so that stochastic effects dominate the dynamics. To capture the stochastic nature, we simplify our model further.

7.6.5   A stochastic model for measles outbreaks

For the formulation of a simple stochastic model, we first observe that outbreaks take place on a relatively short timescale, in the order of weeks or months. Demographic processes in humans generally take place at much slower timescales. We can therefore assume that at the timescale of a measles outbreak, birth and death do not have a major impact on the disease dynamics, and we set birth and death rates to zero. Similarly, the process of vaccination only changes the fraction of susceptible hosts relatively slowly and we also set the vaccination rate to zero. Vaccination enters the simplified model through the fraction initially immunized, c.

If the reproductive number is smaller than one the disease disappears from the population and hardly ever infects a substantial part of it. As a further simplification, we assume that the population is very large, and that the presence of the disease does not have an impact on the fraction of susceptible hosts. The fraction of susceptible hosts is constant and remains at $1 - c$ throughout the outbreak. A second consequence of this assumption is that we only need to keep track of the number of infected people. Under these assumptions, the disease dynamics reduces to a linear birth-and-death process (see Table 7.4), as introduced in Section 3.2. Its birth rate is $b = \beta(1 - c) = \gamma R_c$ and the death rate is $d = \gamma$. It follows from Equation (3.10) that

$$\mathbb{E}[I(t)] = M(t) = e^{\gamma(R_c - 1)t}, \tag{7.83}$$

if the process started from one single case. This can, of course, also be derived directly from the differential equation for the mean, much as done in Section 3.2.

Table 7.4 Transition rates for the simplified continuous-time Markov process.

Event	Type of transition	Rate
Infection	$I \rightarrow I + 1$	$\beta(1 - c)I$
Recovery	$I \rightarrow I - 1$	γI

The differential equation argument also applies to the variance of Markov branching processes $V(t) = \text{Var}[I(t)]$. In the present case

$$\frac{dV}{dt} = \gamma(R_c + 1)M - 2\gamma(1 - R_c)V , \qquad (7.84)$$

which yields exact solutions that, if $R_c < 1$, increase initially, but in the limit of large t decrease as $e^{-2\gamma(1-R_c)t}$, grow linearly in the critical case $R_c = 1$, and, if $R_c > 1$, in the limit of large t increase as $e^{\gamma(R_c-1)t}$. This is in analogy with the asymptotic formulas for more general branching processes in Section 3.2, and exactly as in the discrete Markov (i.e., Galton–Watson) case (Section 2.2). We see that only in the subcritical situations does the variance vanish with time and, indeed, if the reproductive number is close to one, realizations are not at all close to the mean (Figure 7.19). To investigate the behavior near criticality we derive the distribution of outbreak sizes.

7.6.6 The size distribution of outbreaks

In the subepidemic case, outbreaks are sparked by the introduction of the disease into the population. To derive the distribution of their size, we apply the branching process interpretation used to generate the realizations in Figures 7.18 and 7.19. In stochastic simulations the system remains unchanged most of the time and only changes because of transitions: jumps in which the numbers in the model classes change by one. The probability of change per unit time is constant between jumps and therefore the times between successive jumps are distributed exponentially, with a mean equal to the reciprocal of the total rate of leaving the state the system is in. For our simulations the time to the next jump is drawn from an exponential distribution. The probability of leaving the initial state and ending in a particular state is given by the rate that corresponds to this transition, divided by the total rate of leaving the initial state. This procedure, which follows directly from the probabilistic structure of the process, is known as the Gillespie algorithm (Gillespie 1976, 1978; Feistel 1977). It is a computationally efficient way to generate realizations of continuous-time Markov processes.

Let us apply this procedure to the system if it has I infected individuals. The rate of leaving that state is given by $(b + d)I = \gamma(R_c + 1)I$, as previously introduced, and the average waiting time is $1/(\gamma(R_c + 1)I)$. Upon leaving this state the system can either go to the state $I + 1$ if the most recent event was an infection or to the state $I - 1$ if it was a recovery. The next event is an infection with probability $\frac{b}{b+d} = \frac{R_c}{R_c+1}$ and a recovery with probability $\frac{d}{b+d} = \frac{1}{R_c+1}$ (Table 7.5). Note that these probabilities do not depend on the number of infected individuals.

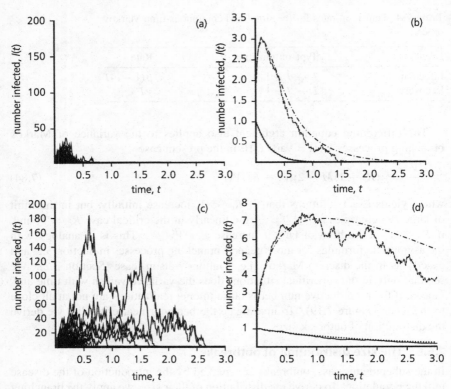

Figure 7.19 The left panels show the different realizations and the right panels show the observed ensemble mean (continuous) and variances (dotted) and predicted ensemble mean (dashed and dotted) and predicted ensemble variance (dashed). For (a) and (b) 10 000 simulations were used, for (c) and (d) 500 simulations. The lower curves almost coincide, which indicates a good approximation. Parameters: (a, b) $\nu = 0.132$, (b, c) $\nu = 0.1215$.

Table 7.5 Transition rates for the simplified event based discrete-time Markov process.

Event	Type of transition	Probability
Infection	$I \to I + 1$	$\frac{R_c}{R_c+1}$
Recovery	$I \to I - 1$	$\frac{1}{R_c+1}$

To calculate the distribution of the number of cases that result from a single introduction we need not know when these cases occurred. We therefore discard the part of the algorithm in which the waiting time is computed, and follow the process from event to event, events being infection or recovery. This transforms the continuous-time branching process into a discrete-time branching process with constant transition probabilities, a simple random walk.

The number of events that have passed is indicated by the variable T. Let $p(I, T)$ be the probability that there are I infected individuals after T events. The algorithm outlined gives the transition probabilities from event to event. If $I > 1$, this probability satisfies the recursion

$$p(I, T+1) = \frac{R_c}{R_c + 1} p(I - 1, T) + \frac{1}{R_c + 1} p(I + 1, T), \tag{7.85}$$

and otherwise

$$p(1, T+1) = \frac{1}{R_c + 1} p(2, T) \tag{7.86}$$

and

$$p(0, T+1) = p(0, T) + \frac{1}{R_c + 1} p(1, T); \tag{7.87}$$

at $T = 0$ all probabilities are 0 except $p(1, 0) = 1$.

By transforming to a discrete branching process we have lost some information (when events happen), but we gain a tremendous simplification in that the transition probabilities have become constants. The recursion above has the solution

$$p(I, T) = T!I \frac{(R_c + 1)^{-T} R_c^{\frac{1}{2}(T+I-1)}}{(\frac{1}{2}(T + I + 1))!(\frac{1}{2}(T - I + 1))!} \tag{7.88}$$

if $T - I - 1 > 0$ and if $T - I - 1$ is even, and $p(I, T) = 0$ otherwise. It is easy to check that this solution is correct by substituting it into Equations (7.85)–(7.87).

We use this to find the probability that an outbreak with a total of x cases occurs. Suppose that the outbreak stops after T events, when the individual infected last recovers. Therefore, at $T - 1$ the number of infected individuals has to be one and $(T - 1)/2$ of the past events must have been infections (otherwise the process cannot have returned to $I = 1$). The total number of cases in the outbreak is $x = 1 + (T - 1)/2$ (half of the events must have been infections, one infection is added to account for the infected individual present at $T = 0$), from which follows $T = 2x - 1$. Let $q(x)$ be the probability that the final size of the outbreak is x. For the outbreak to stop after $2x - 1$ events the last event has to be a recovery and hence this probability is

$$q(x) = \frac{1}{R_c + 1} p(1, 2x - 2) = \frac{R_c^{x-1}}{(R_c + 1)^{2x-1}} \frac{(2x - 2)!}{x!(x - 1)!}. \tag{7.89}$$

Alternatively, the moment generating function of the outbreak size distribution of the embedded Galton–Watson process could be derived, as described in Section 3.1 (see Section 7.6.9). Although this method is systematic it has the disadvantage that the probability distribution is found in terms of a series that is not easy to interpret. We therefore prefer the method given above, as it leads directly to the probability distribution. In case the moments rather than the distribution are of interest, the generating function procedure is readily applicable.

Often the probability of a certain outbreak size or larger is of practical use. This probability is given by $\sum_{x=n}^{\infty} q(x)$. Before deriving an expression for this quantity we first observe that if the reproductive number is smaller than one the disease always disappears from the population and hence $\sum_{x=1}^{\infty} q(x) = 1$. However, if $R_c > 1$, the disease can disappear by chance from the population. This happens with probability $\sum_{x=1}^{\infty} q(x) = 1/R_c$. In our model the disease goes to infinity with probability $1 - 1/R_c$, which corresponds to an outbreak that affects a large fraction of the population in a large finite population. Using this we find that the probability of an outbreak of a certain size or larger is given by

$$\sum_{x=n}^{\infty} q(x) = \max\left(0, 1 - \frac{1}{R_c}\right)$$
$$+ \frac{(4R_c)^{n-1}}{(1+R_c)^{2n-1}} \frac{\Gamma(n-\frac{1}{2})}{\sqrt{\pi}n!} \, {}_2F_1\left(1, n-\frac{1}{2}, n+1, \frac{4R_c}{(1+R_c)^2}\right) \quad (7.90)$$

where ${}_2F_1$ is the hypergeometric function (Abramowitz and Stegun 1964).

7.6.7 Measles outbreaks in small islands

Historical records of measles outbreaks in small islands show tremendous variation in outbreak size: most are small, but sometimes a substantial part of the island population becomes infected. It has been shown that the size distribution of the outbreaks can be described by a power law (Rhodes and Anderson 1996). Power laws are fingerprints of critical systems (Stanley 1971; Jensen 1998), and it has been suggested that the dynamic behavior of measles in small islands is an example of criticality (Rhodes *et al.* 1997). This has been supported by individual-based models in which this behavior was replicated (Rhodes and Anderson 1996; Rhodes *et al.* 1997).

Another characteristic property of critical systems is the divergence of the variance (Stanley 1971; Yeomans 1992). This phenomenon is shown in Figure 7.19 for reproductive numbers close to one. In the critical case the frequency distribution of events can be described by a power law. Figure 7.20a illustrates how the distribution of outbreak sizes approaches a straight line in a log–log plot if the reproductive number goes to one. Indeed, in the limit of the reproductive number tending to one, we find

$$\lim_{R_c \to 1} q(x) = 2^{1-2x} \frac{(2x-2)!}{x!(x-1)!} \approx \frac{x^{-3/2}}{2\sqrt{\pi}}, \quad (7.91)$$

and the probability to find an outbreak of size n or larger tends to

$$\lim_{R_c \to 1} \sum_{x=n}^{\infty} q(x) = 2^{2-2n} \frac{(2n-2)!}{((n-1)!)^2} \approx \frac{n^{-1/2}}{\sqrt{\pi}}, \quad (7.92)$$

where we used Stirling's formula for the approximation and hence it holds for large n (Stollenwerk and Jansen 2003). We have thus found that our branching process model predicts a power law in the frequency of outbreaks with an exponent of

$-3/2$, and the frequency of an outbreak of a particular minimum size with exponent $-1/2$. This is close to the value found in the data for the Faeroe Islands, which was -0.27 ± 0.014 (95% confidence interval; Rhodes *et al.* 1997). For whooping cough (pertussis) and mumps similar exponents were found. This suggests that the reproductive number in these small island populations for all these diseases is close to one. In these populations the number of susceptible hosts builds up between outbreaks, which increases the reproductive number. Every outbreak immunizes a part of the population and thus reduces the reproductive number. This process keeps the reproductive number, on average, at unity. The deviation in the exponent from $-1/2$ in the Faeroe data could result from additional effects, such as spatial structure (Jensen 1998).

The occurrence of power laws in the outbreaks of measles has been explained previously by the spatial structure in the population. Here we show that power laws arise in a simple branching process without any spatial structure. This fact has long been known (see, e.g., Harris 1963; Jensen 1998), but previously has received little attention in the biological literature. Simple branching processes provide a simple and parsimonious explanation for the occurrence of power laws in epidemiological data.

7.6.8 The basic reproductive number following the MMR scare

The reduced vaccine uptake in the UK after the MMR scare has coincided with a large number of measles outbreaks. These outbreaks can be reconstructed from epidemiological data by grouping all the cases that have had epidemiological contact. This requires a detailed investigation and the resultant clusters are, to an extent, subjective. The distribution of the outbreak sizes can be used to infer important epidemiological information, in particular it can be used to estimate the reproductive number (De Serres *et al.* 2000; Farrington *et al.* 2003). We use data on outbreak size to show how the reproductive number in the UK population has changed in response to the MMR scare.

Using the outbreak size distribution derived in this chapter, a maximum likelihood estimate can be found. The rationale behind likelihood estimates is that one tries to identify the most likely estimate for R_c given a set of observed outbreak sizes. To do so, consider a set of n observed outbreaks of size $(x_1, \ldots, x_n)$. The likelihood of the data (i.e., the probability of observing these data given that the reproductive number is R_c) is proportional to

$$\prod_{i=1}^{n} q(x_i) = \frac{R_c^{mn-n}}{(R_c + 1)^{2mn-n}} \prod_{i=1}^{n} \frac{(2x_i - 2)!}{x_i!(x_i - 1)!} , \qquad (7.93)$$

where we use $m = \frac{1}{n}\sum_{i=1}^{n} x_i$ to denote the mean outbreak size. We want to know for what value of R_c the likelihood is maximized. By differentiation with respect to R_c we find that the maximum likelihood is found for

$$(m - 1)\frac{1}{\hat{R}_c} - (2m - 1)\frac{1}{\hat{R}_c + 1} = 0 , \qquad (7.94)$$

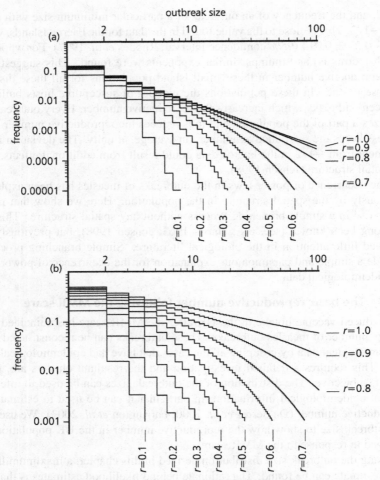

Figure 7.20 (a) The distribution of outbreak sizes. The distribution approaches a straight line with a gradient of $-3/2$ in a log–log plot if the reproductive number goes to one. (b) The distribution of the probability of an outbreak of a certain size or larger [$P(x \geq n)$]. This distribution approaches a straight line with a gradient of $-1/2$ in a log–log plot if the reproductive number goes to one.

and hence for $\hat{R}_c = 1 - 1/m$. Note that the predicted mean outbreak size is given by

$$\sum_{x=1}^{\infty} xq(x) = \frac{1}{1 - R_c} \text{ if } R_c < 1. \tag{7.95}$$

The resultant estimates for the UK were $\hat{R}_c = 0.35$ for the period 1995–1998 and $\hat{R}_c = 0.70$ for 1999–2002, which indicates a clear increase.

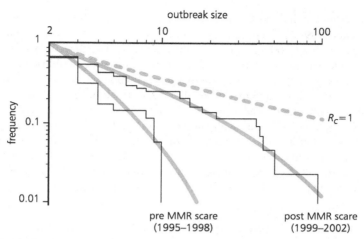

Figure 7.21 The frequency of an outbreak of measles of a particular size or larger in the UK in the years 1995–1998 and 1999–2002. Isolated cases and Steiner outbreaks are excluded. The continuous lines are the expected distributions. The dashed line is the distribution for $R_c = 1$. Modified from Jansen *et al.* (2003).

So far we have used all the reported cases of measles in the UK in the years 1995–2002, grouped over outbreaks. In such data sets isolated cases tend to be misrepresented for several reasons. First, one can expect that the introducing individuals would normally spend less than their entire infectious period in the country they are importing the infection into. This results in an under-representation of isolated cases. A more serious source of error is that sometimes epidemiological analysis fails to connect infections to the outbreak they are part of. This tends to over-represent isolated cases. Indeed, for the data 1999–2002 the model that used all data points can be rejected through a Kolmogorov–Smirnov test.

A more reliable estimate of the reproductive number is found by excluding isolated cases. This can be done as follows. The probability of an isolated case is $q(1) = \frac{1}{R_c+1}$. The distribution of outbreaks of size $x \geq 2$ is given by $q(x)/(1 - q(1))$. We now find the maximum likelihood estimator to be $\hat{R}_c = 1 - 2/m$. The resultant estimates are 0.47 for the period 1995–1998 and 0.82 for 1999–2002. These agree well with observed data (Figure 7.21). The conclusion that the reproductive number has increased was further corroborated by a bootstrap argument in Jansen *et al.* (2003).

Clearly, the reproductive number increased after the reduction in vaccine uptake decreased following the MMR scare (Jansen *et al.* 2003). Vaccine uptake has reached the lowest point for 10 years. The consequence of this is a further accrual of unvaccinated individuals and, inevitably, a further increase in the reproductive number. If the vaccine uptake does not increase, this will eventually lead to a re-emergence of measles as an endemic disease in the UK. An indication that the measles epidemiology in the UK is approaching the critical point at which the reproductive number equals one is provided by the distribution of outbreak sizes.

By comparing the distribution of outbreak sizes before 1999 with the distribution for the years 1999–2002 a progression toward criticality can be seen (Figure 7.21).

Currently, about 200 cases of measles are reported in the UK per year. The risk of a child actually suffering measles complications is negligible because of the low incidence. If the reproductive number increases above unity, this situation will change and unvaccinated individuals will have a substantial chance of contracting the disease. Although the risk of serious complications is small, this risk certainly outweighs the risk associated with vaccination (Carabin *et al.* 2002). To put this risk in context, before mass vaccination was introduced measles caused about 100 deaths per year in the UK (Gay *et al.* 1995). The decision not to vaccinate one's child so as to avoid a perceived risk can have the ironical consequence that the child is exposed to a much larger risk if this behavior is taken up by the population at large.

7.6.9 The moment generating function of the embedded process

Consider the embedded Galton–Watson process of Section 3.1, which defines the outbreak initiator as the ancestor and all those directly infected by the initiator as its direct offspring, which thus constitute the first generation. Those infected by members of the latter make up the second generation, and so forth. If we write Z_n for the number of members of the nth generation thus defined and $Z_0 = 1$ (i.e., one initiator) this constitutes a Galton–Watson process, whose reproduction distribution is easily seen to be geometric with the parameter $R_c/(R_c + 1)$. Since an infection occurs with probability $\frac{R_c}{R_c+1}$, the probability of infecting k or more individuals during the individual's period of infection is

$$\mathbb{P}(\xi \geq k) = \left(\frac{R_c}{R_c + 1} \right)^k . \tag{7.96}$$

The mean number of individuals directly infected by any single person is thus $\mathbb{E}[\xi] = R_c$, and the size of the outbreak is the accumulated total size of the embedded Galton–Watson process,

$$Y = \sum_{k=0}^{\infty} Z_k . \tag{7.97}$$

We note immediately that Y is finite if $R_c \leq 1$, the expected total size of the outbreak being

$$\mathbb{E}[Y] = \begin{cases} \sum_{k=0}^{\infty} R_c^k = 1/(1 - R_c), & \text{if } R_c < 1 , \\ \infty, & \text{if } R_c = 1 . \end{cases} \tag{7.98}$$

If $R_c > 1$, Y is finite if and only if the embedded Galton–Watson process dies out. Since the reproduction generating function is

$$f(s) = \frac{1}{1 + R_c(1 - s)} , \tag{7.99}$$

the probability Q of this, being as always the smallest non-negative root of $Q = f(Q)$, is $1/R_c$ for $R_c > 1$ (see Section 5.3).

To find the distribution of the outbreak size, note that $Y_n = Z_0 + Z_1 + \cdots + Z_n$, starting from one individual, has a generating function, $h_n(s) = \mathbb{E}[s^{Y_n}]$, that satisfies the recursive relation

$$h_{n+1}(s) = \mathbb{E}[s\mathbb{E}[s^{Z_1 + \cdots + Z_{n+1}}|Z_1]] = \mathbb{E}[s\mathbb{E}[s^{1+\cdots+Z_n}]^{Z_1}] = sf(h_n(s)) \,. \quad (7.100)$$

Passage to the limit $n \to \infty$ yields a functional equation, $h(s) = sf(h(s))$, which can be solved in the present case, since it reduces to a second-degree equation.

The generating function then yields the probability distribution after an inverse transformation as a series. See, e.g., Jagers 1975, pp. 39 ff; this and other textbooks also contain the intriguing consequence of the so-called Ballot theorem, that the total size of a branching process or, in this case, an infection outbreak satisfies

$$\mathbb{P}(Y = k) = \mathbb{P}(Z_1 = k - 1|Z_0 = k)/k \,, \quad (7.101)$$

from which the form of the outbreak size probability distributions, in theory, can be found (Jagers 1975, p. 42).

For a detailed description of the use of branching processes in epidemiology see Farrington *et al.* (2003).

Acknowledgments We gratefully acknowledge financial support of The Wellcome Trust (grant no. 063134).

7.7 Metapopulations

M. Gyllenberg

7.7.1 Introduction

Most population models, both deterministic and stochastic, assume that all individuals of the population live in the same habitat and interact homogeneously with each other. Models of this type have been used successfully to describe, explain, and predict the local dynamics of one or several interacting species.

Natural populations of most species have a spatial structure, with several geographically distributed habitat patches that can support local populations. Such a population of populations is called a *metapopulation*. Local populations in a metapopulation are connected by migration. A local population may become extinct while the metapopulation persists. An empty patch may be colonized by migrants from other patches. Extinction and recolonization are the essential features of metapopulation dynamics. Hanski and Gilpin (1991) even characterized the study of metapopulation dynamics as the study of conditions under which these two processes are in balance and the consequences of this balance to associated processes.

Many important questions in ecology, genetics, and evolution require the metapopulation concept to be analyzed appropriately. For instance, conservation biology is an important area in which metapopulation dynamics plays a prominent role [see the book by Hanski (1999) and the books edited by Gilpin and Hanski (1991) and Hanski and Gilpin (1997) for many other examples].

The simplest deterministic patch model, one in which the time evolution of the fraction $p(t)$ of occupied patches is modeled, was introduced by Levins (1969, 1970) and has the form

$$\frac{dp(t)}{dt} = cp(t)(1 - p(t)) - ep(t) \,. \tag{7.102}$$

Here, c and e are the colonization and extinction parameters, respectively. A model very closely related to Equation (7.102) is the *mainland–island* model,

$$\frac{dp(t)}{dt} = c(1 - p(t)) - ep(t) \,, \tag{7.103}$$

which is a single-species version of a model by MacArthur and Wilson (1967). In Equation (7.103), colonization is assumed to occur as a result of migration from the mainland and the colonization rate is thus directly proportional to the fraction of empty patches, whereas in Equation (7.102) empty patches are colonized by migrants from the occupied patches and so the colonization rate is assumed to be proportional to the product of the fractions of empty and occupied patches.

Equations (7.102) and (7.103) make several simplifying assumptions. First of all, since they are deterministic, it is assumed tacitly that the number of patches is infinite. Second, all patches are assumed to be identical, whereas in nature there is always variation in patch size and quality. Third, the spatial arrangement of the habitat patches is completely ignored: an isolated patch located far away from the mainland or the other patches has the same probability of being colonized as a patch close to the mainland or in the middle of a cluster of patches. Moreover, in the Levins model [Equation (7.102)] the colonization rate depends only on the fraction of occupied patches and not on the specific patches that are occupied. It is clear that in real life an empty patch surrounded by nearby occupied patches is more likely to be colonized than a patch with neighboring patches that are all empty. Finally, local dynamics is ignored and the models therefore neglect the effects of migration upon local dynamics.

To relax Levins' assumptions that all patches are identical and that local dynamics is not affected by migration one has to turn to the *structured* metapopulation models treated in Section 4.3. These models are completely analogous to "ordinary" structured population models (Metz and Diekmann 1986; Diekmann *et al.* 1993, 1998, 2001), in which local populations play the role of individuals and the metapopulation corresponds to the population. This conforms to the general view in this book that "individual" is understood, in a broad sense, to include, among other things, local populations (see Section 1.3).

When modeling structured metapopulations, the starting point is to model local dynamics. Local populations grow or decline as a consequence of reproduction, death, emigration, and immigration. Patches may change in size and quality and may be destroyed (Gyllenberg and Hanski 1997). These processes at the local level correspond to processes such as growth and aging at the individual level in "ordinary" population dynamics. We refer to them as *local state development*. There is also an obvious analog of reproduction: empty patches may be recolonized by migrants that arrive from extant local populations, and thus give rise to or, if you wish, give birth to a new local population, and so new patches may be formed.

Gyllenberg *et al.* (1997) modeled local state development and survival as a Markov process with local extinction and/or patch destruction as the transition to an absorbing state. The "birth" process (formation of new populations and patches) was modeled by prescribing a *colonization kernel* to keep track of how many, and of what local state, new populations will be "born" to a local population of a given local state during the course of time. One should keep in mind that the transition probabilities of the Markov process that describes local state development and the colonization kernel depend on the state of the metapopulation. This is clear since, for instance, the immigration rate, which affects local dynamics, depends on the size of the metapopulation and, since only empty patches can be colonized, no new populations are born if all patches happen to be occupied. In a fully stochastic model the colonization kernel should be a probability that depends on the state of the metapopulation. However, Gyllenberg *et al.* (1997) made the model deterministic by letting the colonization kernel express the expectation of the number of offspring with state at birth in a given set. This enabled a comparatively easy treatment of non-linear feedback mechanisms (Diekmann *et al.* 2001), which seem intractable in a fully stochastic model.

As they are deterministic, the structured metapopulation models described above are still based on the assumption of a very large number (effectively infinite) of patches. They also neglect the explicit spatial structure of the habitat patches and information about which patches are actually occupied. Another drawback is that models which take the effect of immigration upon local dynamics into account predict that there cannot be any empty patches. This is because in a deterministic model there is a continuous non-zero influx of individuals into every patch, so if a population becomes extinct through a local catastrophe, the patch is recolonized instantaneously by this stream of immigrants. To arrive at a consistent deterministic model that allows empty patches one has to model colonization in a non-mechanistic way, which seems hard to justify biologically (Gyllenberg *et al.* 1997).

Empty patches do occur in nature. When a migrant arrives at an empty patch it has to adapt to the new environment, survive for a sufficiently long time to be able to found a new clan, and so on. This strongly suggests that successful colonization should be modeled as a *stochastic* event. In this section we present a simple, stochastic metapopulation model that is spatially explicit with a finite number of patches. To keep the exposition simple we neglect local dynamics, but this could be incorporated easily in the model. We illustrate the usefulness of the model by addressing three biologically relevant questions.

The first question relates to conservation biology and viability of metapopulations. When will a small metapopulation grow initially (the supercritical case)? If it does not grow (subcritical case) the metapopulation rapidly becomes extinct, but even in the supercritical case the metapopulation eventually becomes extinct. In such cases the viability can be measured by the expected time to extinction. In this section, we present a formula for the distribution of extinction times in a certain limiting case.

A concept of fundamental importance in stochastic metapopulation models is that of *incidence (of occupancy)*. The incidence of a given patch is defined by Hanski (1994; see also Gilpin and Diamond 1981) as the stationary probability that the patch is occupied. *Incidence functions* describe how the incidence depends on patch size or some other patch characteristic (Diamond 1975). Later, we provide formulas for the incidence in different cases.

Finally, the *distribution* of a species is measured by the expected number of occupied patches. In this section we derive a formula for the probability distribution of the number of occupied patches.

We do not present any proofs. Full proofs of all the results can be found in Gyllenberg and Silvestrov (1994).

7.7.2   The model

We consider a collection of n patches that at the discrete-time instants $t = 0, 1, 2, \ldots$ can be either occupied or empty. The state of patch i at time t is given by the random indicator variable $\eta_i(t)$, which takes on the value 1 if patch i is occupied and 0 if patch i is empty at time t. The state of the metapopulation is described by the vector random process with discrete time $\bar{\eta}(t) = (\eta_1(t), \ldots, \eta_n(t)), t = 0, 1, \ldots$. The state space of the process $\bar{\eta}(t)$ is $\bar{X} = \{\bar{x} = (x_1, \ldots, x_n) : x_i \in \{0, 1\}\}$. It has 2^n states. The state $\bar{0} = (0, 0, \ldots, 0)$ corresponds to metapopulation extinction.

The local dynamics is modeled by preassigning the n by n *interaction matrix* $Q = (q_{ji})$. Here, $q_{ii}, i = 1, 2, \ldots, n$ is the probability that, in the absence of migration, the population that inhabits patch i becomes extinct in one time step, q_{ji} is the probability that patch i is *not* colonized in one time step by a migrant that originates from patch j. Typically, q_{ji} depends on at least the distance between the patches i and j and the area of patch j. Since q_{ji} are probabilities we have $0 \leq q_{ji} \leq 1$. We assume that the local extinction processes and the colonization attempts from different local populations are all independent. As a consequence of this independence the conditional probabilities $q_i(\bar{x})$ for patch i to be empty at moment $t + 1$ under the condition that at moment t the metapopulation is in a state $\bar{x} = (x_1, \ldots, x_n)$ are given by the product

$$q_i(\bar{x}) = \prod_{j=1}^{n} q_{ji}^{x_j}, \quad i = 1, 2, \ldots, n, \tag{7.104}$$

where we have used the convention $0^0 = 1$.

Notice that our model incorporates the *rescue effect* (a decreasing extinction rate with increasing fraction of occupied patches). The overall extinction probability of the local population that inhabits patch i may be considerably less than the "internal" extinction probability q_{ii} if there are many large occupied patches in the vicinity (many small q_{ji}).

Having described the local patch dynamics, we can deduce the law that governs the time evolution of the process $\bar{\eta}(t)$ that gives the state of the metapopulation. The process $\bar{\eta}(t)$ is a homogeneous Markov chain with state space $\bar{X}$ and transition probabilities

$$P(\bar{x}, \bar{y}) = \prod_{i=1}^{n} q_i(\bar{x})^{1-y_i} (1 - q_i(\bar{x}))^{y_i} , \quad \bar{x}, \bar{y} \in \bar{X} . \tag{7.105}$$

Note that the process $\bar{\eta}(t)$ is determined completely by the interaction matrix Q.

Throughout the section, we assume that the interaction matrix satisfies the conditions:

- A1. $q_{ji} > 0, \ j \neq i$;
- A2. $q_{ii} < 1, i \in \{1, \ldots, n\}$;
- A3. For each pair (j, i) of patches, $j, i \in \{1, \ldots, n\}$, there exist an integer m and a chain of indices $j = i_0, \ldots, i_m = i$ such that $\prod_{k=1}^{m}(1 - q_{i_{k-1} i_k}) > 0$.

Condition A1 means that no local population is able to colonize another patch in one time step with probability 1. Condition A2 means that even in the absence of migration (rescue effect) no local population has an extinction probability of 1. Finally, condition A3 means that every local population is able to colonize any other patch either directly or through a chain of patches (stepping-stone dispersal).

7.7.3 The pure mainland–island case

We consider a constellation of a mainland (patch number 1) and $n - 1$ islands. In the pure mainland–island model, we assume that the mainland is inhabited initially, that the mainland population never becomes extinct, and that there is no migration between the islands. However, an empty island is colonized by migrants from the mainland with a certain probability in the next time step. These assumptions mean that the islands are completely decoupled and that the full system with 2^n states is reduced to $n - 1$ independent Markov chains with only two states (empty or occupied).

Mathematically, the mainland–island assumption amounts to

$$q_{11} = 0; \quad q_{1i} > 0, \ i = 2, 3, \ldots, n; \quad q_{ji} = 1, \ j \neq 1, i \neq j . \tag{7.106}$$

The probability that patch i transits from state 1 (occupied) to 0 (empty) is then given by

$$e_i = q_{ii} q_{1i} , \tag{7.107}$$

which is the product of the probability that the population becomes extinct and is not recolonized immediately by migrants from the mainland. Similarly, the colonization probability of patch i is

$$c_i = 1 - q_{1i} . \tag{7.108}$$

The transition matrix of the two-state Markov chain that describes the dynamics
of patch i is thus $\begin{pmatrix} 1 - c_i & c_i \\ e_i & 1 - e_i \end{pmatrix}$.

The incidence of patch i (the stationary probability that patch i is occupied) is
therefore

$$J_i = \frac{c_i}{c_i + e_i} = \frac{1 - q_{1i}}{1 - q_{1i} + q_{ii}q_{1i}}, \quad 1 = 2, 3, \ldots, n . \tag{7.109}$$

The incidence of the mainland is, of course,

$$J_1 = 1 . \tag{7.110}$$

The stationary probability that precisely k patches are occupied is

$$p(k) = \sum J_{i_1} J_{i_2} \cdots J_{i_{k-1}} \left(1 - J_{i_k}\right) \left(1 - J_{i_{k+2}}\right) \cdots \left(1 - J_{i_n}\right) , \tag{7.111}$$

where the sum has $\binom{n-1}{k-1}$ terms that represent different ways to choose the
indices $\{i_1, i_2, \ldots, i_{k-1}\}$ from $\{2, 3, \ldots, n\}$. Note that we do not have to bother
about the index 1 because $J_1 = 1$ by Equation (7.110).

To illustrate the concepts introduced in this section, consider a simple example
with only three patches, numbered 1, 2, and 3. We return to modified versions of
this example in later sections.

Patch number 1 represents the mainland. We assume that patch number 2 is a
small island and therefore prone to extinction, but situated close to the mainland
and therefore frequently rescued or recolonized by migrants from the mainland.
This is formalized by choosing q_{22} large, say, $q_{22} = 0.9$, and q_{12} small, say, $q_{12} = 0.1$. Patch number 3 is a large island and therefore has a small value of q_{33}. We take
$q_{33} = 0.1$, but we assume it is located far away from the mainland and therefore
q_{13} is large, $q_{13} = 0.8$. The interaction matrix for this system is thus given by

$$Q = \begin{pmatrix} 0 & 0.1 & 0.8 \\ 0 & 0.9 & 0 \\ 0 & 0 & 0.1 \end{pmatrix} . \tag{7.112}$$

Using Equation (7.109) we find that

$$J_2 \approx 0.9, \quad J_3 \approx 0.7 . \tag{7.113}$$

This shows clearly the rescue effect. The small island with a high "natural" extinction probability has a higher incidence than the large island because of the frequent
immigration from the mainland.

Next we calculate the stationary probabilities $p(k)$ of exactly k patches being
occupied for $k = 1, 2, 3$. This also illustrates the content of formula (7.111). If
only one patch is occupied it is necessarily the mainland, which forces the two
other patches to be empty. This happens with probability

$$p(1) = (1 - J_2)(1 - J_3) \approx 0.03 . \tag{7.114}$$

If precisely two patches are occupied, one of the islands is empty and the other is occupied. There are two ways in which this can happen. Adding up the probabilities of these two events gives

$$p(2) = J_2(1 - J_3) + J_3(1 - J_2) \approx 0.32 . \tag{7.115}$$

Finally, the probability that all three patches are occupied is

$$p(3) = J_2 J_3 \approx 0.65 . \tag{7.116}$$

Observe that $p(1) + p(2) + p(3) = 1$, as it should.

7.7.4 The mainland–island model with migration between islands

In this subsection we consider the case in which we still have a mainland (patch 1) in which the population does not go extinct, but now we allow for a possible migration between the islands and also from the islands to the mainland. The condition given by Equation (7.106) is therefore replaced by

$$q_{11} = 0; \quad q_{ii} > 0, \quad i = 2, 3, \dots, n . \tag{7.117}$$

Denote by D_1 the set of states $\bar{x} = (1, x_2, \dots, x_n)$ with the first coordinate equal to 1 and arbitrary values of the other coordinates and let D_0 be the set of states $\bar{x} = (0, x_2, \dots, x_n)$ with the first coordinate equal to 0 and $x_j \neq 0$ for at least one $j = 2, 3, \dots, n$. The state space $\bar{X}$ is thus decomposed into a disjoint union of D_1, D_0, and $\{\bar{0}\}$.

If Condition (7.117) holds, then starting from any state $\bar{x} \in D_1$ the system does not exit D_1 and does not reach the absorbing state $\bar{0}$ (i.e., the metapopulation does not become extinct). We call this an ergodic condition and the set D_1 the *ergodic class*.

We denote by $h(\bar{x}, B)$ the probability that the metapopulation state reaches the subset B of the metapopulation state space, given that the initial state is $\bar{x}$. The term h is called the *hitting probability*. We now have

$$h(\bar{x}, D_1) + h(\bar{x}, \bar{0}) = 1 \tag{7.118}$$

for $\bar{x} \in D_0$. Therefore, starting from any state $\bar{x} \in D_0$ the system exits D_0 in a finite time with probability 1, and either enters the ergodic class D_1 or the absorbing state $\bar{0}$. We call the set D_0 the *transient class*.

An initial state in the ergodic class means that the mainland is inhabited initially, which is, of course, the normal situation that corresponds to the very idea of a mainland. However, when migration from the islands to the mainland is possible, the metapopulation may persist even if the mainland is not inhabited initially.

If the mainland is inhabited initially, the metapopulation approaches a stationary distribution π as time tends to infinity. For all the initial states $\bar{x} \in D_1$

$$\lim_{t \to \infty} P_{\bar{x}}\{\bar{\eta}(t) = \bar{y}\} = \begin{cases} \pi(\bar{y}) & \text{if } \bar{y} \in D_1 , \\ 0 & \text{if } \bar{y} \in D_0 \bigcup\{\bar{0}\} . \end{cases} \tag{7.119}$$

Here, and in the sequel, $P_{\bar{x}}(B)$ stands for the probability that the event B occurs, given that the process started at $\bar{x}$, that is, $P_{\bar{x}}(B) = P(B|\bar{\eta}(0) = \bar{x}$.

In mathematical terms the stationary distribution π is the left eigenvector (normalized to a probability distribution) that corresponds to the dominant eigenvalue 1 of the transition matrix given by Equation (7.105) restricted to the ergodic class D_1. Explicitly,

$$\pi(\bar{y}) = \sum_{\bar{x} \in D_1} \pi(\bar{x})P(\bar{x}, \bar{y}), \ \bar{y} \in D_1 , \quad \sum_{\bar{x} \in D_1} \pi(\bar{x}) = 1 . \tag{7.120}$$

If the mainland is not initially inhabited ($\bar{x} \in D_0$), sooner or later it becomes inhabited with a probability $h(\bar{x}, D_1)$. How will this happen? Either the mainland becomes inhabited in one time step or the system remains in the transient class for at least the first time step and then transits to the ergodic class at some later time. Translating this consistency relation from the verbal description to mathematical language, one finds that $h(\bar{x}, D_1)$ must satisfy the following system of linear equations

$$h(\bar{x}, D_1) = \sum_{\bar{z} \in D_1} P(\bar{x}, \bar{z}) + \sum_{\bar{y} \in D_0} P(\bar{x}, \bar{y})h(\bar{y}, D_1) , \ \bar{x} \in D_0 . \tag{7.121}$$

The probability $h(\bar{x}, D_1)$ is obtained as the unique solution of Equation (7.121).

Using the hitting probability $h(\bar{x}, D_1)$, we can write down a formula that describes the long-term behavior of a metapopulation when initially the mainland is not inhabited. For all initial states $\bar{x} \in D_0$

$$\lim_{t \to \infty} P_{\bar{x}}\{\bar{\eta}(t) = \bar{y}\} = \begin{cases} h(\bar{x}, D_1)\pi(\bar{y}) & \text{if } \bar{y} \in D_1 , \\ 0 & \text{if } \bar{y} \in D_0 , \\ h(\bar{x}, \bar{0}) & \text{if } \bar{y} = \bar{0} , \end{cases} \tag{7.122}$$

where the stationary probabilities $\pi(\bar{y})$ and the probabilities $h(\bar{x}, D_1)$ are given by Equations (7.120) and (7.121), respectively. The interpretation of Equation (7.122) is clear. To end up in state $\bar{y} \in D_1$, the metapopulation state first has to hit the set D_1 [which happens with probability $h(\bar{x}, D_1)$] and, once there, it remains in state $\bar{y}$ with the stationary probability $\pi(\bar{y})$. At this point the need to treat the initial conditions with uninhabited mainland may seem strange, but it is essential to understand the pure metapopulation model without a mainland (Section 7.7.6).

We can now derive expressions for biologically important quantities. We are content here with expressions for the incidence of the ith patch (the stationary probability that this patch is occupied) and the stationary distribution of the number of occupied patches. These quantities are defined by

$$J_i = \sum_{y_i=1} \pi(\bar{y}) \tag{7.123}$$

Table 7.6 The probabilities $q_i(\bar{x})$ that patch number i is empty given the metapopulation state $\bar{x}$ at the previous time step for the metapopulation characterized by the interaction matrix Q in Equation (7.127).

State	$\bar{x}$	$q_1(\bar{x})$	$q_2(\bar{x})$	$q_3(\bar{x})$
1	(0,0,0)	1	1	1
2	(0,0,1)	0.95	0.8	0.1
3	(0,1,0)	0.95	0.9	0.95
4	(0,1,1)	0.9025	0.72	0.095
5	(1,0,0)	0	0.1	0.8
6	(1,0,1)	0	0.08	0.08
7	(1,1,0)	0	0.09	0.76
8	(1,1,1)	0	0.072	0.076

and

$$p(k) = \sum_{y_1 + \cdots + y_n = k} \pi(\bar{y}) \,, \tag{7.124}$$

respectively. The summation in Equation (7.123) is taken over all metapopulation states $\bar{y}$ with the ith coordinate equal to 1 and the summation in Equation (7.124) over all states $\bar{y}$ with the sum of the coordinates equal to k. We have

$$\lim_{t \to \infty} P_{\bar{x}} \{\eta_i(t) = 1\} = \begin{cases} J_i & \text{if } \bar{x} \in D_1 \,, \\ h(\bar{x}, D_1) J_i & \text{if } \bar{x} \in D_0 \,, \end{cases} \tag{7.125}$$

$$\lim_{t \to \infty} P_{\bar{x}} \{\eta_1(t) + \cdots + \eta_n(t) = k\} = \begin{cases} p(k) & \text{if } \bar{x} \in D_1 \,, \\ h(\bar{x}, D_1) p(k) & \text{if } \bar{x} \in D_0 \,. \end{cases} \tag{7.126}$$

We now illustrate the theory developed in this section with an example. Since we now allow for migration between the islands we modify the matrix Q in Equation (7.112) by replacing the zero entries with positive entries (the entry q_{11} remains, of course, zero because the first patch is the mainland),

$$Q = \begin{pmatrix} 0 & 0.1 & 0.8 \\ 0.95 & 0.9 & 0.95 \\ 0.95 & 0.8 & 0.1 \end{pmatrix}. \tag{7.127}$$

The choice of the large values $q_{21} = q_{23} = 0.95$ reflects that even if patch number 2 is close both to the mainland and to the other island it has, because of its smallness, a low probability of colonizing the other patches. The large island (patch number 3) has a low probability of colonizing the mainland (because of the long distance involved) and a high probability of colonizing the other island. To compute the transition matrix we first use Equation (7.104) to calculate the probabilities of the patches being empty, given the previous state of the metapopulation. We number the eight metapopulation states according to the binary ordering, that is, $(0, 0, 0)$, $(0, 0, 1)$, $(0, 1, 0)$, ... , $(1, 1, 1)$. We present the result in Table 7.6.

We now obtain the transition matrix P by plugging the values of $q_i(\bar{x})$ in Table 7.6 into Equation (7.105),

$$
P = \begin{pmatrix}
1 & 0 & 0 & 0 & 0 & 0 & 0 & 0 \\
0.076 & 0.684 & 0.019 & 0.171 & 0.004 & 0.036 & 0.001 & 0.009 \\
0.812 & 0.043 & 0.090 & 0.005 & 0.043 & 0.002 & 0.005 & 0.000 \\
0.062 & 0.588 & 0.024 & 0.229 & 0.006 & 0.063 & 0.003 & 0.025 \\
0 & 0 & 0 & 0 & 0.08 & 0.02 & 0.72 & 0.18 \\
0 & 0 & 0 & 0 & 0.006 & 0.074 & 0.074 & 0.846 \\
0 & 0 & 0 & 0 & 0.068 & 0.022 & 0.692 & 0.218 \\
0 & 0 & 0 & 0 & 0.005 & 0.067 & 0.071 & 0.857
\end{pmatrix} . \tag{7.128}
$$

According to our numbering of the metapopulation states, the state 1 is the absorbing state that corresponds to metapopulation extinction, $D_0 = \{2, 3, 4\}$ is the transient class, and $D_1 = \{5, 6, 7, 8\}$ is the ergodic class. The stationary distribution π is therefore the normalized left eigenvector that corresponds to the eigenvalue 1 of the 4×4 submatrix in the lower right corner of P. An easy (numerical) calculation gives

$$
\pi = (0.02 \quad 0.06 \quad 0.22 \quad 0.70) . \tag{7.129}
$$

Note that in Equation (7.129), the first entry is the stationary probability of state number 5, the second of state number 6, and so on. This is consistent with our convention, Equation (7.119), of defining $\pi(\bar{y})$ only for states $\bar{y}$ in the ergodic class.

We now turn our attention to the incidence and the stationary distribution of the number of occupied patches as defined by Equations (7.123) and (7.124), respectively. As patch number 2 is occupied in states 7 and 8, we obtain

$$
J_2 = \pi(7) + \pi(8) = 0.22 + 0.70 = 0.92 , \tag{7.130}
$$

and for patch number 3, in the same manner,

$$
J_3 = \pi(6) + \pi(8) = 0.06 + 0.70 = 0.76 . \tag{7.131}
$$

Comparison of these incidence values with those given by Equation (7.113) for the pure mainland–island model shows a slight increase which, of course, results from the added rescue effect caused by immigration from the other island.

State 5 is the only ergodic state with precisely one patch occupied (the mainland, of course). Therefore

$$
p(1) = \pi(5) = 0.02 . \tag{7.132}
$$

Similarly, we obtain

$$
p(2) = \pi(6) + \pi(7) = 0.06 + 0.22 = 0.28 \tag{7.133}
$$

and

$$
p(3) = \pi(8) = 0.70 . \tag{7.134}
$$

Finally, we calculate the hitting probabilities $h(\bar{x}, D_1)$ for the three states $\bar{x}$ in the transient class D_0. Let h be the column vector that contains the hitting probabilities

$$h = \begin{pmatrix} h(2, D_1) \\ h(3, D_1) \\ h(4, D_1) \end{pmatrix} \tag{7.135}$$

and let P_0 be the 3×3 submatrix obtained by taking the rows and columns 2 to 4 from P. P_0 describes transitions within the transient class. Let f be the column vector with components $f(\bar{x}) = \sum_{\bar{z}=5}^{8} P(\bar{x}, \bar{z})$ for $\bar{x} = 2, 3, 4$. $f(\bar{x})$ is the probability that the metapopulation enters the ergodic class in one time step starting from state $\bar{x}$. The System (7.121) now takes the form

$$(I - P_0)h = f , \tag{7.136}$$

where I is the identity matrix (ones in the diagonal, zeros elsewhere). The solution of Equation (7.136) is

$$h(2, D_1) = 0.40, \quad h(3, D_1) = 0.08, \quad h(4, D_1) = 0.43 . \tag{7.137}$$

7.7.5 Pure metapopulation case and quasi-stationary distribution

We now assume that there is no mainland, that is, all local populations have a positive probability of extinction in one time step,

$$q_{ii} > 0, \quad i = 1, 2, \ldots, n . \tag{7.138}$$

We first consider the question as to under what condition a set of empty patches can be invaded. Such a condition should be formulated in terms of the interaction matrix Q, which completely determines the dynamics of the metapopulation. The interaction matrix contains information about, on the one hand, the patches (such as their mutual distances and their quality) and, on the other hand, the species (such as its migration and colonization capacities).

A set of empty patches can be invaded if a "typical" local population put into an otherwise population-free constellation of patches initially leads to a growing metapopulation. This corresponds to the supercritical case of branching processes. This point of view actually has more to it than a mere analogy. Indeed, the initial phase of the metapopulation dynamics can be viewed as a multi-type branching process, in which the local populations play the roles of individuals and the type is simply the number of the patch the population inhabits. A local population is assumed to live for exactly one time step (if it survives, we say that it gives birth to a new local population of the same type). If migrants from the local population succeed in colonizing patch number i, the local population has given birth to a new local population of type i. Using the model ingredient Q, we can formulate the reproduction process by saying that the probability that a local population of type j gives birth to a local population of type i in an otherwise population-free environment is $1 - q_{ji}$. The reproduction kernel mentioned in Section 7.7.1 reduces to a matrix (because there are only finitely many types) and takes the form

$$1 - Q \,, \tag{7.139}$$

where 1 is the matrix with all entries equal to 1. The dominant eigenvalue R_0 is called the *basic reproduction ratio*. The corresponding left eigenvector β is the stationary type-distribution of "new-born" populations. This gives us an exact mathematical definition of what is meant by a "typical" local population: it is a population sampled from β. As a consequence, R_0 can be interpreted as the expected number of new local populations produced by one typical population in an otherwise population-free environment. Therefore, invasion takes place if and only if $R_0 > 1$.

Since there is a finite number of patches, this means that sooner or later all extant populations simultaneously become extinct and the whole metapopulation is wiped out. Mathematically speaking, the metapopulation becomes extinct with probability 1. There is no ergodic class and the transient class D_0 now consists of all states except the absorption state $\bar{0}$, which corresponds to metapopulation extinction. Instead of Equation (7.118) we have

$$h(\bar{x}, \bar{0}) = 1 \tag{7.140}$$

for all initial states $\bar{x} \in \bar{X}$

$$\lim_{t \to \infty} P_{\bar{x}} \{\bar{\eta}(t) = \bar{y}\} = \begin{cases} 0 & \text{if } \bar{y} \in \bar{X} \setminus \{\bar{0}\} \,, \\ 1 & \text{if } \bar{y} = \bar{0} \,. \end{cases} \tag{7.141}$$

In the pure metapopulation case there is, of course, no stationary distribution except the trivial one that corresponds to metapopulation extinction. However, we can define a so-called *quasi-stationary* distribution (Darroch and Seneta 1965), which is the stationary distribution conditioned on the metapopulation not becoming extinct (see also Sections 6.8 and 6.9).

In this setting, the quasi-stationary distribution π is the left eigenvector (normalized to a probability distribution) that corresponds to the dominant eigenvalue p of the transition matrix given by Equation (7.105) restricted to the transient class D_0,

$$\pi(\bar{y}) = p \sum_{\bar{x} \in D_0} \pi(\bar{x}) P(\bar{x}, \bar{y}), \quad \bar{y} \in D_0, \quad \sum_{\bar{x} \in D_0} \pi(\bar{x}) = 1 \,. \tag{7.142}$$

The dominant eigenvalue p has a clear-cut interpretation: it is the probability that a metapopulation *sampled from the quasi-stationary distribution* does not become extinct in one time step.

Suppose that the metapopulation does not become extinct, so the state distribution approaches the quasi-stationary distribution π. It therefore makes sense to view the dynamics of the metapopulation as a two-state Markov process, the states being metapopulation extinction $\bar{0}$ and the quasi-stationary distribution π. The transition matrix of this process is $\begin{pmatrix} 1 & 0 \\ 1-p & p \end{pmatrix}$.

We next calculate the distribution of the time to extinction. Let the stochastic variable T denote the time to extinction, given that the state of the metapopulation is sampled from the quasi-stationary distribution. Then

$$P(T = 1) = 1 - p \,, \tag{7.143}$$

$$P(T = 2) = p(1 - p) \,, \tag{7.144}$$

and, in general, we obtain the distribution of the extinction time

$$P(T = k) = p^{k-1}(1 - p) \,. \tag{7.145}$$

From Equation (7.145) we obtain the expected extinction time

$$E(T) = \sum_{k=1}^{\infty} k P(T = k) = \frac{1}{1 - p} \,. \tag{7.146}$$

Equation (7.146) gives an exact expression for the expected time to extinction, provided the metapopulation is initially at the quasi-stationary distribution. However, the calculation of p (the eigenvalue of a $2^n - 1$ by $2^n - 1$ matrix) becomes computationally prohibitive as the number of patches n grows. We refer to Etienne and Heesterbeek (2001) for examples of how Equation (7.146) can be applied to reach practical conclusions about, for instance, how changes in the connectivity of patches (changing q_{ji}) affect the viability of the metapopulation.

Let us again return to our simple three-patch model. As we are here considering a pure metapopulation model, patch number 1 is no longer a mainland, but has a positive extinction probability $\varepsilon > 0$. The interaction matrix becomes

$$Q = \begin{pmatrix} \varepsilon & 0.1 & 0.8 \\ 0.95 & 0.9 & 0.95 \\ 0.95 & 0.8 & 0.1 \end{pmatrix} . \tag{7.147}$$

The entries in Table 7.6 remain unchanged with the exception of the four zeros in the $q_1(\bar{x})$ column, which change to ε, 0.95ε, 0.95ε, and 0.9025ε, respectively. The first four rows of the transition matrix P do not change, and the other entries, of course, depend on ε. We calculated numerically the dominant eigenvalue p of the 7×7 submatrix that corresponds to the transient class for $0 < \varepsilon < 0.5$, and used Equation (7.146) to compute the expected time to extinction. The result is presented in Figure 7.22.

7.7.6 The quasi-mainland–island model

In Section 7.7.5 we considered the case of a pure metapopulation model and, among other things, derived an exact formula for the expected time to extinction, provided that the initial metapopulation state is sampled from the quasi-stationary distribution. That formula is of practical importance only if the time to extinction is long enough for the metapopulation to settle down to the quasi-stationary distribution.

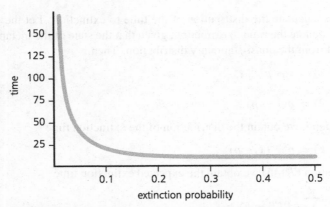

Figure 7.22 The expected time to metapopulation extinction as a function of ε for the model defined by Equation (7.147).

One situation in which it is intuitively clear that the metapopulation survives for a "sufficiently long time" is when there is one patch with a very low (ε, say) extinction probability. In this subsection we consider this case and derive exact formulas for biologically important quantities in the limiting case when ε tends to zero. These formulas are given in terms of the stationary distribution π for the mainland–island model with migration between islands, that is, π is obtained from Equation (7.120). Throughout this subsection π has this meaning. That we have used the same symbol for the quasi-stationary distribution of Section 7.7.5 is not accidental. We return to the relation between the two uses of π at the end of this subsection.

We assume that patch number 1 has extinction probability $q_{11}^{(\varepsilon)} = \varepsilon$, and call this patch the *quasi-mainland*. All other entries q_{ij} of the interaction matrix are as in Section 7.7.4, that is, we replace Condition (7.117) by

$$q_{11}^{(\varepsilon)} = \varepsilon; \quad q_{ii}^{(\varepsilon)} = q_{ii} > 0, \ i = 2, 3, \ldots, n, \quad q_{ij}^{(\varepsilon)} = q_{ij}, \quad \varepsilon > 0 . \quad (7.148)$$

The corresponding interaction matrix is denoted by $Q^{(\varepsilon)}$ and the resultant Markov process by $\bar{\eta}^{(\varepsilon)}$.

Much more general perturbations of the mainland–island model with migration between islands (or, for that matter, of general ergodic interaction matrices) are treated in detail by Gyllenberg and Silvestrov (1994, 1997, 1999, 2000).

The quantities

$$\hat{q}_i(\bar{x}) = q_{1i} \prod_{j \neq 1} q_{ji}^{x_j}, \ i \neq 1, \quad \hat{q}_1(\bar{x}) = \hat{q}_{11} \prod_{j \neq 1} q_{j1}^{x_j} \quad (7.149)$$

have clear-cut interpretations. $\hat{q}_i(\bar{x})$ is the probability that patch i is not colonized in the next time step, given that the quasi-mainland is inhabited. $\hat{q}_1(\bar{x})$ is the probability that the quasi-mainland is not colonized in the next time step. We denote the transition probability matrix that corresponds to the interaction matrix $\hat{Q}$ by $\hat{P}$, that is,

$$\hat{P}(\bar{x}, \bar{y}) = \hat{q}_1(\bar{x}) \prod_{i \neq 1} \hat{q}_i(\bar{x})^{1-y_i} \left(1 - \hat{q}_i(\bar{x})\right)^{y_i} . \tag{7.150}$$

Finally, we define the constant λ as the probability of extinction when the initial metapopulation state is sampled from the stationary distribution π of the mainland–island model [π is obtained from Equation (7.120)] and the metapopulation evolves according to the process determined by $\hat{P}$,

$$\lambda = \sum_{\bar{x} \in D_1} \pi(\bar{x}) \left(\hat{P}(\bar{x}, \bar{0}) + \sum_{\bar{y} \in D_0} \hat{P}(\bar{x}, \bar{y}) h(\bar{y}, \bar{0}) \right) . \tag{7.151}$$

We are interested in the long-run behavior of the metapopulation. The exact notion of "long run" turns out to be crucial for the results. As long as the quasi-mainland remains occupied we can expect the system to behave as in the ergodic situation, that is, as if we had a mainland–island system with migration between the islands. When the local population on the quasi-mainland becomes extinct, the quasi-mainland can either be recolonized or the whole metapopulation becomes extinct. The number of such recolonizations of the quasi-mainland before metapopulation extinction is a random variable with a distribution that resembles the geometrical distribution. As the expected lifetime of the quasi-mainland population is of the order $1/\varepsilon$, the expected lifetime of the metapopulation is of the same order and we anticipate that the asymptotic behavior depends on whether a "long time" is considerably less than or greater than $1/\varepsilon$.

We next investigate the behavior of the t-step transition probabilities $P_{\bar{x}}\{\bar{\eta}^{(\varepsilon)}(t) = \bar{y}\}$ as $\varepsilon \to 0$ and $t \to \infty$ simultaneously. To make the above intuitive ideas precise we assume that time $t = t_\varepsilon$ is also a function of the parameter ε, so as to satisfy the condition

$$t_\varepsilon \to \infty \text{ as } \varepsilon \to 0, \text{ and } \varepsilon t_\varepsilon \to s \text{ as } \varepsilon \to 0, \text{ where } 0 \leq s \leq \infty . \tag{7.152}$$

We have the important result in Theorem 7.11.

Theorem 7.11 *(i) For any initial state $\bar{x} \in D_1$*

$$\lim_{\varepsilon \to 0} P_{\bar{x}}\left\{\bar{\eta}^{(\varepsilon)}(t_\varepsilon) = \bar{y}\right\} = \begin{cases} e^{-\lambda s} \pi(\bar{y}) & \text{if } \bar{y} \in D_1 , \\ 0 & \text{if } \bar{y} \in D_0 , \\ 1 - e^{-\lambda s} & \text{if } \bar{y} = \bar{0} . \end{cases} \tag{7.153}$$

(ii) For any initial state $\bar{x} \in D_0$

$$\lim_{\varepsilon \to 0} P_{\bar{x}}\left\{\bar{\eta}^{(\varepsilon)}(t_\varepsilon) = \bar{y}\right\} = \begin{cases} e^{-\lambda s} h(\bar{x}, D_1) \pi(\bar{y}) & \text{if } \bar{y} \in D_1 , \\ 0 & \text{if } \bar{y} \in D_0 , \\ 1 - e^{-\lambda s} h(\bar{x}, D_1) & \text{if } \bar{y} = \bar{0} . \end{cases} \tag{7.154}$$

From Theorem 7.11 we immediately obtain expressions for the incidence and the asymptotic distribution of the number of occupied patches.

Corollary 7.1 *For any initial state $\bar{x} \in D_1$ we have*

$$\lim_{\varepsilon \to 0} P_{\bar{x}} \left\{ \eta_i^{(\varepsilon)}(t_\varepsilon) = 1 \right\} = e^{-\lambda s} J_i, \quad i \in \{1, \dots, n\}, \tag{7.155}$$

and

$$\lim_{\varepsilon \to 0} P_{\bar{x}} \left\{ \eta_1^{(\varepsilon)}(t_\varepsilon) + \cdots + \eta_n^{(\varepsilon)}(t_\varepsilon) = k \right\} = \begin{cases} e^{-\lambda s} p(k), & \text{if } k \in \{1, \dots, n\}, \\ 1 - e^{-\lambda s} & \text{if } k = 0, \end{cases}$$

$$\tag{7.156}$$

where J_i and $p(k)$ are given by Equations (7.123) and (7.124), respectively.

Theorem 7.11 calls for some explanation. As expected, the parameter s has a decisive influence on the asymptotic behavior.

When $s = 0$ the time t_ε grows slowly compared with $1/\varepsilon$, that is, t_ε is asymptotically considerably less (on a different timescale) than the expected lifetime of the quasi-mainland population. We therefore expect the metapopulation to behave as a mainland–island system with migration between the patches. Indeed, in this case the factor $e^{-\lambda s}$ takes on the value 1 and the limiting expressions in Equations (7.153) and (7.154) coincide with the limiting expressions in Equations (7.119) and (7.122), respectively.

When $s = \infty$ the time t_ε is also on a different timescale from the expected lifetime of the quasi-mainland population, but now it is much larger. The factor $e^{-\lambda s}$ vanishes and consequently the limiting expressions in Equations (7.153) and (7.154) coincide with the limiting expression in Equation (7.141) for the pure metapopulation model.

The intermediate quasi-stationary case $0 < s < \infty$ is the most interesting. It corresponds to the situation in which the rates of growth of t_ε and $1/\varepsilon$ are of the same order; in other words, extinction of the quasi-mainland population happens on the timescale of t_ε. The asymptotic behavior of the metapopulation therefore exhibits components typical of both the mainland–island (with migration between islands) and the pure metapopulation situations. This is reflected in the formulas through the factor $e^{-\lambda s}$.

Next we consider the random variable $T^{(\varepsilon)}$, which by definition is the metapopulation extinction time for the system in which the quasi-mainland population has extinction probability ε. It is clear that $T^{(\varepsilon)} \to \infty$ as $\varepsilon \to 0$. We show that the order of this growth is ε^{-1}. Actually, we do more and show that the metapopulation extinction time normalized by ε has a limiting distribution. We also give an explicit formula for this distribution.

Theorem 7.12, which gives an expression for the limiting distribution of the metapopulation extinction time, is a corollary of Theorem 7.11.

Theorem 7.12 *Let λ be given by Equation (7.151). Then for all $u \in (0, \infty)$*

$$\lim_{\varepsilon \to 0} P_{\bar{x}} \left\{ \varepsilon T^{(\varepsilon)} > u \right\} = \begin{cases} e^{-\lambda u} & \text{if } \bar{x} \in D_1 , \\ e^{-\lambda u} h(\bar{x}, D_1) & \text{if } \bar{x} \in D_0 . \end{cases} \tag{7.157}$$

Notice that in the case $u = 0$, $\bar{x} \in D_1$, the right-hand side of Equation (7.157) is trivially equal to 1.

When the initial state $\bar{x}$ is in the ergodic class D_1, the limiting distribution is exponential. When $\bar{x} \in D_0$, it is the distribution of the random variable $\chi_{\bar{x}} \alpha_\lambda$, which is a product of two independent random variables $\chi_{\bar{x}}$, which takes on the values 1 and 0 with probabilities $h(\bar{x}, D_1)$ and $h(\bar{x}, \bar{0}) = 1 - h(\bar{x}, D_1)$, respectively, and α_λ, which is exponentially distributed with parameter λ.

In Section 7.7.5 we introduced the quasi-stationary distribution as the stationary distribution conditioned on that extinction has not yet taken place. In the present subsection we have derived formulas for the long-term behavior of the metapopulation using the stationary distribution of the corresponding mainland–island system with migration between the islands, together with a limiting procedure in which one patch tends toward a mainland (more precisely, the extinction probability of its population tends to zero). We denoted the quasi-stationary distribution of the pure metapopulation model and the stationary distribution of the mainland–island model with migration between the islands by the same symbol π. This suggests that the two distributions are, in some sense, the same. We close this section by giving an exact meaning of this sameness.

Taking $u = s$ [the parameter in Condition (7.152)] in Theorem 7.12 and recalling that asymptotically $t_\varepsilon = s/\varepsilon$, we see from Equation (7.157) that the correction factors $e^{-\lambda s}$ and $e^{-\lambda s} h(\bar{x}, D_1)$, which occur in Equations (7.153) and (7.154), respectively, are the probabilities that metapopulation extinction has not yet taken place. This observation suggests that the probabilities $\pi(\bar{y})$ may be interpreted as conditional stationary probabilities, given that metapopulation extinction has not yet occurred. To formulate this intuitive idea in a precise way, we introduce the conditional probabilities

$$P_{\bar{x}} \left\{ \bar{\eta}^{(\varepsilon)}(t_\varepsilon) = \bar{y} | T^{(\varepsilon)} > t_\varepsilon \right\} = \frac{P_{\bar{x}} \left\{ \bar{\eta}^{(\varepsilon)}(t_\varepsilon) = \bar{y}, \ T^{(\varepsilon)} > t_\varepsilon \right\}}{P_{\bar{x}} \left\{ T^{(\varepsilon)} > t_\varepsilon \right\}} . \tag{7.158}$$

We then have Theorem 7.13.

Theorem 7.13 *Let $0 < s < \infty$. Then for any initial state $\bar{x} \in D_1 \bigcup D_0$*

$$\lim_{\varepsilon \to 0} P_{\bar{x}} \left\{ \bar{\eta}^{(\varepsilon)}(t_\varepsilon) = \bar{y} | T^{(\varepsilon)} > t_\varepsilon \right\} = \begin{cases} \pi(\bar{y}) & \text{if } \bar{y} \in D_1, \\ 0 & \text{if } \bar{y} \in D_0. \end{cases} \tag{7.159}$$

If $\bar{x} \in D_1$, then Equation (7.159) holds in the case $s = 0$, too.

From Theorem 7.13, we immediately obtain the characterization in Corollary 7.2 of the incidence and the distribution of the number of occupied patches under the condition that the metapopulation has not yet gone extinct.

Corollary 7.2 *Let* $0 < s < \infty$. *Then for any initial state* $\bar{x} \in D_1 \bigcup D_0$

$$\lim_{\varepsilon \to 0} P_{\bar{x}} \left\{ \eta_i(t_\varepsilon) = 1 | T^{(\varepsilon)} > t_\varepsilon \right\} = J_i, \quad i \in \{1, \ldots, n\}, \tag{7.160}$$

and

$$\lim_{\varepsilon \to 0} P_{\bar{x}} \left\{ \eta_1^{(\varepsilon)}(t_\varepsilon) + \cdots + \eta_n^{(\varepsilon)}(t_\varepsilon) = k | T^{(\varepsilon)} > t_\varepsilon \right\} = p(k), \quad k \in \{1, \ldots, n\}, \tag{7.161}$$

where J_i *and* $p(k)$ *are given by Equations (7.123) and (7.124), respectively.*

7.8   Multi-type Branching Processes and Adaptive Dynamics of Structured Populations

M. Durinx and Johan A.J. Metz

7.8.1   Introduction

Adaptive dynamics deals with the consequences of the repeated establishment of rare mutants in environments set by large equilibrium populations of residents. It studies which mutants can potentially invade, which successful invasions lead to the demise of the original residents, and what the evolutionary outcome could be of a prolonged series of such substitution events (Metz *et al.* 1996). The main assumptions are:

- *Rare mutations.* The community dynamics settles on an attractor between mutation events, and hence the ecological and evolutionary timescales are separated.
- *Initially rare mutants.* The well-mixed resident populations have a large size, while the mutant's population starts up from a single mutant. The assumed magnitude of the resident population makes its dynamics deterministic, whereas the rarity of the invading mutant induces a strong stochastic effect.
- *Small mutational steps.* To allow sensible topological and geometrical inferences, mutants must be similar to one of the residents.

In this section, we restrict the postulate of deterministic resident dynamics even further, to the case in which its attractors are fixed points. Combined with the timescale supposition, this allows the residents to be regarded as stationary on the ecological timescale.

The main attraction of this restrictive set of assumptions is that it allows the construction of a mathematically consistent framework within which to study the possible patterns of evolutionary outcomes, based on a precise if not always equally realistic foundation of mechanistic biological reasoning.

The traits under evolutionary control through mutation and selection must be thought of as parameters that govern the life history of individuals. Each resident population consists of a large number of individuals that share a characteristic trait. Several such populations, with different traits or trait values, make up the community of residents. In this environment, a single newborn individual may have an advantageous mutation, which gives rise to an initially small mutant population. In the long run, the presence of the newcomers affects the fitness and thereby, possibly, the persistence of the resident populations; this interplay lets evolution shape the species of the community.

Study of the evolutionary dynamics can be divided broadly into two categories. First, away from special points called singularities (as defined in Section 7.8.2), directional selection acts. Then the dictum *"invasion implies fixation"* holds, which means that any persistent mutant drives its ancestral resident to extinction. This property is proved in Geritz *et al.* (2002), given the assumption that the population dynamic attractors are sufficiently well-behaved. Through this mechanism, the amount of diversity remains the same, since the emergence of a successful new type implies the disappearance of an older type. One way of failing to be well-behaved is to have a qualitative change in the pattern of the population dynamics (known in dynamic systems theory as a bifurcation). The appearance of a mutant may then lead to the demise of not only the resident that spawned it, but also of other resident types, and thereby reduce the diversity present in the community. The more extreme cases involve evolutionary suicide, in which a sequence of substitution events can drive the entire community to extinction (e.g., Gyllenberg and Parvinen 2001).

Second, at singularities the attractiveness of adaptive dynamics as an evolutionary framework is revealed. One of the categories a singularity can belong to is that of the classic, but inappropriately named, *evolutionarily stable strategy* (ESS) known from evolutionary game theory (Maynard Smith 1982). However, there are other naturally occurring types of singularities. Foremost is the *branching point*, an attracting singularity in the proximity of which the population dynamics exerts disruptive selection. This selection acts such that a newly established mutant does not send its progenitor to kingdom come. Instead, the two coexisting types diverge further and further, so that after a short time they are on opposite sides of the singularity. When plotting the resident strategies against evolutionary time, the "branches" that gave the singularity its name become clear. This splitting of genetic lines has an obvious appeal as a model for (the initiation of) speciation. For a deeper understanding, we refer to the book on adaptive speciation (Dieckmann *et al.* 2004) in this series.

In this section, we consider evolution at a distance from singularities only. Under such a regime of directional selection, the rate of trait substitutions can be estimated as Dieckmann and Law (1996) did, formulating the *"canonical equation of adaptive dynamics."* We show here how Durinx *et al.* (unpublished) extend this equation to physiologically structured populations. In particular, we sketch how the speed of evolution is assessed from the underlying branching process of the invasion dynamics.

7.8.2 Invasion fitness

The main tool of adaptive dynamics is the *invasion fitness function*. By definition this is the long-term average per capita growth rate of a rare type (the *invader*) in an equilibrium community of a given set of types (the *residents*). Thus, it is the Malthusian parameter of the invader. Any resident typecast in the role of invader always has zero invasion fitness, since, on average, it neither grows nor diminishes in abundance. An invader with a negative fitness does not gain a foothold in the given community, whereas a positive fitness implies a positive *probability* of establishment. As this concerns a stochastic process with a very small number of invaders, even a positive average growth rate does not prevent extinction in a fair number of cases: establishment must be studied as the outcome of a branching process.

The assumption of small mutational steps means the *mutants* differ only slightly from one of the resident types. The tacit biological suggestion is that any mutation with a large effect is detrimental because of pleiotropy, which thus guarantees a truncation of the effective mutation distribution.

When a mutant has a positive invasion fitness, but because of stochasticity its attempt at establishment fails, this is not the end of the process. Evolution can bide its time and a later occurrence of a similar mutation may become established with better luck. In the simplest setting, the probability of success of individual invasion attempts affects only the speed of evolution rather than its endpoint. However, in polymorphic populations or higher dimensional strategy spaces, the speed, direction, and outcome can all be affected.

The *strategy*, *trait value*, or simply *trait* of an individual denotes its particular set of values for the parameters that are under evolutionary control; this can be a vector or just a single scalar value. A *polymorphic* community has individuals with different traits. We only consider cases in which the number of strategies present is finite. If we lump together all individuals with an identical strategy, they are referred to collectively as a *type*, *population*, or *species*. Clearly, this last term is very loose at this point, and does not imply any well-defined concept of biological species. For simplicity, we consider populations to be clonal. The term *community* refers to the collection of all resident populations.

We denote a resident's strategy as X and an invader's as Y. Hence, a community with N different strategies present consists of the types $1, 2, \ldots, N$, distinguished by their strategies $X_1, X_2, \ldots, X_N$. The community as a whole is indicated by $\mathbb{X} := \{X_1, X_2, \ldots, X_N\}$.

The invasion fitness function as defined above is referred to as the s-function; this choice of notation underlines its heritage as a conceptual extension of the selection coefficient of population genetics. A *monomorphic s-function* $s_X(Y)$ describes the invasion fitness of a mutant with type Y in an environment set by a population of X-type residents. In the case of a community of N populations, we similarly speak of a *polymorphic s-function* and denote it by $s_{X_1, X_2, \ldots, X_N}(Y)$ or $s_{\mathbb{X}}(Y)$ to show which N strategies are present.

A further function of central concern is the *invasion gradient*. The invasion gradient at a given strategy X_i is the derivative of s in the mutant direction at that trait value: $\partial s_{\mathbb{X}}(Y)/\partial Y|_{Y=X_i}$. Trait values for which the invasion gradient is zero are called *evolutionarily singular strategies*; these are particularly interesting as they are possible evolutionary endpoints (attracting, non-invadable points), or sources of diversity (branching points).

For mathematical reasons, the existence of a unique, fixed-point attractor for the community as a whole is usually assumed. We restrict ourselves in this section to attractors composed of positive equilibrium densities for the N types that make up the community. The basic assumption of rare mutations implies that between two mutation events the population dynamics settles at its attractor. So in an N-species community at equilibrium in which there is a mutation in the ith species, a mutant $Y \approx X_i$ is introduced. After some time, the community again reaches an equilibrium state. The N strategies that make up this new attractor depend on whether the mutant has disappeared or driven its ancestor to extinction: we deal in this section only with situations away from singularities, so that "invasion implies fixation" holds.

Time is always scaled such that it reflects slow, evolutionary time and not the much faster ecological timescale. Then the population dynamics is so rapid that the community always seems to be at its attractor, at densities determined by the unique equilibrium for the set of strategies present. In this way, the N strategies present fully describe the community at each point in time.

7.8.3 The deterministic path

The canonical equation is a first-order prediction for the speed of trait evolution. This estimate of the rate at which strategy substitutions happen is valid under a regime of directional selection; that is to say, away from evolutionarily singular strategies, so that the fitness gradient $\partial s_{X_1,X_2,\dots,X_N}(Y)/\partial Y$ is non-zero.

As given in the assumptions, the resident community is very large and hence can be described by a deterministic model. Furthermore, we can regard the population dynamics as stationary because of the assumption that it has a fixed-point attractor. However, the appearance of mutants and their eventual success or failure at establishment are both inherently stochastic processes. This makes the trait values themselves stochastic and time dependent. As we are dealing with a Markov process, the community has a probability $\Pi(\mathbb{X}, t)$ of being in state $\mathbb{X}$ at time $t > 0$. This probability can be calculated from the state at $t = 0$, together with all transition rates $\pi(B, A)$ from state A into B.

As we consider rare mutations here, any transition must be a mutation that affects a single strategy vector, which simplifies $\pi(A, B)$: all the action arises from the terms $\pi_0(X_i', X_i, \mathbb{X})$, which are the rates at which the ith species (known by its strategy X_i) in community $\mathbb{X}$ is replaced by one with strategy X_i'.

Application of the Kolmogorov backward equations to the right-hand side of $\frac{d}{dt}\mathbb{E}[\mathbb{X}] := \int \mathbb{X} \frac{\partial \Pi(\mathbb{X},t)}{\partial t} \, d\mathbb{X}$ shows, after some algebra, that the expected rate of evolutionary change is

$$\tfrac{d}{dt}\mathbb{E}[\mathbb{X}] = \mathbb{E}[J(\mathbb{X})] , \qquad (7.162)$$

where $J(\mathbb{X})$ is an $N{\times}N$ matrix with as the ith column $\int (X_i' - X_i)\pi_0(X_i', X_i, \mathbb{X}) dX_i'$. The unique solution that satisfies a given initial condition is called the *mean path*, starting from that given state. If J happens to be a linear function of $\mathbb{X}$, or if the distribution of $\mathbb{X}$ is concentrated at a point, we have the self-contained equation

$$\tfrac{d}{dt}\mathbb{E}[\mathbb{X}] = J(\mathbb{E}[\mathbb{X}]) . \qquad (7.163)$$

The solution to this simplified problem is called the *deterministic path*. The validity of this approximation has been argued, based on considerations in van Kampen (1981) and on simulations, in Dieckmann and Law (1996). It hinges on whether the solution to Equation (7.162) is dominated by the first-order part of the equation or not. Intuitively, one can expect this to happen, as the path proceeds, in time, by very many very small steps. In such a case some effect similar to the Law of Large Numbers leads to a concentrated distribution of $\mathbb{X}$ and hence to the applicability of the approximation (7.163). This heuristic idea is explored in a mathematically rigorous fashion in Champagnat *et al.* (2001), based on convergence theorems in Ethier and Kurtz (1986).

We follow the assumption that a deterministic approximation is valid, and henceforward focus on the attendant path. To lighten the notation, we write X_i instead of $\mathbb{E}[X_i]$ for each resident and similarly for $\mathbb{X}$, as these are the expected values from here onward (not so for invaders Y or mutants X_i'). Our concrete aim is thus to derive analytical expressions for the right-hand side of

$$\frac{d}{dt}X_i = \int (X_i' - X_i)\pi_0(X_i', X_i, \mathbb{X}) \, dX_i' , \qquad (7.164)$$

for all species i in the community. The first step is to separate the factors that make up the transition probability π_0. Seeing that mutation and selection are independent processes, it must be the product of the appearance rate of mutants with their establishment probability,

$$\pi_0(X_i', X_i, \mathbb{X})$$

$$= \text{(rate of mutations } X_i \to X_i' \text{ given } \mathbb{X}) \text{ (establishment chance)}$$

$$= \overbrace{\text{(birth rate of } X_i \text{ types) (mutation chance } X_i \to X_i')}\ \overbrace{P(X_i', \mathbb{X})}$$

$$= \overbrace{\lambda(X_i, \mathbb{X}) \quad \hat{n}_i}\quad \overbrace{\mu(X_i) \quad \mathfrak{M}(X_i' - X_i, X_i)}\quad P(X_i', \mathbb{X}) . \quad (7.165)$$

At this point we stress that above we are treating expected values, but with simplified notation; $\hat{n}_i$ denotes the equilibrium density of the ith species. The chance P of establishment warrants a separate computation, but the other factors are intuitive. The verbal reasoning is that the production rate of mutants that descend from an X_i-type parent is simply the total production of X_i-type offspring times the mutation chance. The production rate of X_i strategists is the per capita birth rate λ times the equilibrium density $\hat{n}_i \equiv \hat{n}(X_i; X_1, X_2, \ldots, X_N)$ in the given community. Which and how many mutants appear (as a fraction of the newborns) is the product of the mutation probability μ per birth event with the mutational distribution $\mathfrak{M}(V, X_i)$, that is, the probability that the mutation changes a trait value from X_i to $X_i + V$.

Under very general conditions, the per capita birth rate in a closed, critical system is the inverse of the expected life span. The argument is called the *microcosm principle* in Mollison (1995) and runs as follows: in a large population that ergodically fluctuates around its attracting density,

$$\mathbb{E}[\text{density}] = \mathbb{E}[\text{influx of individuals per area}]\, \mathbb{E}[\text{duration of stay}]$$
$$= \mathbb{E}[\text{per capita birth rate}]\, \mathbb{E}[\text{density}]\, \mathbb{E}[\text{life span}]$$
$$\hat{n}_i = \lambda(X_i, \mathbb{X})\, \hat{n}_i \, \mathbb{E}[\text{life span}] \, . \tag{7.166}$$

Hence the life expectancy

$$T \equiv T(X_i, \mathbb{X}) = \lambda^{-1}(X_i, \mathbb{X}) \tag{7.167}$$

for any resident. This assumes closed populations, so that influx is only caused by reproduction, and an individual's stay is ended by death only.

The chance of establishment P depends on the underlying population model. As presented by Dieckmann and Law (1996), the canonical equation traditionally considers unstructured populations in continuous time. In this case, a linear birth-and-death process determines the fate of the mutants. We, however, want to allow populations to be structured (e.g., by size or by sexual differences). For this, we introduce the concepts of structured population models as far as are needed to estimate P.

7.8.4 Physiologically structured population models

The two best-known classes of demographic models are probably Lotka–Volterra and stage- (or age-) structured matrix models. Both have their shortcomings: matrix models can deal only with finite numbers of discrete stages, and time must be discrete too. In Lotka–Volterra models all the individuals are born equal and their death rates are independent of age or reproductive state; the whole population is essentially a soup of identical creatures.

We therefore consider the class of physiologically structured population models, which has both Lotka–Volterra and matrix models as subclasses. Section 4.3 gives a more detailed, mathematical description of this class, but the biological considerations that shape the formalism are

■ *Size does matter.* A large fish may happily eat a smaller conspecific, but will refrain from attacking a healthy individual of similar size; a healthy fat insect will lay more eggs, and a big baby has a head start in life.
Furthermore, reaching a given size may trigger a stage transition. A life cycle diagram for any insect shows that these are important events. Thinking further along these lines, this life cycle may depend equally drastically on the sex of

the individual or similar characteristics. Populations could also live on several patches in which the resource availability differs.

All these features together form the *states* an individual can be in, whether they are described as continuous (like size or age), or discrete (like stage or sex). We attempt to show later on that we can focus on only the *birth states*, those states that individuals could possibly start their life in. Examples could be the size of a plant seed, the mass of the yolk in an egg, or the gender or morph of an individual.

■ *Conditional linearity.* A key insight to disentangle the interactions between individuals in a community is the *separation of individual and environment*. One has to conceive of a formal environment that influences individuals and vice versa through a feedback loop. The implicit definition is that, given an environment, all individuals are independent.

The environmental condition must contain all the information to predict what will happen at the next moment to each individual. In other words, given the environment, an individual's fate is a Markov process (see Section 2.5). The feedback loop is closed by the simple fact that each individual is accounted for when calculating the condition of the environment. The reader is referred to the discussion in Sections 2.9 and 5.10 to relate this concept of environment to the view in branching processes.

For example, if gender matters and competition acts differently within age classes rather than across them, then the environment has components that show densities and sex ratios in each class separately. Furthermore, the environment logically also has components that describe relevant external parameters, such as temperature, influx of resources, or harvesting, which may have their own dynamics.

The idea that such an environment can be constructed is justified in large populations: any two individuals, being exceedingly rare as a proportion of the total population, will experience the same world populated by the same types, states, and quantities of "others," even if their expected reaction to it may differ, depending on the individual state and strategy. This decoupling of individual and environment is a helpful step in the systematic understanding of structured population models (see Metz and Diekmann 1986; Diekmann *et al.* 2003) and linearizes the equations when the environment is given as a function of time. To derive the correct description of the environment may involve a lot of work in specific situations, but most models are presented in such a way that the derivation is trivial.

At this point, it is important to distinguish strategies and birth states: both are parameters an individual starts its life with, and some states, like sex, might not change during an individual's lifetime. One requires a priori that the life history of all individuals with a given strategy can be sketched in one indecomposable life-cycle graph. If not, isolated groups could be separated into species by adding a parameter to the strategy. This consideration provides a strict distinction between traits and birth states in models in which individuals with different strategies cannot reproduce together, which we have guaranteed trivially by assuming clonal reproduction. An intuitive example of the distinction between strategy and birth state is seed size in plants (Geritz *et al.* 1999). Part of a plant's strategy may be the decision to allocate its energy reserve into many small seeds with a low survival chance, few large seeds with a good chance, or a given mixture of these; while a plant's birth state can be the size it has as a seed.

That birth states suffice to study invasion demographics follows from uncoupling the feedback loop. If it is assumed that the environment is given, the probability of being in each possible state at each later moment can be calculated for any given newborn. Similarly, for an individual in a given state, the chance of having any number of newborns in any birth state can be computed for any later point in time. By combining these for any given newborn, the environment-dependent probabilities of having any amount of offspring in any birth state at any later moment can be established. From this generational viewpoint, birth states are the only things that have to be kept track of. This is an important observation that forms the basis of multi-type branching processes.

Furthermore, for a deterministic population to be at equilibrium must mean that in each generation there are born an exact quantity of young in each birth state, such that the individuals have, on average, precisely one offspring over their lifetime, and the new generation has the same distribution of birth states as the former. In short, we do not have to keep track of offspring production in all possible states; it suffices to follow lifetime offspring production of the individuals in all possible birth states.

We denote by $I \equiv I(X_1, X_2, \ldots, X_N)$ the environment as set by the resident community and consider a given invader with strategy Y. For any structured population model, there must necessarily exist a *reproduction kernel*, which we formulate here as a matrix function with entries $(\Lambda(Y, I, a))_{hj}$, which is the expected number of Y-type invaders born in state j to an invader newly born in state h, before it reaches age a if the environment is in state I. Here we remark that $\Lambda(Y, I, da)$ corresponds to $\mu(r, ds \times da)$ as introduced in Section 3.3, and that the connection to the mean matrix is seen from

$$\mathbb{E}[\xi_{hj}] = \Lambda(Y, I, \infty)_{hj} , \tag{7.168}$$

where the strategy and environment must be the same on both sides of the equality. We conform here to branching processes notation, insofar that Λ is usually defined as its transpose in structured populations literature.

According to the definition we give, the (generally unique) solution for ψ of Lotka's equation

$$\text{Dominant Eigenvalue of } \left[\int_0^\infty e^{-\psi a} \Lambda(Y, I, da) \right] = 1 \tag{7.169}$$

is the invasion fitness $s_{X_1,X_2,\ldots,X_N}(Y)$. Alternatively, it is called the instantaneous growth rate r in life-history theory, or the Malthusian parameter α in branching processes.

If $\psi = 0$ is fixed, the integral on the left-hand side corresponds to the mean matrix as used throughout this book, and its dominant eigenvalue is the lifetime reproductive output R_0, which is denoted by ρ in branching processes. The notation R_0 is traditional in both life-history theory and epidemiology, where it is the expected number of secondary infections caused by an infective individual [see the snappily titled thesis of Hans Heesterbeek (1992)].

Durinx *et al.* (unpublished) show, by a Taylor expansion of the left-hand side around $\psi = 0$, that for any mutant Y, we have the relationship

$$s_{\mathbb{X}}(Y) = \frac{\ln R_0}{\beta(0)} + o(\ln R_0) \,, \tag{7.170}$$

where the *average age at giving birth* $\beta(0)$ is that of the mutant's progenitor, calculated from the reproduction kernel as

$$\beta(0) \equiv \beta(X_i, I) = u(0)^{\mathsf{T}} \left[\int_0^{\infty} a \Lambda(X_i, I, \, da) \right] v(0) \,, \tag{7.171}$$

with $u(0)$ the "stable type distribution" of the ith resident, which in our context is its *stable birth state distribution*, and $v(0)$ its reproductive value.

7.8.5 Establishment probability

The first paragraph on invasion fitness verbally states that, under very general conditions, a positive chance of establishment is equivalent to a positive s-value,

$$P(Y, \mathbb{X}) > 0 \Leftrightarrow s_{\mathbb{X}}(Y) > 0 \,. \tag{7.172}$$

We must now seek a quantitative relationship between these terms.

Consider an invader that differs but slightly from one of the resident species; each of these residents has zero growth rate in the community, as it is assumed to be stationary. Hence the d-type branching process that describes the fate of this mutant is slightly supercritical, $R_0 = 1 + \varepsilon$ for a small $\varepsilon > 0$. As explained in Section 5.6, Athreya (1993) proved that, under very general conditions, the establishment probability of a single X_i-type mutant with birth state h can therefore be approximated as

$$P_h(X_i', \mathbb{X}) = 2\frac{R_0 - 1}{B(\varepsilon)} v(\varepsilon)_h + o(\varepsilon) = 2\frac{\ln R_0}{B(\varepsilon)} v(\varepsilon)_h + o(\varepsilon) \,, \tag{7.173}$$

where $v(\varepsilon)$ denotes the mutant's reproductive value. This complies with our earlier notation as the ith resident naturally has $\varepsilon = 0$, and the same applies to the parameters $u(\varepsilon)$ and $\xi(\varepsilon)_{hj}$. The factor $B(\varepsilon)$ may be interpreted as a variance (see Durinx *et al.*, unpublished),

$$B(\varepsilon) = \sum_j u(\varepsilon)_j \mathrm{Var}[\sum_l v(\varepsilon)_l \xi(\varepsilon)_{lj}] + o(1) \,. \tag{7.174}$$

The initial mutant has a probability $u(0)_h$ of being born in state h, since $u(0)$ is the stationary offspring distribution of its parent. As the eigenvectors for residents and mutants differ at most by the order ε, we can here approximate $v(\varepsilon)_h$ by $v(0)_h$, and so forth, and show that

$$P(X_i', \mathbb{X}) = \sum_h u(0)_h P_h(X_i', \mathbb{X})$$

$$= 2 \sum_h u(0)_h v(0)_h \frac{\ln R_0}{\sum_j u(0)_j \text{Var}[\sum_l v(0)_l \xi(0)_{lj}]} + o(\varepsilon)$$

$$= \frac{2 \ln R_0}{\sum_j u(0)_j \text{Var}[\sum_l v(0)_l \xi(0)_{lj}]} + o(\varepsilon) , \qquad (7.175)$$

since the scalar product of the eigenvectors sums to 1 (see Section 2.3).

This means we can wrap up the formulation of the establishment probability by substituting the relation found between s and R_0 [Equation (7.170)] and, finally, by linearly approximating $s_{\mathbb{X}}(X_i')$ as close to $s_{\mathbb{X}}(X_i) = 0$,

$$P(X_i', \mathbb{X}) = \frac{2\beta(0) s_{\mathbb{X}}(X_i')}{\sum_j u(0)_j \text{Var}[\sum_l v(0)_l \xi(0)_{lj}]} + o(\varepsilon) \qquad (7.176)$$

$$= \frac{2\beta(0)}{\sum_j u(0)_j \text{Var}[\sum_l v(0)_l \xi(0)_{lj}]} (X_i' - X_i)^{\top} \frac{\partial S_{\mathbb{X}}(X_i)^{\top}}{\partial Y} + o(\varepsilon) . \quad (7.177)$$

Note that we use values derived for X_i for all but the mutation step $(X_i' - X_i)$. More importantly, the above approximation only holds if it returns a positive value, as we started by assuming $\varepsilon > 0$.

7.8.6 The canonical equation for structured populations

The calculated transition rates [Equation (7.165)], combined with the microcosm principle [Equation (7.167)] and the estimated chance of establishment [Equation (7.177)], show that the evolutionary movement along the deterministic path given by Equation (7.164) is generated by

$$\frac{d}{dt} X_i = \int (X_i' - X_i) \frac{\hat{n}_i \mu(X_i)}{T} \mathfrak{M}(X_i' - X_i, X_i) P(X_i', \mathbb{X}) \, dX_i'$$

$$= \frac{\beta \hat{n}_i \mu(X_i)}{T \sum_j u_j \text{Var}[\sum_l v_l \xi_{lj}]} \int V \mathfrak{M}(V, X_i) V^{\top} \frac{\partial S_{\mathbb{X}}(X_i)^{\top}}{\partial Y} \, dV + o(\varepsilon)$$

$$= \frac{\beta \hat{n}_i \mu(X_i)}{T \sum_j u_j \text{Var}[\sum_l v_l \xi_{lj}]} \mathbb{M}(X_i) \frac{\partial S_{\mathbb{X}}(X_i)^{\top}}{\partial Y} + o(\varepsilon) \qquad (7.178)$$

with $\mathbb{M}(X_i) := \int V \mathfrak{M}(V, X_i) V^{\top} \, dV$ the *mutational covariance matrix at* X_i.

The second remark after Equation (7.177) explains the disappearance of factor 2 in the first equality above: for each strategy V that returns a positive value for $V^{\top} \frac{\partial S_{\mathbb{X}}(X_i)^{\top}}{\partial Y}$, the strategy $-V$ returns a negative value, and vice versa since we are away from singular strategies. Hence, we have to replace exactly half of the estimates by zero. The value under the integral is not influenced by this, as V and $-V$ return the same value under the second integral if the distribution $\mathfrak{M}$ is symmetrical.

This allows us to finalize the canonical equation for structured populations as the first-order prediction

$$\frac{d}{dt} X_i \approx \frac{\beta}{T} \frac{\hat{n}_i \, \mu(X_i)}{\sum_j u_j \text{Var}[\sum_l v_l \xi_{lj}]} \mathbb{M}(X_i) \frac{\partial S_{\mathbb{X}}(X_i)^{\top}}{\partial Y} \, . \tag{7.179}$$

Keep in mind that these are all expected values for each of the parameters and strategies, and most parameters that relate to a species i depend also on the other strategies in the community $\mathbb{X}$.

7.8.7 Discussion

A fundamental open problem in evolutionary biology is the development of a straightforward, systematic way to study long-term evolutionary trajectories. Over the years, some of the issues have been addressed.

How selection can change morphological traits was first described by Lande (1979), based on the breeder's equation as found in animal sciences. In what has become known in evolutionary biology as Lande's equation, we have a formula that is remarkably similar to the canonical equation. The major difference is that the population density does not appear as a factor (the other differences amount to a different interpretation of the parameters). What it essentially describes is how a population changes through selection on standing genetic variation. Such variation accumulates when a species' strategy is close to the evolutionary optimum in a stable environment, especially if the optimum is relatively weak. Selection on the diversity then occurs when external environmental parameters change. A typical example of this is the introduction of a wild population into a laboratory setup that has directional selection applied, which accounts for the good fit of Lande's equation with laboratory data. With none of the wild variants optimal for the laboratory, the winner of the laboratory race does not represent an evolutionary endpoint. After the initial relatively rapid modification of the traits in reaction to external changes, further innovations and long-term evolution must come from mutations. Haldane (1927) was the first to realize this and to argue that mutation-limited evolution is slower than initially suspected, because many advantageous mutants fail to become established. The canonical equation builds on these ideas to derive a quantitative relation between the factors involved and, in particular, how the ecology determines the selective pressures.

It is clear that the assumptions we have worked with amount to a crude oversimplification, but the question is important: how can we link ecology with paleontology? The Modern Synthesis went no further than to show that the two are compatible. Lande's equation and the canonical equation are the best (being the only) tools we have so far with which to reason about the connection.

This section illustrates how branching processes underlie mutation-limited evolution and hence their fundamental importance to adaptive dynamics theory. The calculations are meant to be heuristic and biological, at the cost of mathematical precision and exhaustiveness. For a more mathematical treatment of the canonical equation, consult Champagnat *et al.* (2001). The restriction to finitely many possible birth states means we can fall back on established theory, but limits the applicability. To overcome this requires an extension of the theorem of Section 5.6.2 (Athreya 1993) to branching processes with infinitely many types. If the resident attractor is not a fixed point, but a limit cycle, every individual can be assigned a birth state that depends on where in the cycle the individual is born. In discrete time this allows the attractor to be treated analogously to a fixed point, so that our analysis immediately applies. In continuous time the suggested extension of Athreya's proof is applied similarly.

Analytically, no extension of the canonical equation to non-periodic attractors has been found yet. The first heuristic explorations for ergodically fluctuating environments with invader dynamics that follow a linear birth-and-death process suggest that the canonical equation is robust against such extension. Dieckmann (personal communication) shows, by approximating the process as formulated in Kendall (1948), that the establishment probability is approximately proportional to the fitness [as in Equation (7.177)], so that a similar result holds.

The analysis as presented applies to spatial models with finite numbers of patches, if the residents are locally sufficiently numerous and well mixed. The patch an individual inhabits is expressed as a component of its state. For some more complicated spatial models an equation similar to the canonical equation may well apply. The crucial part is that the chance of success on invading must scale linearly with changes in strategy. This is an area in which more research is badly needed.

We have assumed an unbiased mutation distribution. Mutation bias arises from the non-linearity of the genotype/phenotype mapping, and becomes prominent when high mutational variance is combined with a highly curved mapping. However, since we assume small mutational steps, we follow the biological literature in neglecting this effect. Champagnat *et al.* (2001) discuss the relevance of bias in the context of the canonical equation.

A far more complicated issue underlies timescale separation. There are several latent limits: large resident populations (or the limit $\frac{1}{\Omega} \to 0$ where Ω is the system size), small mutations (or $\epsilon \to 0$), and rare mutations (or $\mu(X_i)\Omega \to 0$). These limits are not a priori commutative, and so, depending on the order in which the details of mechanistic, individual-based parameters are scaled away, a different limit process is obtained. An initial discussion of these issues, in particular of the necessity to stay away from singularities, is given in Metz *et al.* (1996).

Acknowledgments Michel Durinx is supported by the Dutch National Science Organization (NWO) through PhD grant 809.34.002. Both authors received support from the European Research Training Network *ModLife* and from the NWO through the Dutch–Hungarian exchange grant 048.011.039.

We thank Peter Jagers, Patsy Haccou, and Tom van Dooren for their patience and unrelenting criticism.

Appendix

A.1 Expectation and Variance

A.1.1 Expectation

Let X be a random variable that can take on a finite number of values $x_1, \dots, x_n$. The *expectation* of X is defined as

$$\mathbb{E}[X] = \sum_{k=1}^{n} x_k \mathbb{P}(X = x_k) . \tag{A.1}$$

If X takes on a countable number of values $x_1, x_2, \dots, x_n, \dots$, the expectation of X is

$$\mathbb{E}[X] = \sum_{k=1}^{\infty} x_k \mathbb{P}(X = x_k) , \tag{A.2}$$

provided that the sum exists and is finite, that is,

$$\sum_{k=1}^{\infty} |x_k| \, \mathbb{P}(X = x_k) < \infty . \tag{A.3}$$

The latter condition is added to make the value of the infinite sum uniquely defined. In most applications X only takes values $0, 1, \dots$, so we obtain

$$\mathbb{E}[X] = \sum_{k=0}^{\infty} k \mathbb{P}(X = k) . \tag{A.4}$$

A random variable X is called *discrete* if it can take a finite or a countable number of values. It is called *continuous* if there exists a non-negative function $p(y)$, $y \in (-\infty, \infty)$, such that for any number x

$$\mathbb{P}(X \leq x) = \int_{-\infty}^{x} p(y) \, dy . \tag{A.5}$$

The function $p(x)$ is called the density of X. Clearly,

$$p(x) \geq 0 \text{ and } \int_{-\infty}^{+\infty} p(y) \, dy = 1 . \tag{A.6}$$

If X is a continuous random variable, the expectation of X is

$$\mathbb{E}[X] = \int_{-\infty}^{\infty} x p(x) \, dx , \tag{A.7}$$

provided that

$$\int_{-\infty}^{\infty} |x|\, p(x)\, dx < \infty .$$ (A.8)

From these definitions it is not difficult to derive the following properties:

1. If c is a constant then $\mathbb{E}[c] = c$, $\mathbb{E}[cX] = c\mathbb{E}[X]$.
2. $\mathbb{E}[X + Y] = \mathbb{E}[X] + \mathbb{E}[Y]$.
3. If X is a non-negative random variable ($X \geq 0$), then $\mathbb{E}[X] \geq 0$.

If X takes only integer values $0, 1, 2, \ldots$, then $\mathbb{E}[X]$ can also be calculated by the formula

$$\mathbb{E}[X] = \sum_{k=1}^{\infty} \mathbb{P}(X \geq k) .$$ (A.9)

Indeed, in this case

$$
\begin{aligned}
\mathbb{E}[X] &= \sum_{k=0}^{\infty} k\mathbb{P}(X = k) = \sum_{k=1}^{\infty} k\,(\mathbb{P}(X \geq k) - \mathbb{P}(X \geq k+1)) \\
&= \sum_{k=1}^{\infty} k\mathbb{P}(X \geq k) - \sum_{k=1}^{\infty} k\mathbb{P}(X \geq k+1) \\
&= \sum_{k=1}^{\infty} k\mathbb{P}(X \geq k) - \sum_{k=1}^{\infty} (k-1)\,\mathbb{P}(X \geq k) \\
&= \sum_{k=1}^{\infty} \mathbb{P}(X \geq k) .
\end{aligned}
$$ (A.10)

A.1.2 Variance

The *variance* of X is defined as

$$\mathrm{Var}[X] = \mathbb{E}\left[(X - \mathbb{E}[X])^2\right] = \sum_{k}(x_k - \mathbb{E}[X])^2 \mathbb{P}(X = x_k) ,$$ (A.11)

if X is a discrete random variable, and as

$$\mathrm{Var}[X] = \mathbb{E}\left[(X - \mathbb{E}[X])^2\right] = \int_{-\infty}^{\infty} (x - \mathbb{E}[X])^2\, p(x)\, dx ,$$ (A.12)

if X is continuous.

It is not difficult to check that

$$\mathrm{Var}[X] = \mathbb{E}\left[X^2\right] - (\mathbb{E}[X])^2$$ (A.13)

and that

$$\mathrm{Var}[cX] = c^2 \mathrm{Var}[X] .$$ (A.14)

The *covariance* between two random variables X and Y is defined as

$$\text{Cov}[X, Y] = \mathbb{E}[(X - \mathbb{E}[X])(Y - \mathbb{E}[Y])] = \mathbb{E}[XY] - \mathbb{E}[X]\mathbb{E}[Y] . \quad (A.15)$$

Hence $\text{Cov}[X, X] = \text{Var}[X]$ and

$$\text{Var}[X + Y] = \text{Var}[X] + \text{Var}[Y] + 2\text{Cov}[X, Y] . \quad (A.16)$$

Two random variables X and Y are said to be *independent* if for all numbers x and y,

$$\mathbb{P}(X \leq x, Y \leq y) = \mathbb{P}(X \leq x)\mathbb{P}(Y \leq y) . \quad (A.17)$$

The covariance of independent random variables vanishes (but the converse is not generally true). Hence, for independent variables, variance is additive,

$$\text{Var}[X + Y] = \text{Var}[X] + \text{Var}[Y] . \quad (A.18)$$

The *correlation coefficient* R between X and Y is defined as

$$R = \frac{\text{Cov}[X, Y]}{\sqrt{\text{Var}[X]\text{Var}[Y]}} . \quad (A.19)$$

One can check that $R = 0$ if X and Y are independent, $R = 1$ if $X = cY$, for some $c > 0$, and $R = -1$ if $X = cY$, $c < 0$.

A.2 Useful Equalities and Inequalities

A.2.1 A simple inequality

For non-negative integers k and $0 \leq x \leq 1$,

$$1 - (1 - x)^k \geq kx - \frac{k(k - 1)}{2}x^2 \quad (A.20)$$

or, equivalently,

$$kx - (1 - (1 - x)^k) \leq \frac{k(k - 1)}{2}x^2 . \quad (A.21)$$

Indeed, for $k = 0$ and $k = 1$ this is obvious, whereas for $k \geq 2$ it follows from

$$kx - (1 - (1 - x)^k) = k(k - 1)\int_0^x du \int_0^u (1 - y)^{k-2} \, dy$$

$$\leq k(k - 1)\int_0^x du \int_0^u dy = \frac{k(k - 1)}{2}x^2 . \quad (A.22)$$

A.2.2 Indicators

The *indicator* 1_A of an event A is a function that equals 1 if the event A occurs and 0 otherwise. More generally, for any set A, like, for instance, an interval, $A = [a, b]$,

$$1_A(x) = \begin{cases} 1 & \text{if } x \in A , \\ 0 & \text{otherwise} . \end{cases} \quad (A.23)$$

For a fixed event A, $1_A(u)$ is thus viewed as a function of the outcome u. It is not difficult to check the following:

1. $1_{A \cap B} = 1_A \times 1_B$.
2. If $\bar{A}$ is the event complementary to A (i.e., $\bar{A}$ occurs if and only if A does not occur), then $1_{\bar{A}} = 1 - 1_A$.
3. Let $B = A_1 \cup A_2 \cup \cdots \cup A_n$ be the union of the events $A_1, A_2, \ldots, A_n$, that is, the event that occurs if and only if at least one of the events $A_1, A_2, \ldots, A_n$ occurs. The event $\bar{B}$, which is complementary to B, is the intersection of the events $\bar{A}_1, \bar{A}_2, \ldots, \bar{A}_n$, that is $\bar{B} = \bar{A}_1 \cap \bar{A}_2 \cap \cdots \cap \bar{A}_n$. It occurs if and only if the events $\bar{A}_1, \bar{A}_2, \ldots, \bar{A}_n$ occur simultaneously.

Clearly,

$$\mathbb{E}[1_A] = 0 \times \mathbb{P}(1_A = 0) + 1 \times \mathbb{P}(1_A = 1)$$
$$= \mathbb{P}(1_A = 1) = \mathbb{P}(A) \,. \tag{A.24}$$

A.2.3 Inclusion–exclusion

The equality

$$\mathbb{P}(A \cup B) = \mathbb{P}(A) + \mathbb{P}(B) - \mathbb{P}(AB) \tag{A.25}$$

has an extension to unions of an arbitrary sequence $A_1, A_2, \ldots, A_n$ of events. (We write AB for $A \cap B$, etc.)

$$\mathbb{P}(A_1 \cup A_2 \cup \cdots \cup A_n) = \mathbb{P}(A_1) + \mathbb{P}(A_2) + \cdots + \mathbb{P}(A_n)$$
$$- \mathbb{P}(A_1 A_2) - \mathbb{P}(A_1 A_3) - \cdots - \mathbb{P}(A_1 A_n) - \mathbb{P}(A_1 A_2)$$
$$- \cdots - \mathbb{P}(A_2 A_n) - \cdots - \mathbb{P}(A_{n-1} A_n) + \mathbb{P}(A_1 A_2 A_3) + \mathbb{P}(A_1 A_2 A_4)$$
$$+ \cdots + \mathbb{P}(A_1 A_2 A_n) + \mathbb{P}(A_1 A_3 A_4) + \cdots + \mathbb{P}(A_{n-2} A_{n-1} A_n)$$
$$- \cdots + \cdots - \cdots + (-1)^{n-1} \mathbb{P}(A_1 A_2 \ldots A_n) \,. \tag{A.26}$$

In particular,

$$\mathbb{P}(A_1 \cup A_2 \cup A_3) = \mathbb{P}(A_1) + \mathbb{P}(A_2) + \mathbb{P}(A_3) - \mathbb{P}(A_1 A_2)$$
$$- \mathbb{P}(A_1 A_3) - \mathbb{P}(A_2 A_3) + \mathbb{P}(A_1 A_2 A_3) \,. \tag{A.27}$$

Proof. For simplicity we prove only the last case ($n = 3$).

$$1_{A \cup B \cup C} = 1 - 1_{\overline{A \cup B \cup C}} = 1 - 1_{\bar{A} \bar{B} \bar{C}}$$
$$= 1 - 1_{\bar{A}} 1_{\bar{B}} 1_{\bar{C}} = 1 - (1 - 1_A)(1 - 1_B)(1 - 1_C)$$
$$= 1_A + 1_B + 1_C - 1_A 1_B - 1_A 1_C - 1_B 1_C + 1_A 1_B 1_C$$
$$= 1_A + 1_B + 1_C - 1_{AB} - 1_{AC} - 1_{BC} + 1_{ABC} \,. \tag{A.28}$$

The expectation of this is the desired formula.

$$\diamond \ \diamond \ \diamond$$

A.2.4   Classic inequalities

In the text we make repeated use of the following inequalities, which we state here without proof. Proofs can be found in any probability theory textbook.

1. *Markov's inequality.* Let X be a non-negative random variable, then for any $x > 0$,

$$\mathbb{P}(X \geq x) \leq \mathbb{E}[X]/x . \tag{A.29}$$

2. *Chebyshev's inequality.* Let Y be a random variable with finite variance, then for any $\epsilon > 0$,

$$\mathbb{P}(|Y - \mathbb{E}[Y]| > \epsilon) \leq \text{Var}[Y]/\epsilon^2 . \tag{A.30}$$

3. *Lyapunov's inequality.* Let X be a non-negative random variable with

$$\mathbb{E}[X^2] < \infty ,$$

then

$$\mathbb{P}(X > 0) \geq (\mathbb{E}[X])^2/\mathbb{E}[X^2] . \tag{A.31}$$

4. *Jensen's inequality.* Let $u(x)$ be a function defined for $x \in D \subseteq (-\infty, +\infty)$. Assume that D is a finite or infinite interval and let $u(x)$ be differentiable. The function is called *concave* if for any fixed $m \in D$ and all $x \in D$

$$u(x) \leq u(m) + u'(m)(x - m) , \tag{A.32}$$

and *convex* if for any fixed $m \in D$ and all $x \in D$

$$u(x) \geq u(m) + u'(m)(x - m) . \tag{A.33}$$

This means that a graph of the function lies under (concave) or over (convex) its tangent at any point m. If it has a second derivative the following simple criterion works: if $u''(x) > 0$ for all $x \in D$, then $u(x)$ is convex in D; if $u''(x) < 0$, then $u(x)$ is concave in D.

Let X be a random variable with values in D. If $u(x)$ is concave in D, then

$$\mathbb{E}[u(X)] \leq u(\mathbb{E}[X]) , \tag{A.34}$$

and if it is convex in D, then

$$\mathbb{E}[u(X)] \geq u(\mathbb{E}[X]) . \tag{A.35}$$

Examples. The function $\ln x, x > 0$, is concave. Therefore, for any positive random variable X,

$$\mathbb{E}[\ln X] \leq \ln \mathbb{E}[X] . \tag{A.36}$$

The function e^x is convex for $x \in D = (-\infty, +\infty)$ and, therefore, for any random variable X,

$$\mathbb{E}\left[e^X\right] \geq e^{\mathbb{E}[X]} . \tag{A.37}$$

◇ ◇ ◇

A.3 Conditioning

A.3.1 Conditional expectation

Let X and Y be two discrete random variables, the latter with the possible values $y_1, \ldots, y_n, \ldots$, and the former with a well-defined expectation. (Recall that this means $\mathbb{E}[|X|] < \infty$.) The *conditional expectation* of X, given a fixed value $Y = y_j$, is defined as

$$\mathbb{E}\left[X \mid Y = y_j\right] = \sum_k x_k \mathbb{P}(X = x_k | Y = y_j)$$

$$= \frac{\sum_k x_k \mathbb{P}(X = x_k; Y = y_j)}{\mathbb{P}(Y = y_j)} . \tag{A.38}$$

The conditional expectation of X given Y, $\mathbb{E}[X|Y]$, is a random variable defined to equal $\mathbb{E}[X|Y = y_j]$ when $Y = y_j$. It is thus a function of Y, as the name indicates. If we take the expectation of the conditional expectation,

$$\mathbb{E}\left[\mathbb{E}\left[X \mid Y\right]\right] = \sum_j \mathbb{E}[X \mid Y = y_j]\mathbb{P}(Y = y_j)$$

$$= \sum_j \sum_k x_k \mathbb{P}(X = x_k; Y = y_j)$$

$$= \sum_k x_k \sum_j \mathbb{P}(X = x_k; Y = y_j)$$

$$= \sum_k x_k \mathbb{P}(X = x_k) = \mathbb{E}\left[X\right] , \tag{A.39}$$

we obtain the original expectation. This is called the *rule of double expectation*,

$$\mathbb{E}[\mathbb{E}[X|Y]] = \mathbb{E}[X] . \tag{A.40}$$

It was derived here for discrete variables, but in fact Equation (A.40) is valid for *any* random variables X and Y, such that $\mathbb{E}[X]$ is well defined.

A.3.2 Wald's equation

Let $\xi_1, \ldots, \xi_n, \ldots$ be a sequence of independent identically distributed random variables with $m = \mathbb{E}[\xi_1] < \infty$, and let Y be a non-negative integer-valued random variable, independent of all ξ_n. Set $X = \xi_1 + \xi_2 + \cdots + \xi_Y$. Then,

$$\mathbb{E}\left[X\right] = m E\left[Y\right] . \tag{A.41}$$

This follows immediately from

$$\mathbb{E}\left[X\right] = \mathbb{E}\left[\mathbb{E}\left[X|Y\right]\right] = \mathbb{E}\left[mY\right] , \tag{A.42}$$

by the additivity of expectation and independence of ξ_i and Y.

A.3.3   Conditional variance

The *conditional variance* of X given $Y = y_j$ is

$$\text{Var}[X|Y = y_j] = \mathbb{E}[(X - \mathbb{E}[X|Y = y_j])^2 | Y = y_j] . \tag{A.43}$$

Again, the conditional variance of X given Y is a random variable, usually denoted $\text{Var}[X|Y]$, such that if $Y = y_j$ it equals $\text{Var}[X|Y = y_j]$. In particular,

$$\text{Var}[X|Y] = \mathbb{E}[X^2|Y] - (\mathbb{E}[X|Y])^2 , \tag{A.44}$$

and clearly

$$\begin{aligned}
\text{Var}[X] &= \mathbb{E}[X^2] - (\mathbb{E}[X])^2 = \mathbb{E}[\mathbb{E}[X^2|Y]] - (\mathbb{E}[\mathbb{E}[X|Y]])^2 \\
&= \mathbb{E}[(\mathbb{E}[X|Y])^2] - (\mathbb{E}[\mathbb{E}[X|Y]])^2 + \mathbb{E}[\mathbb{E}[X^2|Y]] \\
&\quad - \mathbb{E}[(\mathbb{E}[X|Y])^2] \\
&= \text{Var}[\mathbb{E}[X|Y]] + \mathbb{E}[\text{Var}[X|Y]] . \tag{A.45}
\end{aligned}$$

This proves the *variance partitioning* formula,

$$\text{Var}[X] = \text{Var}[\mathbb{E}[X|Y]] + \mathbb{E}[\text{Var}[X|Y]] . \tag{A.46}$$

The following is a useful application of this to sums of a random number of random variables. Let $X = \xi_1 + \xi_2 + \cdots + \xi_Y$, where Y is a non-negative integer-valued random variable with finite variance and independent of $\xi_1, \ldots, \xi_n, \ldots$, which are in their turn independent and identically distributed among themselves. Write $m = \mathbb{E}[\xi_1] < \infty$ and $\sigma^2 = \text{Var}[\xi_1] < \infty$, then

$$\text{Var}[X] = \sigma^2 \mathbb{E}[Y] + m^2 \text{Var}[Y] . \tag{A.47}$$

Proof. Since ξ_i are independent and identically distributed, and independent of Y,

$$\begin{aligned}
\text{Var}[X|Y] &= \text{Var}[\xi_1 + \xi_2 + \cdots + \xi_Y | Y] \\
&= Y\text{Var}[\xi_1] = Y\sigma^2 , \tag{A.48}
\end{aligned}$$

and therefore,

$$\mathbb{E}[\text{Var}[X|Y]] = \sigma^2 \mathbb{E}[Y] . \tag{A.49}$$

Since $\mathbb{E}[X|Y] = mY$,

$$\text{Var}[\mathbb{E}[X|Y]] = \text{Var}[mY] = m^2 \text{Var}[Y] . \tag{A.50}$$

A.4   Distributions and Their Transforms

A.4.1   Probability generating functions

Let X be a non-negative integer-valued random variable with

$$p_k = \mathbb{P}(X = k), \ k = 0, 1, 2, \ldots .$$

The function

$$f(s) = \mathbb{E}\left[s^X\right] = \sum_{k=0}^{\infty} p_k s^k, \ 0 \le s \le 1, \tag{A.51}$$

is called the *generating function* of X. It has the following properties:

1.

$$f(1) = \sum_{k=0}^{\infty} p_k = 1. \tag{A.52}$$

2. If $\sum_{k\ge 1}^{\infty} k p_k < \infty$, then

$$f'(1) = \sum_{k=0}^{\infty} k p_k = \mathbb{E}\left[X\right]. \tag{A.53}$$

3. If $\sum_{k\ge 1}^{\infty} k^2 p_k < \infty$, then

$$f''(1) = \sum_{k=0}^{\infty} k(k-1) p_k = \mathbb{E}\left[X(X-1)\right] \tag{A.54}$$

and

$$\mathrm{Var}\left[X\right] = \mathbb{E}\left[X^2\right] - (\mathbb{E}\left[X\right])^2 = \mathbb{E}\left[X(X-1)\right] + \mathbb{E}\left[X\right] - (\mathbb{E}\left[X\right])^2$$
$$= f''(1) + f'(1) - \left(f'(1)\right)^2. \tag{A.55}$$

4. If the random variables X_1 and X_2 are independent with generating functions $f_1(s)$ and $f_2(s)$, then the generating function of their sum is

$$\mathbb{E}[s^{X_1+X_2}] = f_1(s) f_2(s). \tag{A.56}$$

5. If $X_1, X_2, \ldots, X_n, \ldots$ are independent and identically distributed non-negative and integer-valued random variables with the generating function $f(s)$, and Y is an integer-valued random variable with $g(s) = \mathbb{E}[s^Y]$, independent of X, then

$$\mathbb{E}\left[s^{X_1+X_2+\cdots+X_Y}\right] = \mathbb{E}\left[(\mathbb{E}\left[s^{X_1}\right])^Y\right] = g(f(s)). \tag{A.57}$$

A.4.2 Some basic distributions and their generating functions

Binary events or the Bernoulli law. If X is a random variable that is either zero or one, and $\mathbb{P}(X = 1) = p, 0 \le p \le 1, q = 1 - p$, then

$$f(s) = q s^0 + p s^1 = q + p s = 1 - p + p s. \tag{A.58}$$

Binomial distributions. The number of successes X in n independent trials with success probability p follows the *binomial* distribution with parameters n and p,

$$\mathbb{P}(X = k) = \binom{n}{k} p^k (1-p)^{n-k}, \ k = 0, 1, \ldots, n. \tag{A.59}$$

This is often expressed by saying that X is Binomial(n, p) or Bin(n, p). The generating function is

$$f(s) = \mathbb{E}\left[s^X\right] = \sum_{k=0}^{n} \binom{n}{k} p^k (1 - p)^{n-k} s^k = (1 - p + ps)^n . \tag{A.60}$$

Clearly, X is the sum of n independent Bernoulli random variables.

Geometric distributions. The number of trials X until the first success in a sequence of independent trials each with the success probability p has the *geometric* distribution with parameter p [is Geometric(p) or Geo(p)],

$$\mathbb{P}(X = k) = qp^k, k = 0, 1, 2, \dots ; \quad q = 1 - p . \tag{A.61}$$

The generating function is

$$f(s) = \mathbb{E}\left[s^X\right] = \sum_{k=0}^{\infty} qp^k s^k = \frac{q}{1 - ps} . \tag{A.62}$$

Poisson distributions. A random variable Y has a *Poisson* distribution with parameter $m \geq 0$ [or X is Poisson(m) or Poi(m)] if

$$\mathbb{P}(Y = k) = \frac{m^k}{k!} e^{-m}, k = 0, 1, 2, \dots . \tag{A.63}$$

The generating function is

$$f(s) = \mathbb{E}\left[s^Y\right] = \sum_{k=0}^{\infty} \frac{m^k}{k!} e^{-m} s^k = e^{-m} \sum_{k=0}^{\infty} \frac{(ms)^k}{k!} = e^{m(s-1)} . \tag{A.64}$$

The Poisson distribution is sometimes referred to as the "Law of Rare Events." The reason for this is that numbers of unlikely events sparked by many independent sources tend to follow a Poisson distribution. Thus, the Bernoulli$(n, m/n)$ distribution converges to Poisson(m) as $n \to \infty$.

In its turn, the Poisson distribution is conserved if events are filtered (are independently deleted with a probability $1 - p$). Let $Z = X_1 + X_2 + \cdots + X_Y$, where $X_i, i = 1, 2, \dots$ are independent identically distributed random variables such that

$$q = \mathbb{P}(X_i = 0), \quad p = \mathbb{P}(X_i = 1), \quad p + q = 1 . \tag{A.65}$$

If Y has a Poisson(m) distribution, then Z is Poisson(pm) distributed.

Proof. By Equations (A.57), (A.64), and (A.58), we have

$$\mathbb{E}\left[s^Z\right] = \mathbb{E}\left[s^{X_1+X_2+\cdots+X_Y}\right] = \mathbb{E}\left[(\mathbb{E}\left[s^{X_1}\right])^Y\right] = \mathbb{E}\left[(q + ps)^Y\right]$$
$$= e^{m(q+ps-1)} = e^{m(ps-p)} = e^{pm(s-1)} , \tag{A.66}$$

as desired. (As is tacitly assumed, not only does a distribution determine its generating function, but the reverse holds as well.)

$$\diamond \diamond \diamond$$

A.4.3 Multivariate generating functions

Now consider a d-dimensional vector $\mathbf{X} = (X_1, \dots, X_d)$ with non-negative integer-valued random components. The generating function of $\mathbf{X}$ is then defined as

$$f(\mathbf{s}) = \mathbb{E}[\mathbf{s}^{\mathbf{X}}] = f(s_1, \dots, s_d) = \mathbb{E}[s_1^{X_1} \cdots s_d^{X_d}]$$

$$= \sum_{k_1=0}^{\infty} \cdots \sum_{k_d=0}^{\infty} \mathbb{P}(X_1 = k_1, \dots, X_d = k_d) s_1^{k_1} \cdots s_d^{k_d} . \tag{A.67}$$

For instance, let n independent experiments be conducted, each of which may result in one of d possible outcomes, the probability of the ith outcome being $p_i, i = 1, 2, \dots, d$. Let X_i be the number of trials that result in the ith outcome. The distribution of the random vector $\mathbf{X} = (X_1, \dots, X_d)$ is called a Multinomial$(n, p_1, \dots, p_d)$ distribution. One can show that

$$\mathbb{P}(X_1 = k_1, \dots, X_d = k_d) = \binom{n}{k_1, \dots, k_d} p_1^{k_1} \cdots p_d^{k_d}$$

$$= \frac{n!}{k_1! \dots k_d!} p_1^{k_1} \cdots p_d^{k_d} . \tag{A.68}$$

The generating function of this distribution is

$$f(\mathbf{s}) = f(s_1, \dots, s_d) = (p_1 s_1 + p_2 s_2 + \cdots + p_d s_d)^n . \tag{A.69}$$

A.4.4 Laplace transforms

Let X be a discrete non-negative random variable that takes on values $x_0, x_1, x_2, \dots$ with probabilities $p_0, p_1, p_2, \dots$, respectively. The function

$$\phi(u) = \mathbb{E}\left[e^{-uX}\right] = \sum_k \mathbb{P}(X = x_k) e^{-ux_k} = \sum_k p_k e^{-ux_k}, \ u \geq 0 , \tag{A.70}$$

is called the *Laplace* transform of X. Clearly, if X is integer-valued, then $\phi(u) = f(e^{-u})$. For instance, if X is Poisson(m) distributed, then

$$\phi(u) = f(e^{-u}) = \exp\{m\left(e^{-u} - 1\right)\} . \tag{A.71}$$

If $X \geq 0$ is a continuous random variable with density $p(x)$, its Laplace transform is defined to be

$$\phi(u) = \mathbb{E}\left[e^{-uX}\right] = \int_0^{\infty} e^{-ux} p(x) \, dx, \ u \geq 0 . \tag{A.72}$$

Since, in this case,

$$\int_0^{\infty} e^{-ux} p(x) \, dx \leq \int_0^{\infty} p(x) \, dx = 1 , \tag{A.73}$$

the Laplace transform is well defined for all $u \geq 0$.

Like the generating function, the Laplace transform determines the distribution function of a random variable uniquely.

A.4.5   Some distributions and Laplace transforms

Exponential distributions.   A random variable X is distributed exponentially with parameter $\lambda > 0$, or $X \sim \text{Exp}(\lambda)$, if its density is

$$p(x) = \lambda e^{-\lambda x}, x \ge 0, \qquad \text{and} \qquad p(x) = 0, x < 0. \tag{A.74}$$

Then, the Laplace transform of X is

$$\phi(u) = \lambda \int_0^\infty e^{-ux} e^{-\lambda x}\, dx = \frac{\lambda}{u + \lambda}. \tag{A.75}$$

As pointed out in the main text, exponential distributions are often used for life spans. This has to be carried out with some care though, as having exponentially distributed life spans is equivalent to the property of *no aging*.

If $T \sim \text{Exp}(\lambda)$, then for any $t, u > 0$, by the definition of conditional probability,

$$\mathbb{P}(T > t + u \mid T > t) = \frac{\mathbb{P}(T > t + u, T > t)}{\mathbb{P}(T > t)} = \frac{\mathbb{P}(T > t + u)}{\mathbb{P}(T > t)}$$

$$= \frac{e^{-\lambda(t+u)}}{e^{-\lambda u}} = e^{-\lambda t} = \mathbb{P}(T > t), \tag{A.76}$$

so that the chances of a t-year-old surviving for another u years are the same as those for a newborn.

Conversely, if $\mathbb{P}(T > t + u \mid T > t) = \mathbb{P}(T > u)$ for any $t, u > 0$, then (as above)

$$\mathbb{P}(T > t + u) = \mathbb{P}(T > t)\mathbb{P}(T > u). \tag{A.77}$$

Thus,

$$\mathbb{P}(T > t + u) - \mathbb{P}(T > t) = -(1 - \mathbb{P}(T > u))\mathbb{P}(T > t). \tag{A.78}$$

If we write $S(t) = \mathbb{P}(T > t)$ for the survival function, and divide the equation by $u \to 0$, it follows that

$$S'(t) = S'(0)S(t). \tag{A.79}$$

The only feasible solution of this is $S(t) = e^{-\lambda t}$ with $\lambda = -S'(0)$.

Gamma distributions.   A non-negative random variable X has the *Gamma* distribution with parameters (a, λ) [or $X \sim \Gamma(a, \lambda)$] if

$$p(x) = \frac{\lambda^a x^{a-1}}{\Gamma(a)} e^{-\lambda x}, x \ge 0, \qquad \text{and} \qquad p(x) = 0, x < 0, \tag{A.80}$$

where

$$\Gamma(a) = \int_0^\infty x^{a-1} e^{-x}\, dx \tag{A.81}$$

is the gamma function. Observe that $\Gamma(n + 1) = n! = 1 \times 2 \times \cdots \times (n - 1) \times n$ for positive integers n.

The Laplace transform of X is

$$\phi(u) = \frac{\lambda^a}{\Gamma(a)} \int_0^\infty x^{a-1} e^{-ux} e^{-\lambda x} \, dx = \frac{\lambda^a}{(u+\lambda)^a} \, . \tag{A.82}$$

A.4.6 Generalizations of transforms

Definitions of expectation, variance, and transforms can be extended to random variables that are partially discrete and partially continuous. We do not go into detail (see Box 3.1, though).

Here we give an example. We say that a random variable Y is a *mixture* of the independent random variables X_1 and X_2 if

$$Y = I X_1 + (1 - I) X_2 \, , \tag{A.83}$$

where I is a Bernoulli random variable that is independent of X_1 and X_2. If Y is a mixture of X_1 and X_2 and $p = \mathbb{P}(I = 1)$, the Laplace transforms of the random variables involved are related by

$$\mathbb{E}[e^{-uY}] = (1 - p)\mathbb{E}[e^{-uX_1}] + p\mathbb{E}[e^{-uX_2}] \, . \tag{A.84}$$

For example, if $\mathbb{P}(X_1 = 0) = 1$ and X_2 is $\text{Exp}(\lambda)$, then

$$\mathbb{E}[e^{-uY}] = q + p\frac{\lambda}{u+\lambda}, \qquad q = 1 - p \, . \tag{A.85}$$

The Laplace transform of an *arbitrary* non-negative random variable X is defined to be

$$\phi(u) = \mathbb{P}(X = 0) + u \int_0^\infty e^{-ux} \mathbb{P}(X \le x) \, dx, \ u \ge 0 \, . \tag{A.86}$$

One can check, by direct integration for discrete random variables or by integration by parts for continuous random variables, that this definition is consistent with those given earlier [Equations (A.70) and (A.72)].

If X is a continuous (not necessarily non-negative) random variable such that the integral

$$\int_{-\infty}^\infty e^{-ux} p(x) \, dx \tag{A.87}$$

is convergent for all u that belong to a domain $D \subseteq (-\infty, +\infty)$, the Laplace transform can be defined by Equation (A.72) for $u \in D$. For instance, if X is distributed normally with parameters μ and σ^2, that is, it has the density

$$p(x) = \frac{1}{\sigma\sqrt{2\pi}} e^{-\frac{(x-\mu)^2}{2\sigma^2}} \, , \tag{A.88}$$

often written $X \sim N(\mu, \sigma^2)$, then

$$\phi(u) = \frac{1}{\sigma\sqrt{2\pi}} \int_{-\infty}^\infty e^{-ux} e^{-\frac{(x-\mu)^2}{2\sigma^2}} \, dx = e^{\frac{u^2\sigma^2}{2} - u\mu} \tag{A.89}$$

is convergent and thus well defined for all real values of u.

A.5   Convergence

A.5.1   Convergence forms

Probability theory contains several kinds of convergence of sequences of random variables. The most important for this book are convergences in distribution, in probability, in mean square, and with probability one, also called almost sure (a.s.) convergence.

Let $X_1, \ldots, X_n, \ldots$ be a sequence of random variables. The sequence is said to converge *in distribution* to a random variable X if

$$\lim_{n \to \infty} \mathbb{P}(X_n \le x) = \mathbb{P}(X \le x) \tag{A.90}$$

at all points x where the function $G(x) = \mathbb{P}(X \le x)$ is continuous.

The sequence $X_1, \ldots, X_n, \ldots$ is said to converge *in probability* to X if, for any $\varepsilon > 0$,

$$\lim_{n \to \infty} \mathbb{P}(|X_n - X| > \varepsilon) = 0 \,, \tag{A.91}$$

in mean square if

$$\lim_{n \to \infty} \mathbb{E}[(X_n - X)^2] = 0 \,, \tag{A.92}$$

and *with probability one* if

$$\mathbb{P}(\lim_{n \to \infty} X_n = X) = 1 \,. \tag{A.93}$$

A.5.2   Fatou's lemma

Fatou's lemma is a fundamental relation in probability and integration theory. In this book it is also crucial for obtaining bounds to fence in entities that are important, but difficult to calculate.

Lemma A.1 *Let Y and Z be two non-negative random variables with $\mathbb{E}[Y] < \infty$ and $\mathbb{E}[Z] < \infty$, and let $X_1, \ldots, X_n, \ldots$ be a sequence of random variables. If $-Y \le X_n$ for all n or $X_n \le Z$ for all n then, respectively,*

$$\mathbb{E}[\liminf_{n \to \infty} X_n] \le \liminf_{n \to \infty} \mathbb{E}[X_n] \tag{A.94}$$

and

$$\limsup_{n \to \infty} \mathbb{E}[X_n] \le \mathbb{E}[\limsup_{n \to \infty} X_n] \,. \tag{A.95}$$

A.5.3   Martingales

Let $Y_1, \ldots, Y_n, \ldots$ be a sequence of random variables with $\mathbb{E}[|Y_n|] < \infty$ for all n. This sequence is called a *martingale* if, for each $n = 1, 2, \ldots,$

$$\mathbb{E}[Y_{n+1} | Y_n, Y_{n-1}, \ldots, Y_1] = Y_n \,. \tag{A.96}$$

If $\mathbb{E}[Y_{n+1}|Y_n, Y_{n-1}, \ldots, Y_1] \leq Y_n$, the sequence is called a *supermartingale*. (Supermartingales thus tend to decrease – the reason for the name is found in mathematical analysis and is an analogy to so-called superharmonic functions.)

Example. If $X_1, X_2, \ldots$ are independent random variables with $\mathbb{E}[X_i] = 0$ and $Y_n = X_1 + X_2 + \cdots + X_n$, then

$$\mathbb{E}[Y_{n+1}|Y_n, Y_{n-1}, \ldots, Y_1]$$
$$= \mathbb{E}[X_1 + X_2 + \cdots + X_n + X_{n+1}|Y_n, Y_{n-1}, \ldots, Y_1]$$
$$= \mathbb{E}[Y_n + X_{n+1}|Y_n, Y_{n-1}, \ldots, Y_1]$$
$$= \mathbb{E}[Y_n|Y_n, Y_{n-1}, \ldots, Y_1] + \mathbb{E}[X_{n+1}|Y_n, Y_{n-1}, \ldots, Y_1]. \tag{A.97}$$

Clearly,

$$\mathbb{E}[Y_n|Y_n, Y_{n-1}, \ldots, Y_1] = Y_n, \tag{A.98}$$

since if we know exactly the value of Y_n, its expected value is, of course, Y_n. However, since X_{n+1} is independent of $Y_n, Y_{n-1}, \ldots, Y_1$,

$$\mathbb{E}[X_{n+1}|Y_n, Y_{n-1}, \ldots, Y_1] = \mathbb{E}[X_{n+1}] = 0. \tag{A.99}$$

Hence,

$$\mathbb{E}[Y_{n+1}|Y_n, Y_{n-1}, \ldots, Y_1] = Y_n \tag{A.100}$$

and, consequently, the sequence in question is a martingale.

◇ ◇ ◇

Theorem A.1, the famous theorem by Doob, is important in theory as well as in applications.

Theorem A.1 *If $Y_1, Y_2, \ldots, Y_n, \ldots$ is a martingale (or supermartingale) of nonnegative elements and there exists a constant $C < \infty$ such that $\mathbb{E}[Y_n] < C$ for all n, then there exists a random variable $Y < \infty$ such that $\lim_{n\to\infty} Y_n = Y$ with probability one.*

A.5.4 Martingales and branching processes

Martingales and Doob's theorem have many uses in branching processes. The most illustrious yields Malthusian growth of supercritical Galton–Watson processes. Let Z_n be the number of individuals in the nth generation. As usual, write m for the expected number of offspring. Then

$$\mathbb{E}[Z_{n+1}|Z_n] = mZ_n, \quad n = 0, 1, \ldots. \tag{A.101}$$

(Since the process is Markov, we need only condition upon the last preceding element.) Thus, if $m = 1$ (the process is critical), Z_n is a martingale.

Now, write $W_n = m^{-n}Z_n$. Then, by properties of the conditional expectation,

$$\mathbb{E}[W_{n+1}|W_n] = \mathbb{E}[m^{-(n+1)}Z_{n+1}|m^{-n}Z_n]$$
$$= m^{-(n+1)}\mathbb{E}[Z_{n+1}|m^{-n}Z_n]. \tag{A.102}$$

Clearly,

$$\mathbb{E}\left[Z_{n+1}|m^{-n}Z_n\right] = \mathbb{E}\left[Z_{n+1}|Z_n\right] , \tag{A.103}$$

since conditioning on $m^{-n}Z_n$ is the same as on Z_n – they determine each other. Hence, by Equation (A.101),

$$\mathbb{E}\left[W_{n+1}|W_n\right] = m^{-(n+1)}\mathbb{E}\left[Z_{n+1}|Z_n\right] = m^{-(n+1)}m\mathbb{E}\left[Z_{n+1}|Z_n\right]$$
$$= m^{-n}Z_n = W_n . \tag{A.104}$$

Therefore, W_n is a martingale. Since $\mathbb{E}[W_n] = 1$, $\lim_{n\to\infty} W_n = W$ exists and is finite for *any* Galton–Watson branching process with $m < \infty$.

Note, however, that since subcritical and critical Galton–Watson branching processes die out, $W = 0$ for these cases.

Theorem A.2 (by Kesten and Stigum) shows that norming is accurate in most cases (i.e., that dividing by m^n does not ultimately annihilate Z_n, except in rather degenerate cases).

Theorem A.2 *Let Z_n be a supercritical branching process with $\mathbb{E}[Z_1] < \infty$ and $W = \lim_{n\to\infty} m^{-n}Z_n$, then:*

1. *if $\mathbb{E}\left[Z_1 \log (1 + Z_1)\right] < \infty$, then*

$$EW = 1 \text{ and } \mathbb{P}(W = 0) = \lim_{n\to\infty} \mathbb{P}(Z_n = 0) ; \tag{A.105}$$

2. *if $\mathbb{E}\left[Z_1 \log (1 + Z_1)\right] = \infty$, then $\mathbb{P}(W = 0) = 1$.*

Note that a finite reproduction variance implies case (1) above.

A.5.5   Stationary and independent identically distributed random variables

Let $X_1, \dots, X_n, \dots$ be a sequence of independent, identically distributed random variables with $a = \mathbb{E}[X_i] \in (-\infty, \infty)$.

The *Law of Large Numbers* states that for any $\varepsilon > 0$

$$\lim_{n\to\infty} \mathbb{P}\left(\left|\frac{X_1 + \cdots + X_n}{n} - a\right| > \varepsilon\right) = 0 . \tag{A.106}$$

The *Strong Law of Large Numbers* says that

$$\mathbb{P}\left(\lim_{n\to\infty} \frac{X_1 + \cdots + X_n}{n} = a\right) = 1 . \tag{A.107}$$

If $\sigma^2 = \text{Var}[X_i] \in (0, +\infty)$, then the *Central Limit Theorem* states that as $n \to \infty$,

$$\mathbb{P}\left(\frac{X_1 + \cdots + X_n - na}{\sigma\sqrt{n}} < x\right) \to \frac{1}{\sigma\sqrt{2\pi}} \int_{-\infty}^{x} e^{-y^2/2} \, dy , \tag{A.108}$$

uniformly in $x \in (-\infty, \infty)$.

The Law of Large Numbers and the Strong Law of Large Numbers are not only valid for sequences of independent identically distributed random variables. Below we describe some important classes of random variables for which these relations are preserved.

Consider a sequence of random variables that is infinite in both directions, $\dots, X_{-2}, X_{-1}, X_0, X_1, X_2, \dots$. The sequence is called *strongly stationary* if for any integers n, m, and t_j $(1 \le j \le n)$,

$$\mathbb{P}\left(X_{t_i} < x_1, \dots, X_{t_n} < x_n\right) = \mathbb{P}\left(X_{t_i+m} < x_1, \dots, X_{t_n+m} < x_n\right) . \tag{A.109}$$

This means that all n-tuples of observations have the same distribution, wherever they are located in the sequence. It is said to be *weakly stationary* if all variables have the same expectation and variance and $\text{Cov}\left[X_i, X_j\right]$ depends only on $|j - i|$. In words, the covariance between two variables is determined by the distance between their indices. The correlation coefficient between X_t and X_{t+m} is thus independent of t and equals

$$R(m) = \frac{\text{Cov}\,[X_0, X_m]}{\sqrt{\text{Var}\,[X_0]}\sqrt{\text{Var}\,[X_m]}} . \tag{A.110}$$

For weakly stationary sequences with $R(m) \to 0$ as $m \to \infty$, the Law of Large Numbers, Equation (A.106), holds and for strongly stationary sequences that satisfy this condition the Strong Law of Large Numbers, Equation (A.107), holds.

A.6 The Perron–Frobenius Theorem

A matrix M is called irreducible if there is an integer n_0 such that all the elements of the matrix M^{n_0} are positive.

Theorem A.3 *If M is an irreducible matrix with non-negative elements, then it has a unique positive eigenvalue ρ that is greater in absolute value than any other eigenvalue. All elements of the left and right eigenvectors $u = (u_1, \dots, u_d)^T$ and $v = (v_1, \dots, v_d)^T$ that correspond to ρ can be chosen positive and such that*

$$\sum_{k=1}^{d} u_k = 1, \quad \sum_{k=1}^{d} u_k v_k = 1 . \tag{A.111}$$

Then the eigenvectors are unique. In addition,

$$M^n = \rho^n A + B^n , \tag{A.112}$$

where $A = (v_i u_j)_{i,j=1}^{d}$ and B are matrices that satisfy the conditions:

1. *$AB = BA = 0$;*
2. *there are constants $\rho_1 \in (0, \rho)$ and $C > 0$ such that none of the elements of the matrix B^n exceeds $C\rho_1^n$.*

References

Page numbers of reference citations in this volume are given in square brackets.

Abramowitz M & Stegun I (1964). *Handbook of Mathematical Functions.* New York, NY, USA: Dover   [*244*]

Aldous D (1999). Deterministic and stochastic models for coalescence (aggregation, coagulation): A review of the mean-field theory for probabilists. *Bernoulli* **5**:3–48   [*200*]

Alexandersson M (2000). An application of general branching processes to a cell cycle model with two uncoupled subcycles and unequal cell division. *International Journal of Applied Mathematics and Computer Sciences* **10**:131–145   [*222, 225*]

Allee WC (1931). *Animal Aggregations: A Study in General Sociology.* Chicago, IL, USA: University of Chicago Press   [*89*]

Anderson RM & May RM (1991). *Infectious Diseases of Humans.* Oxford, UK: Oxford University Press   [*236–237*]

Alsmeyer G & Rösler U (1996). The bisexual Galton–Watson process with promiscuous mating: Extinction probabilities in the supercritical case. *Annals of Applied Probability* **6**:922–939   [*141, 143–144*]

Alsmeyer G & Rösler U (2002). Asexual versus promiscuous bisexual Galton–Watson processes: The extinction probability ratio. *Annals of Applied Probability* **12**:125–142   [*141, 143*]

Arino O & Kimmel M (1993). Comparison of approaches to modeling of cell population dynamics. *SIAM Journal of Applied Mathematics* **53**:1480–1504   [*78*]

Arino O, Kimmel M & Webb GF (1995). Mathematical modeling of the loss of telomere sequences. *Journal of Theoretical Biology* **177**:45–57   [*226, 228, 231*]

Arino O, Sanchez E & Webb GF (1997). Polynomial growth dynamics of telomere loss in a heterogeneous cell population. *Dynamics of Continuous, Discrete and Impulsive Systems* **3**:262–282   [*230*]

Asmussen S (1987). *Applied Probability and Queues.* New York, NY, USA: John Wiley & Sons   [*229*]

Asmussen S & Hering H (1983). *Branching Processes.* Boston, MA, USA: Birkhäuser   [*154, 157*]

Athreya KB (1992). Rates of decay for the survival probability of a mutant gene. *Journal of Mathematical Biology* **30**:577–581   [*124*]

Athreya KB (1993). Rates of decay for the survival probability of a mutant gene. II. The multitype case. *Journal of Mathematical Biology* **32**:45–93   [*124, 274, 277*]

Athreya KB & Karlin S (1971a). On branching processes with random environments. I. Extinction probabilities. *Annals of Mathematical Statistics* **42**:1499–1520   [*51, 150*]

Athreya KB & Karlin S (1971b). Branching processes with random environments. II. Limit theorems. *Annals of Mathematical Statistics* **42**:1843–1858   [*51*]

Athreya KB & Ney P (1972). *Branching Processes.* Berlin, Germany: Springer-Verlag   [*27, 51, 123, 150, 154*]

Azevedo RBR & Leroi A (2001). A power law for cells. *Proceedings of the National Academy of Sciences of the USA* **98**:5699–5704   [*20, 74*]

Azevedo RBR, Cunha AS, Emmons SW & Leroi AM (2000). The demise of the platonic worm. *Nematology* **2**:71–79   [*20, 74*]

Bagley JH (1986). On the asymptotic properties of a supercritical bisexual Galton–Watson branching process. *Journal of Applied Probability* **23**:820–826   [*178*]

Balding D, Bishop M & Cannings C (2001) *Handbook of Statistical Genetics.* Chichester, UK: John Wiley & Sons   [*202*]

Barbour AD (1976). Quasi-stationary distributions in Markov population processes. *Advances in Applied Probability* **8**:296–314   [*184*]

Bartlett MS (1957). Measles periodicity and community size. *Journal of the Royal Statistical Society Series A* **120**:48–70   [*237*]

Beaumont MA, Zhang W & Balding DJ (2002). Approximate Bayesian computation in population genetics. *Genetics* **162**:2025–2035   [*212*]

Becker N (1974). On parametric estimation

for mortal branching processes. *Biometrika* **61**:393–453 [*168*]

Bell GI & Anderson EC (1967). Cell growth and division I: A mathematical model with applications to cell volume distributions in mammalian suspension cultures. *Biophysics Journal* **7**:329–351 [*219*]

Bienaymé IJ (1845). De la loi de multiplication et de la durée des familles. *Société Philomatique de Paris Extraits, Series 5* **10**:37–39 [*114*]

Billingsley P (1968). *Convergence of Probability Measures*. New York, NY, USA: John Wiley & Sons [*203*]

Bhattacharya RN & Waymire EC (1990). *Stochastic Processes with Applications*. New York, NY, USA: John Wiley & Sons [*85*]

Biggins JD & Nadarajah S (1993). Near-constancy of the Harris function in the simple branching process. *Communications in Statistics – Stochastic Models* **9**:435–444 [*144*]

Billingsley P (1979). *Probability and Measure*. New York, NY, USA: John Wiley & Sons [*77*]

Boiko RV (1982). Limit theorems for a branching process with a varying environment which describes the development of a population in a limiting environment. *Ukrainskyj Matematychnyj Zhurnal* **34**:681–687 [*39*]

Bolker BM & Grenfell BT (1993). Chaos and complexity in measles dynamics. *Proceedings of the Royal Society of London B* **251**:75–81 [*237*]

Bolthausen E & Sznitman AS (1998). On Ruelle's probability cascades and an abstract cavity method. *Communications in Mathematical Physics* **197**:247–276 [*207*]

Brakefield PM & Larsen TB (1984). The evolutionary significance of dry and wet season forms in some tropical butterflies. *Biologial Journal of the Linnean Society* **22**:1–12 [*49*]

Bruss FT (1984). A note on extinction criteria for bisexual Galton–Watson processes. *Journal of Applied Probability* **21**:915–919 [*45, 137*]

Bruss FT & Slavtchova-Bojkova M (1999). On waiting times to populate environment and a question of statistical inference. *Journal of Applied Probability* **36**:261–267 [*170*]

Bull J (1983). *Evolution of Sex Determining Mechanisms*. Menlo Park, CA, USA: Benjamin/Cummings [*136*]

Cannings C (1974). The latent roots of certain Markov chains arising in genetics: A new approach. I. Haploid models. *Advances in Applied Probability* **6**:260–290 [*204*]

Carabin H, Edmunds WJ, Kou U, van den Hof S & Nguyen VH (2002). The average cost of measles cases and adverse events following vaccination in industrialised countries. *BMC Public Health* **2**:22 (http://www.biomedcentral.com/1471-2458/2/22) [*237, 248*]

Case TJ (2000). *An Illustrated Guide to Theoretical Ecology*. New York, NY, USA: Oxford University Press [*88*]

Caswell H (2001). *Matrix Population Models: Construction, Analysis, and Interpretation*, Second Edition. Sunderland, MA, USA: Sinauer Associates Inc. [*12, 25, 166–167*]

Caswell H, Fujiwara M & Brault S (1999). Declining survival probability threatens the North Atlantic right whale. *Proceedings of the National Academy of Sciences of the USA* **96**:3308–3313 [*121–122*]

Champagnat N, Ferrière R & Arous GB (2001). The canonical equation of adaptive dynamics: A mathematical view. *Selection* **2**:73–83 [*270, 276–277*]

Chistyakov VP, Kolchin VF & Sevastyanov BA (1978). *Random Allocations*. New York, NY, USA: John Wiley & Sons [*40*]

Cohen D (1966). Optimizing reproduction in a randomly varying environment. *Journal of Theoretical Biology* **12**:119–129 [*51*]

Consumer's Association (2003). MMR vaccine: How effective and how safe. *Drugs and Therapeutics Bulletin* **41**:25–29 [*237*]

Cornish-Bowden A (1979). *Fundamentals of Enzyme Kinetics*. London, UK: Butterworths [*234*]

Daley D (1968). Extinction probabilities for certain bisexual Galton–Watson branching processes. *Zeitschrift für Wahrscheinlichkeitstheorie und verwandte Gebiete* **9**:315–322 [*44, 136*]

Daley D & Vere-Jones D (1988). *Introduction to the Theory of Point Processes*. Berlin, Germany: Springer-Verlag [*67*]

Daley D, Hull DM & Taylor JM (1986). Bisexual Galton–Watson branching processes with superadditive mating functions. *Journal of Applied Probability* **23**:585–600 [*45, 138, 141–142, 144–145*]

Darroch J & Seneta E (1965). On quasi-stationary distributions in absorbing discrete-time finite Markov chains. *Jour-*

nal of Applied Probability **2**:88–100 *[260]*

Dawson DA (1993). Measure-valued Markov processes. In *Lecture Notes in Mathematics*, Vol. 1541, ed Hennequin PL, pp. 1–260. Berlin, Germany: Springer-Verlag *[96]*

De Melker H, Pebody RG, Edmunds WJ, Lévy-Bruhl D, Valle M, Rota MC, Salmaso S, Van Den Hof S, Berbers G, Saliou P, Conyn-Van Spaendonck M, Crovari P, Davidkin I, Gabutti G, Hesketh L, Morgan-Capner P, Plesner AM, Raux M, Tischer A & Miller E (2001). The seroepidemiology of measles in Western Europe. *Epidemiology and Infections* **126**:249–259 *[237]*

De Serres G, Gay NJ & Farrington CP (2000). Epidemiology of transmissable disease after elimination. *American Journal of Epidemiology* **151**:1039–1048 *[245]*

Devaney RL (1989). *An Introduction to Chaotic Dynamical Systems*, Second Edition. Reading, MA, USA: Addison-Wesley *[194]*

Diamond JM (1975). Assembly of species communities. In *Ecology and Evolution of Communities*, eds. Cody ML & Diamond JM, pp. 342–444. Cambridge, MA, USA: Harvard University Press *[252]*

Dieckmann U & Law R (1996). The dynamical theory of coevolution: A derivation from stochastic ecological processes. *Journal of Mathematical Biology* **34**:579–612 *[130, 267, 270–271]*

Dieckmann U, Law R & Metz JAJ (2000). *The Geometry of Ecological Interactions*. Cambridge, UK: Cambridge University Press *[42]*

Dieckmann U, Doebeli M, Metz JAJ & Tautz D (2004). *Adaptive Speciation*. Cambridge, UK: Cambridge University Press *[267]*

Diekmann O (1986). The cell size and semi-groups of linear operators. In *The Dynamics of Physiologically Structured Populations*, eds. Metz JAJ & Diekmann O, pp. 46–77. Lecture Notes in Biomathematics **68**. Berlin, Germany: Springer-Verlag *[78]*

Diekmann O, Heijmans H & Thieme HR (1984). On the stability of the cell size distribution. *Journal of Mathematical Biology* **19**:227–248 *[78, 103]*

Diekmann O, Gyllenberg M, Metz JAJ & Thieme HR (1993). The "cumulative" formulation of (physiologically) structured population models. In *Evolution Equations, Control Theory and Biomathematics*, eds. Clément P & Lumer G, pp. 145–154. New

York, NY, USA: Marcel Dekker *[95, 250]*

Diekmann O, Gyllenberg M, Metz JAJ & Thieme HR (1998). On the formulation and analysis of general deterministic structured population models: I. Linear theory. *Journal of Mathematical Biology* **36**:349–388 *[6, 95, 101–103, 250]*

Diekmann O, Gyllenberg M, Huang H, Kirkilionis M, Metz JAJ & Thieme HR (2001). On the formulation and analysis of general deterministic structured population models: II. Nonlinear theory. *Journal of Mathematical Biology* **43**:157–189 *[6, 95, 100, 105, 250–251]*

Diekmann O, Gyllenberg M & Metz JAJ (2003). Steady-state analysis of structured population models. *Theoretical Population Biology* **63**:309–338 *[272]*

Diggle P (1990). *Time Series. A Biostatistical Introduction*. Oxford, UK: Clarendon Press *[3]*

Donald A & Muhtu V (2002). No evidence that MMR vaccine is associated with autism or bowel disease. *Clinical Evidence* **7**:331–340 *[237]*

Donnelly P & Tavaré S (1995). Coalescents and genealogical structure under neutrality. *Annual Review of Genetics* **29**:401–423 *[202]*

Drepper FR, Engbert R & Stollenwerk N (1994). Non-linear time-series analysis of empirical population-dynamics. *Ecological Modelling* **75**:171–181 *[237]*

Dunn AM, Terry RS & Taneyhill DE (1998). Within-host transmission strategies of transovarial, feminizing parasites of *Gammarus duebeni*. *Parasitology* **117**:21–30 *[31]*

Durinx M, Metz JAJ & Meszéna G. Adaptive dynamics of structured populations. Unpublished *[267, 274]*

Durrett R (1988). *Lectures on Particle Systems and Percolation*. Belmont, CA, USA: Wadsworth *[42]*

Durrett R (1995). *Probability: Theory and Examples*, Second Edition. Belmont, CA, USA: Duxbury Press *[40]*

Dynkin EB (1991). Branching particle systems and superprocesses. *Annals of Probability* **19**:1157–1194 *[96]*

Eshel I (1981). On the survival probability of a slightly advantageous mutant gene with a general distribution of progeny size: A branching process model. *Journal of Mathematical Biology* **12**:355–362 *[124]*

Ethier SN & Kurtz TG (1986). *Markov Processes, Characterization and Convergence.* New York, NY, USA: John Wiley & Sons [270]

Etienne RS & Heesterbeek JAP (2001). Rules of thumb for conservation of metapopulations based on a stochastic winking-patch model. *The American Naturalist* **158**:389–407 [261]

Euler L (1767). Recherches générales sur la mortalité et la multiplication du genre humain. *Histoire de l'Académie Royale des Sciences et Belles-Lettres* **1760**:144–164 [162]

Ewens WJ (1968). Some applications of multitype branching processes in population genetics. *Journal of the Royal Statistical Society* **30B**:164–175 [124]

Ewens WJ (1979). *Mathematical Population Genetics.* Berlin–Heidelberg–New York: Springer-Verlag [200]

Farrington CP, Kanaan MN & Gay NJ (2003). Branching process models of surveillance of infectious diseases controlled by mass vaccination. *Biostatistics* **4**:279–295 [245, 249]

Feistel R (1977). Betrachtung der Realisierung stochastischer Prozesse aus automatentheoretischer Sicht. *Wissenschaftliche Zeitschrift der Wilhelm-Pieck-Universität in Rostock* **26**:663–670 [241]

Feller W (1951). Diffusion processes in genetics. In *Proceedings of the Second Berkeley Symposium on Mathematical Statistics and Probability, 1950*, ed. Neyman P, pp. 227–246. Berkeley and Los Angeles, CA, USA: University of California Press [84–85]

Feller W (1957). *An Introduction to Probability Theory and Its Applications*, Vol. 1, Second Edition. New York, NY, USA: John Wiley & Sons [40]

Feller W (1966). *An Introduction to Probability Theory and Its Applications.* New York, NY, USA: John Wiley & Sons [73]

Feller W (1971). *An Introduction to Probability Theory and its Applications*, Volume 2. New York, NY, USA: John Wiley & Sons [229]

Ferrari PA, Martinez S & Picco P (1992). Existence of non-trivial quasi-stationary distributions in the birth–death chain. *Advances in Applied Probability* **24**:795–813 [193]

Fisher R (1930). *The Genetical Theory of Natural Selection.* (Second Edition 1958). New York, NY, USA: Dover [200]

Fujimagari T (1976). Controlled Galton–

Watson process and its asymptotic behavior. *Kodai Mathematical Seminar Reports* **27**:11–18 [133]

Galton F & Watson HW (1875). On the probability of the extinction of families. *Journal of the Anthropological Society of London* **4**:138–144 [111–112]

Gay NJ, Hesketh LM, Morgan-Capner P & Miller E (1995). Interpretation of serological surveillance data for measles using mathematical models: Implications for vaccine strategy. *Epidemiology and Infections* **115**:139–156 [248]

Geritz SAH, van der Meijden E & Metz JAJ (1999). Evolutionary dynamics of seed size and seedling competitive ability. *Theoretical Population Biology* **55**:324–343 [273]

Geritz SAH, Gyllenberg M, Jacobs FJA & Parvinen K (2002). Invasion dynamics and attractor inheritance. *Journal of Mathematical Biology* **44**:548–560 [267]

Gillespie DT (1976). A general method for numerically simulating the stochastic time evolution of coupled chemical reactions. *Journal of Computational Physics* **22**:403–434 [241]

Gillespie DT (1978). Monte Carlo simulation of random walks with residence time dependent transition probability rates. *Journal of Computational Physics* **28**:395–407 [241]

Gilpin ME & Diamond JM (1981). Immigration and extinction probabilities for individual species: Relation to incidence functions and species colonization curves. *Proceedings of the National Academy of Sciences of the USA* **78**:392–396 [252]

Gilpin ME & Hanski I (1991). *Metapopulation Dynamics.* New York, NY, USA: Academic Press [250]

González M & Molina M (1996). On the limit behaviour of a superadditive bisexual Galton–Watson branching process. *Journal of Applied Probability* **33**:960–967 [178]

González M & Molina M (1997). On the L^2-convergence of a superadditive bisexual Galton–Watson branching process. *Journal of Applied Probability* **34**:575–582 [178]

Grimmett G (1999). *Percolation.* Heidelberg, Germany: Springer-Verlag [42]

Guttorp P (1991). *Statistical Inference for Branching Processes.* New York, NY, USA: John Wiley & Sons [208]

Guttorp P (1995). *Stochastic Modeling of Scientific Data.* London, UK: Chapman & Hall [208]

Gyllenberg M (1986). The size and scar distributions of the yeast *Saccharomyces cervisiae. Journal of Mathematical Biology* **24**:81–101 [*103*]

Gyllenberg M & Hanski I (1997). Habitat deterioration, habitat destruction and metapopulation persistence in a heterogeneous landscape. *Theoretical Population Biology* **52**:198–215 [*250*]

Gyllenberg M & Parvinen K (2001). Necessary and sufficient conditions for evolutionary suicide. *Bulletin of Mathematical Biology* **63**:981–993 [*267*]

Gyllenberg M & Silvestrov DS (1994). Quasi-stationary distributions of a stochastic metapopulation model. *Journal of Mathematical Biology* **33**:35–70 [*252, 262*]

Gyllenberg M & Silvestrov DS (1997). *Exponential Asymptotics for Perturbed Renewal Equations and Pseudo-stationary Phenomena for Stochastic Systems.* Research Report 3, Umeå, Sweden: Department of Mathematics and Statistics, Umeå University [*262*]

Gyllenberg M & Silvestrov DS (1999). Quasi-stationary phenomena for semi-Markov processes. In *The Advances on Semi-Markov Models: Theory and Applications*, eds. Janssen J & Limnios N, pp. 33–60. Dordrecht, Netherlands: Kluwer [*262*]

Gyllenberg M & Silvestrov DS (2000). Nonlinearly perturbed regenerative processes and pseudo-stationary phenomena for stochastic systems. *Stochastic Processes and Their Applications* **86**:1–27 [*262*]

Gyllenberg M, Högnäs G & Koski T (1994). Population models with environmental stochasticity. *Journal of Mathematical Biology* **32**:93–108 [*196–198*]

Gyllenberg M, Hanski I & Hastings A (1997). Structured metapopulation models. In *Metapopulation Dynamics: Ecology, Genetics and Evolution*, eds. Hanski I & Gilpin M, pp. 93–122. London, UK: Academic Press [*95, 251*]

Haccou P & Iwasa Y (1996). Establishment probability in fluctuating environments: A branching process model. *Theoretical Population Biology* **50**:254–280 [*151–152*]

Haccou P & Meelis E (1992). *Statistical Analysis of Behavioural Data: A Time-Structured Approach.* Oxford, UK: Oxford University Press [*61*]

Haccou P & Vatutin V (2003). Establishment success and extinction risk in autocorrelated environments. *Theoretical Population Biology* **64**:303–314 [*152*]

Haldane JBS (1927). A mathematical theory of natural and artificial selection. V. Selection and mutation. *Proceedings of the Cambridge Philosophical Society* **23**:838–844 [*124–125, 276*]

Halley JM & Iwasa Y (1998). Extinction rate of a population under both demographic and environmental stationarity. *Theoretical Population Biology* **53**:1–15 [*110*]

Hanski I (1994). A practical model of metapopulation dynamics. *Journal of Animal Ecology* **63**:151–162 [*252*]

Hanski I (1999). *Metapopulation Ecology.* Oxford, UK: Oxford University Press [*250*]

Hanski I & Gilpin ME (1991). Metapopulation dynamics: Brief history and conceptual domain. In *Metapopulation Dynamics*, eds. Gilpin ME & Hanski I, pp. 3–16. New York, NY, USA: Academic Press [*249*]

Hanski I & Gilpin M (1997). *Metapopulation Dynamics: Ecology, Genetics and Evolution.* London, UK: Academic Press [*250*]

Hardy K, Spanos S, Becker D, Iannelli P, Winston RML & Stark J (2001). From cell death to embryo arrest: Mathematical models of human preimplantation embryo development. *Proceedings of the National Academy of Sciences* **98**:1655–1660 [*148–149*]

Harris TE (1963). *The Theory of Branching Processes.* Berlin, Germany: Springer-Verlag [*74, 245*]

Hassell MP (1974). Density-dependence in single-species populations. *Journal of Animal Ecology* **44**:283–296 [*191*]

Hassell MP, Lawton JH & May RM (1976). Patterns of dynamical behaviour in single-species populations. *Journal of Animal Ecology* **45**:471–486 [*89*]

Heesterbeek H (1992). R_0. PhD Thesis. Leiden, Netherlands: Leiden University [*274*]

Heesterbeek JAP (2002). A brief history of R_0 and a recipe for its calculation. *Acta Biotheoretica* **50**:375–376 [*236*]

Heyde CC & Seneta E (1977). *Statistical Theory Anticipated, I.J. Bienaymé.* Heidelberg, Germany: Springer-Verlag [*7, 13*]

Hille E & Phillips R (1957). *Functional Analysis and Semi-Groups.* Providence, RI, USA: American Mathematical Society [*140*]

Högnäs G (1997). On the quasi-stationary distribution of a stochastic Ricker model. *Stochastic Processes and Their Applications*

2:243–263   [9, 195]

Höpfner R (1985). On some classes of population size dependent Galton–Watson processes. *Journal of Applied Probability* **22**:25–36   [135, 175]

Hoppe FM (1992). The survival probability of a mutant in a multidimensional population. *Journal of Mathematical Biology* **30**:547–566   [124]

Hudson RR (1991). Gene genealogies and the coalescent process. In *Oxford Surveys in Evolutionary Biology*, eds. Futuyama D & Antonovics J, Vol. 7, pp. 1–44. Oxford, UK: Oxford University Press   [202]

Hull DM (1982). A necessary condition for extinction in those bisexual Galton–Watson branching processes governed by superadditive mating functions. *Journal of Applied Probability* **19**:847–850   [44, 137, 140]

Hull DM (1998). A reconsideration of Galton's problem (using a two-sex population). *Theoretical Population Biology* **54**:105–116   [145]

Hull DM (2001). A reconsideration of Lotka's extinction probability using a bisexual branching process. *Journal of Applied Probability* **38**:776–780   [145]

Iwasa Y (2000). Lattice models and pair approximation in ecology. In *The Geometry of Ecological Interactions*, eds. Dieckmann U, Law R & Metz JAJ, pp. 227–251. Cambridge, UK: Cambridge University Press   [42]

Jagers P (1975). *Branching Processes with Biological Applications*. London: John Wiley & Sons   [7, 71, 74, 132, 154, 233, 249]

Jagers P (1981). How probable is it to be firstborn? *Mathematical Biosciences* **59**:1–15   [80]

Jagers P (1989). General branching processes as Markov fields. *Stochastic Processes and Their Applications* **32**:183–212   [76, 100, 227]

Jagers P (1991). The growth and stabilization of populations. *Statistical Science* **6**:269–283   [80, 161, 200]

Jagers P (1992). Stabilities and instabilities in population dynamics. *Journal of Applied Probability* **29**:770–780   [110, 161]

Jagers P (1997a). Towards dependence in general branching processes. In *Classical and Modern Branching Processes*, eds. Athreya KB & Jagers P, pp. 127–140. New York, NY, USA: Springer-Verlag   [172]

Jagers P (1997b). Coupling and population dependence in branching processes. *Annals of Applied Probability* **7**:281–298   [174]

Jagers P (2001). The deterministic evolution of general branching populations. In *State of the Art in Statistics and Probability Theory: Festschrift for Willem R. van Zwet*, eds. de Gunst M, Klaassen C & van der Vaart A, pp. 384–398. IMS Lecture Notes, Vol. 36. Beechwood, OH, USA: Institute of Mathematical Sciences   [103, 220]

Jagers P & Klebaner FC (2000). Population-size-dependent and age-dependent branching processes. *Stochastic Processes and Their Applications* **87**:235–254   [173]

Jagers P & Klebaner FC (2003). Random variation and concentration effects in PCR. *Journal of Theoretical Biology* **224**:299–304   [94, 235]

Jagers P & Klebaner FC (2004). Branching processes in near-critical random environments. In *Stochastic Methods and their Applications. Papers in Honour of Chris Heyde*, eds. Gani J & Seneta E. *Journal of Applied Probability* **41A**:17–23   [42, 235]

Jagers P & Lu Z-W (2002). Branching processes with random deteriorating environments. *Journal of Applied Probability* **39**:395–401   [52]

Jagers P & Nerman O (1996). The asymptotic composition of supercritical multi-type branching populations. *Séminaire des probabilités. Springer Lecture Notes in Mathematics* 1626, pp. 40–54. Berlin, Germany: Springer-Verlag   [163]

Jagers P & Sagitov S (2000). The growth of general population-size-dependent branching processes year by year. *Journal of Applied Probability* **37**:1–14   [173]

Jagers P, Nerman O & Taib Z (1992). When did Joe's great-grandfather live? Or: On the time-scale of evolution. In *Selected Proceedings of the Sheffield Symposium on Applied Probability*, eds. Basawa I & Taylor RL, pp. 118–126. Monograph Series, IMS Lecture Notes, Volume 18. Beechwood, OH, USA: Institute of Mathematical Studies   [200]

Jansen VAA, Stollenwerk N, Jensen HJ, Ramsay ME, Edmunds WJ & Rhodes CJ (2003). Measles outbreaks in a population with declining vaccine uptake. *Science* **301**:804   [247]

Jensen HJ (1998). *Self-organized Criticality: Emergent Complex Behavior in Physical and Biological Systems*. Cambridge, UK:

Cambridge University Press [244–245]
Jiřina M (1969). On Feller's branching diffusion processes. *Časopis pro pěstování matematiky* **94**:84–90 [85]

Jones RB, Whitney RG & Smith JR (1985). Intramitotic variation in proliferative potential: Stochastic events in cellular aging. *Mechanisms of Ageing and Development* **29**:143–149 [231]

Kalbfleisch JD & Prentice RL (1980). *The Statistical Analysis of Failure Time Data.* New York, NY, USA: John Wiley & Sons [60]

Kallenberg O (1983). *Random Measures.* Berlin, Germany: Akademie-Verlag [67]

Karr A (1986). *Point Processes and Their Statistical Inference.* New York, NY, USA: Marcel Dekker [67]

Keller G, Kersting G & Rösler U (1987). On the asymptotic behaviour of discrete time stochastic growth processes. *Annals of Probability* **15**:305–343 [176]

Kendall DG (1948). On the generalized "birth-and-death" process. *Annals of Mathematical Statistics* **19**:1–15 [277]

Kendall DG (1966). Branching processes since 1873. *Journal of the London Mathematical Society* **41**:385–406 [7]

Kersting G (1992). Asymptotic Γ distribution for stochastic difference equations. *Stochastic Processes and Their Applications* **40**:15–28 [175]

Kersting G (1986). On recurrence and transience of growth models. *Journal of Applied Probability* **23**:614–625 [135]

Kesten H & Stigum B (1966a). A limit theorem for multidimensional Galton–Watson processes. *Annals of Mathematical Statistics* **37**:1211–1223 [27, 154]

Kesten H & Stigum B (1966b). Additional limit theorems for indecomposable multidimensional Galton–Watson processes. *Annals of Mathematical Statistics* **37**:1463–1481 [157]

Kesten H & Stigum B (1967). Limit theorems for decomposable multi-dimensional Galton–Watson processes. *Journal of Mathematical Analysis and Applications* **17**:309–338 [157]

Keyfitz N (1977). *Applied Mathematical Demography.* New York, NY, USA: Wiley-Interscience, Wiley [88]

Kingman JFC (1982a). On the genealogy of large populations. *Journal of Applied Probability* **19A**:27–43 [200, 202–203]

Kingman JFC (1982b). Exchangeability and the evolution of large populations. In *Exchangeability in Probability and Statistics,* eds. Koch G & Spizzichino F, pp. 97–112. Dordrecht, Netherlands: North-Holland [205]

Klebaner FC (1983). Population-size-dependent branching process with linear rate of growth. *Journal of Applied Probability* **20**:242–250 [174]

Klebaner FC (1984). On population-size-dependent branching processes. *Advances in Applied Probability* **16**:30–55 [133, 172, 175]

Klebaner FC (1985). A limit theorem for population-size-dependent branching processes. *Journal of Applied Probability* **22**:48–57 [172]

Klebaner FC (1989a). Stochastic difference equations and generalized gamma distributions. *Annals of Probability* **17**:178–188 [175]

Klebaner FC (1989b). Geometric growth in near-supercritical population-size-dependent multi-type Galton–Watson processes. *Annals of Probability* **17**:1466–1477 [173]

Klebaner FC (1989c). Linear growth in near-critical population-size-dependent multitype Galton–Watson processes. *Journal of Applied Probability* **26**:431–445 [39]

Klebaner FC (1990). Conditions for the unlimited growth in multitype population size dependent Galton–Watson processes. *Bulletin of Mathematical Biology* **52**:527–534 [135]

Klebaner FC (1991). Asymptotic behaviour of near-critical multi-type branching processes. *Journal of Applied Probability* **28**:512–519 [173]

Klebaner FC (1993). Population-dependent branching processes with a threshold. *Stochastic Processes and Their Applications* **46**:115–127 [93, 190]

Klebaner FC & Nerman O (1994). Autoregressive approximation in branching processes with a threshold. *Stochastic Processes and Their Applications* **51**:1–7 [93, 190]

Klebaner FC, Lazar J & Zeitouni O (1998). On the quasi-stationary distribution for some randomly perturbed transformations of an interval. *Annals of Applied Probability* **8**:300–315 [190]

Kooijman SALM (1993). *Dynamic Energy Budgets in Biological Systems.* Cambridge, UK: Cambridge University Press [174]

Küster P (1985). Asymptotic growth of controlled Galton–Watson processes. *Annals of Probability* **13**:1157–1178 [*176*]

Kurtz TG (1983). Gaussian approximations for Markov chains and counting processes. *Bulletin of the Institute of International Statistics* **50**:361–376 (With a discussion in Vol. 3, pp. 237–248. Proceedings of the 44th Session of the International Statistical Institute, Vol. 1, Madrid, 1983) [*88*]

Lalam N, Jacob C & Jagers P (2004). Modelling the PCR amplification process by a size-dependent branching process and estimation of the efficiency. *Advances in Applied Probability* **36**:602–615 [*234–235*]

Lande R (1979). Quantitative genetic analysis of multivariate evolution, applied to brain:body size allometry. *Evolution* **33**:402–416 [*276*]

Larson DD, Spangler EA & Blackburn EH (1987). Dynamics of telomere length variation in *Tetrahymena thermophila*. *Cell* **50**:477–483 [*230*]

Law R & Dieckmann U (1998). Symbiosis through exploitation and the merger of lineages in evolution. *Proceedings of the Royal Society of London B* **265**:1245–1253 [*124, 128–130*]

Lawton JH & May RM (1995). *Extinction Rates*. Oxford, UK: Oxford University Press [*107*]

Levins R (1969). Some demographic and genetic consequences of environmental heterogeneity for biological control. *Bulletin of the Entomological Society of America* **15**:237–240 [*250*]

Levins R (1970). Extinction. In *Some Mathematical Problems in Biology*, ed. Gerstenhaber M, pp. 77–107. Providence, RI, USA: American Mathematical Society [*250*]

Levy MZ, Allsopp RC, Futcher AB, Greider CW & Harley CB (1992). Telomere end-replication problem and cell aging. *Journal of Molecular Biology* **225**:951–960 [*225, 231*]

Lewontin RH & Cohen D (1969). On population growth in randomly varying environments. *Proceedings of the National Academy of Sciences of the USA* **62**:1056–1060 [*51*]

Liggett T (1999). *Stochastic Interacting Systems: Contact, Voter, and Excursion Processes*. New York, NY, USA: Springer-Verlag [*42*]

Lindvall T (1972). Convergence of critical Galton–Watson branching processes. *Journal of Applied Probability* **9**:445–450 [*86*]

Lindvall T (1974). Limit theorems for some functionals of certain Galton–Watson branching processes. *Advances in Applied Probability* **6**:309–321 [*84, 86*]

Lotka AJ (1931a). Population analysis – the extinction of families I. *Washington Academy of Sciences* **21**:337–380 [*145*]

Lotka AJ (1931b). Population analysis – the extinction of families II. *Washington Academy of Sciences* **21**:453–459 [*145*]

Lotka AJ (1931c). The extinction of families. *Journal of the Washington Academy of Sciences* **21**:377–380; 453–459 [*16*]

MacArthur RH & Wilson EO (1967). *The Theory of Island Biogeography*. Princeton, NJ, USA: Princeton University Press [*81, 250*]

Malthus TR (1798). *An Essay on the Principle of Population, as it Affects the Future Improvements of Society, with Remarks on the Speculations of Mr. Godwin, M. Condorcet, and Other Writers*. London, UK: John Murray [*113*]

Markovtsova L, Marjoram P & Tavaré S (2000a). The age of a unique event polymorphism. *Genetics* **156**:401–409 [*211*]

Markovtsova L, Marjoram P & Tavaré S (2000b). The effects of rate variation on ancestral inference in the coalescent. *Genetics* **156**:1427–1436 [*211*]

Martin-Löf A (1966). A limit theorem for the size of the *n*th generation of an age dependent branching process. *Journal of Mathematical Analysis and Applications* **15**:273–279 [*80*]

May RM (1976). Simple mathematical models with very complicated dynamics. *Nature* **261**:459–467 [*89, 194*]

Maynard Smith J (1982). *Evolution and the Theory of Games*. Cambridge, UK: Cambridge University Press [*267*]

Metz JAJ & Diekmann O, eds. (1986). *The Dynamics of Physiologically Structured Populations*. Lecture Notes in Biomathematics **68**. Heidelberg, Germany: Springer-Verlag [*6, 94, 250, 272*]

Metz JAJ, Geritz SAH, Meszéna G, Jacobs FJA & Van Heerwaarden JS (1996). Adaptive dynamics, a geometrical study of the consequences of nearly faithful reproduction. In *Stochastic and Spatial Structures of Dynamical Systems*, eds. Van Strien SJ & Verduyn Lunel SM, pp. 183–231. Amsterdam, Netherlands: North-Holland [*266, 277*]

Mode C (1971). *Multitype Branching Pro-*

cesses. Oxford, UK: Elsevier [*26, 225*]

Möhle M (2000). Total variation distances and rates of convergence for ancestral coalescent processes in exchangeable population models. *Advances in Applied Probability* **32**:1547–1562 [*205*]

Möhle M & Sagitov S (1999). Coalescent patterns in exchangeable diploid population models. *Journal of Mathematical Biology* **47**:337–352 [*205*]

Möhle M & Sagitov S (2001). A classification of coalescent processes for haploid exchangeable population models. *Annals of Probability* **29**:983–993 [*207*]

Mollison D (1995). The structure of epidemic models. In *Epidemic Models, Their Structure and Relation to Data*, ed. Mollison D, pp. 17–33. Cambridge, UK: Cambridge University Press [*271*]

Nilsson M, Rasmussen S, Mayer B & Whitten D (2000). Constructive molecular dynamics lattice gases: 3-D molecular self-assembly. In *New Constructions in Cellular Automata*, eds. Griffeath D & Moore C, pp. 275–290. Oxford, UK: Oxford University Press [*200*]

Novak B & Tyson JJ (1995). Mathematical modeling of the cell division cycle. In *Mathematical Population Dynamics: Analysis of Heterogeneity*, eds. Arino O, Axelrod D & Kimmel M, Vol. 2, pp. 155–169. Winnipeg, Canada: Wuerz Publishing [*78*]

Olofsson P (1997). Branching processes with local dependencies. In *Classical and Modern Branching Processes*, eds. Athreya K & Jagers P, pp. 239–256. New York, NY, USA: Springer-Verlag [*42, 171–172*]

Olofsson P (2000). A branching process model of telomere shortening. *Communications in Statistics – Stochastic Models* **16**:167–177 [*226, 230*]

Olofsson P & Kimmel M (1999). Stochastic models of telomere shortening. *Mathematical Biosciences* **1**:75–92 [*226, 228*]

Pakes AG (1989). Asymptotic results for the extinction time of Markov branching processes allowing immigration. I. Random walk decrements. *Advances in Applied Probability* **21**:243–269 [*119*]

Philippi T & Seger J (1989). Hedging one's evolutionary bets, revisited. *Trends in Ecology and Evolution* **4**:41–44 [*52*]

Pitman J (1999). Coalescents with multiple collisions. *Annals of Probability* **27**:1870–1902 [*207*]

Ramanan K & Zeitouni O (1999). The quasi-stationary distribution for small random perturbations of certain one-dimensional maps. *Stochastic Processes and Their Applications* **84**:25–51 [*196*]

Ramsay ME (2003). The elimination of indigenous measles transmission in England and Wales. *Journal of Infectious Diseases* **187**:S198–S207 [*237*]

Ramsay ME, Gay N, Miller E, White J, Morgan-Capner P & Brown D (1994). The epidemiology of measles in England and Wales: Rationale for the 1994 national vaccination campaign. *CDR Review* **4**:R141–R146 [*237*]

Raup DM (1991). *Extinction*. New York, NY, USA: Norton [*107, 114*]

Rhodes CJ & Anderson RM (1996). Power laws governing epidemics in isolated populations. *Nature* **381**:600–602 [*237, 244*]

Rhodes CJ, Jensen HJ & Anderson RM (1997). On the critical behavior of simple epidemics. *Proceedings of the Royal Society of London B* **264**:1639–1646 [*237, 244–245*]

Richter-Dyn N & Goel NS (1972). On the extinction of a colonizing species. *Theoretical Population Biology* **3**:406–433 [*81*]

Ricker WE (1954). Stock and recruitment. *Journal of the Fisheries Research Board of Canada* **11**:559–623 [*193*]

Rosenberg NA & Nordborg M (2002). Genealogical trees, coalescent theory and the analysis of genetic polymorphisms. *Nature Reviews Genetics* **3**:380–390 [*202*]

Roughgarden J (1979). *Theory of Population Genetics and Evolutionary Ecology: An Introduction*. New York, NY, USA: MacMillan [*23*]

Rubelj I & Vondraček Z (1999). Stochastic mechanism of cellular aging: Abrupt telomere shortening as a model for stochastic nature of cellular aging. *Journal of Theoretical Biology* **197**:425–438 [*231*]

Rudin W (1987). *Real and Complex Analysis*. New York, NY, USA: McGraw-Hill [*77*]

Sagitov S (1999). The general coalescent with asynchronous mergers of ancestral lines. *Journal of Applied Probability* **36**:1116–1125 [*206*]

Sagitov S (2003). Convergence to the coalescent with simultaneous multiple mergers. *Journal of Applied Probability* **40**:839–854 [*206*]

Sato K & Iwasa Y (2000). Pair approximations for lattice-based ecological models. In *The Geometry of Ecological Interactions*, eds.

Dieckmann U, Law R & Metz JAJ, pp. 341–357. Cambridge, UK: Cambridge University Press   [*42*]

Schaffer WM (1974). Optimal reproductive effort in fluctuating environments. *The American Naturalist* **108**:783–790   [*52*]

Schnell S & Mendoza C (1997a). Enzymological considerations for a theoretical description of the quantitative competitive polymerase chain reaction (QC-PCR). *Journal of Theoretical Biology* **184**:433–440   [*94, 234*]

Schnell S & Mendoza C (1997b). Theoretical description of the polymerase chain reaction (QC-PCR). *Journal of Theoretical Biology* **188**:313–318   [*234*]

Schweinsberg J (2000). Coalescents with simultaneous multiple collisions. *Electronic Journal of Probability* **5**:1–50   [*207*]

Schweinsberg J (2003). Coalescent processes obtained from supercritical Galton–Watson processes. *Stochastic Processes and Their Applications* **106**:107–139   [*206, 208*]

Seger J & Brockmann HJ (1988). What is bet-hedging? In *Oxford Surveys in Evolutionary Biology*, Vol. 4, eds. Harvey PH & Partridge L, pp. 182–211. Oxford, UK: Oxford University Press   [*52*]

Seneta E (1981). *Non-negative Matrices and Markov Chains*, Second Edition. New York, NY, USA: Springer-Verlag   [*25*]

Sennerstam R & Strömberg JO (1995). Cell cycle progression: Computer simulation of uncoupled subcycles of DNA replication and cell growth. *Journal of Theoretical Biology* **175**:177–189   [*219–220, 222*]

Sevastyanov BA (1971). *Vetvyaščiesya protsessy (Branching Processes)*. Moscow, Russia: Nauka. [Also available in German, Sewastjanow BA (1974). *Verzweigungsprozesse*. Berlin, Germany: Akademie-Verlag]   [*26, 64*]

Sharkovskii AN, Maistrenko YL & Romanenko EY (1993). *Difference Equations and Their Applications*. Dordrecht, Netherlands: Kluwer   [*90*]

Shields GF, Schmeichen AM, Frazier BL, Redd A, Vovoeda MI, Reed JK & Ward RH (1993). mtDNA sequences suggest a recent evolutionary divergence for Beringian and Northern North American populations. *American Journal of Human Genetics* **53**:549–562   [*211*]

Smith M (1974). *Models in Ecology*. Cambridge, UK: Cambridge University Press   [*89*]

Smith WL & Wilkinson WE (1969). On branching processes in random environments. *Annals of Mathematical Statistics* **40**:814–827   [*51*]

Solomon W (1987). Representation and approximation of large population age distributions using Poisson random measures. *Stochastic Processes and Their Applications* **26**:237–255   [*88*]

Stanley HE (1971). *An Introduction to Phase Transitions and Critical Phenomena*. Oxford, UK: Oxford University Press   [*244*]

Stigler SM (1970). Estimating the age of a Galton–Watson branching process. *Biometrika* **57**:505–512   [*208*]

Stollenwerk N & Jansen VAA (2003). Meningitis, pathogenicity near criticality: The epidemiology of meningococcal disease as a model for accidental pathogens. *Journal of Theoretical Biology* **222**:347–359   [*244*]

Taib Z (1992). *Branching Processes and Neutral Evolution*. Lecture Notes in Biomathematics **93**. Berlin, Germany: Springer-Verlag   [*200*]

Taib Z (1999). A branching process version of the Bell–Anderson cell population model. *Stochastic Models* **15**:719–729   [*220*]

Tan Z (1999). Intramitotic and intraclonal variation in proliferative potential of human diploid cells: Explained by telomere shortening. *Journal of Theoretical Biology* **198**:259–268   [*231*]

Taneyhill DE, Dunn AM & Hatcher MJ (1999). The Galton–Watson branching process as a quantitative tool in parasitology. *Parasitology Today* **15**:159–165   [*31*]

Tavaré S (1984). Line-of-descent and genealogical processes, and their applications in population genetics models. *Theoretical Population Biology* **26**:119–164   [*202*]

Tavaré S, Balding DJ, Griffiths RC & Donnelly P (1997). Inferring coalescence times for molecular sequence data. *Genetics* **145**:505–518   [*211*]

Thompson MT & Stuart HB (1986). *Nonlinear Dynamics and Chaos*. New York, NY, USA: John Wiley & Sons   [*89*]

Tsao J, Yatabe Y, Salovaara R, Järvinen HJ, Mecklin J, Altonen LA, Tavaré S & Shibata D (2000). Genetic reconstruction of individual colorectal tumor histories. *Proceedings of the National Academy of Sciences of the USA* **97**:1236–1241   [*212–216*]

Tyson JJ, Chen KC & Novak B (1997). The eukaryotic cell cycle: Molecules, mechanisms, and mathematical models. In *Case Studies in Mathematical Modeling: Ecology, Physiology, and Cell Biology*, eds. Othmer HG, Adler FR, Lewis MA & Dallon JC, pp. 127–147. Upper Saddle River, NJ, USA: Prentice Hall   *[78]*

Van Kampen NG (1981). *Stochastic Processes in Physics and Chemistry*. Amsterdam, Netherlands: North-Holland   *[270]*

Vellekoop MH & Högnäs G (1997). A unifying framework for chaos and stochastic stability in discrete population models. *Journal of Mathematical Biology* **35**:557–588   *[197–198]*

Von Bortkiewicz L (1898). *Das Gesetz der kleinen Zahlen (The Law of Small Numbers)*. Leipzig, Germany: Teubner   *[17]*

Wakefield AJ, Murch SH, Anthony A, Linnell J, Casson DM, Malik M, Berelowitz M, Dhillon AP, Thomson MA, Harvey P, Valentine A, Davies SE & Walker-Smith JA (1998). Ileal–lymphoid–nodular hyperplasia, non-specific colitis, and pervasive developmental disorder in children. *The Lancet* **351**:637–641   *[237]*

Ward RH, Frazier BL, Dew K & Pääbo S (1991). Extensive mitochondrial diversity within a single Amerindian tribe. *Proceedings of the National Academy of Sciences of the USA* **88**:8720–8724   *[211]*

Watkins J (2000). Consistency and fluctuation theorems for discrete time structured population models having demographic stochasticity. *Journal of Mathematical Biology* **41**:253–271   *[93]*

Watterson GA (1975). On the number of segregating sites in generical models without recombination. *Theoretical Population Biology* **7**:256–276   *[202]*

Weiss G & von Haeseler A (1997). A coalescent approach to the polymerase chain reaction. *Nucleic Acids Research* **25**:3082–3087   *[214]*

Wilkinson WE (1969). On calculating extinction probabilities for branching processes in random environments. *Journal of Applied Probability* **6**:478–492   *[151]*

Williams D (1991). *Probability with Martingales*. Cambridge, UK: Cambridge University Press   *[156]*

Wright S (1931). Evolution in Mendelian populations. *Genetics* **16**:97–159   *[200]*

Yakovlev AY, Mayer-Proschel M & Noble M (1998). A stochastic model of brain cell differentiation in tissue culture. *Journal of Mathematical Biology* **37**:49–60   *[37–38]*

Yeomans JM (1992). *Statistical Mechanics of Phase Transitions*. Oxford, UK: Oxford University Press   *[244]*

Index

The International Institute for Applied Systems Analysis

is an interdisciplinary, nongovernmental research institution founded in 1972 by leading scientific organizations in 12 countries. Situated near Vienna, in the center of Europe, IIASA has been producing valuable scientific research on economic, technological, and environmental issues for nearly three decades.

IIASA was one of the first international institutes to systematically study global issues of environment, technology, and development. IIASA's Governing Council states that the Institute's goal is: *to conduct international and interdisciplinary scientific studies to provide timely and relevant information and options, addressing critical issues of global environmental, economic, and social change, for the benefit of the public, the scientific community, and national and international institutions.* Research is organized around three central themes:

- Energy and Technology;
- Environment and Natural Resources;
- Population and Society.

The Institute now has National Member Organizations in the following countries:

Austria
The Austrian Academy of Sciences

China
National Natural Science
Foundation of China

Czech Republic
The Academy of Sciences of the
Czech Republic

Egypt
Academy of Scientific Research and
Technology (ASRT)

Estonia
Estonian Association for
Systems Analysis

Finland
The Finnish Committee for IIASA

Germany
The Association for the Advancement
of IIASA

Hungary
The Hungarian Committee for Applied
Systems Analysis

Japan
The Japan Committee for IIASA

Netherlands
The Netherlands Organization for
Scientific Research (NWO)

Norway
The Research Council of Norway

Poland
The Polish Academy of Sciences

Russian Federation
The Russian Academy of Sciences

Sweden
The Swedish Research Council for
Environment, Agricultural Sciences
and Spatial Planning (FORMAS)

Ukraine
The Ukrainian Academy of Sciences

United States of America
The National Academy of
Sciences

Printed in the United States
By Bookmasters